4th International Conference on Electronic Devices, Systems, and Applications (ICEDSA 2015)

IOP Conference Series: Materials Science and Engineering
Volume 99

Kuala Lumpur, Malaysia
14 - 15 September 2015

ISBN: 978-1-5108-1636-7
ISSN: 1757-8981

4th International Conference on Electronic Devices, Systems, and Applications (ICEDSA 2015)

IOP Conference Series: Materials Science and Engineering
Volume 99

Kuala Lumpur, Malaysia
14 - 15 September 2015

IOP
Institute of Physics

Printed from e-media with permission by:

Curran Associates, Inc.
57 Morehouse Lane
Red Hook, NY 12571

CURRAN ASSOCIATES INC. proceedings.com

Some format issues inherent in the e-media version may also appear in this print version.

Printed by Curran Associates, Inc. (2016)

Institute of Physics
Dirac House, Temple Back
Bristol BS1 6BE UK

Phone: 44 1 17 929 7481
Fax: 44 1 17 920 0979

techtracking@iop.org

Additional copies of this publication are available from:

Curran Associates, Inc.
57 Morehouse Lane
Red Hook, NY 12571 USA
Phone: 845-758-0400
Fax: 845-758-2634
Email: curran@proceedings.com
Web: www.proceedings.com

Table of contents

Volume 99

4th International Conference on Electronic Devices, Systems and Applications 2015 (ICEDSA)
14–15 September 2015, Kuala Lumpur, Malaysia

Accepted papers received: 20 October 2015
Published online: 19 November 2015

Papers

Fabrication and Nanotechnology 1

Circuits & Systems

Engineering and Technology

IOP Conf. Series: Materials Science and Engineering **99** (2015) 011001 doi:10.1088/1757-899X/99/1/011001

International Conference on Electronic Devices, Systems and Applications (ICEDSA) September 14 – 15, 2015

IOP Conf. Series: Materials Science and Engineering **99** (2015) 011001 doi:10.1088/1757-899X/99/1/011001

ORGANIZING COMMITTEE

General Chair
Sukreen Hana Herman

General Co-Chair
Mousa Mohsen

International Liaison / Secretary
Uzer Mohd Noor

Finance / Registration Chair
Zurita Zulkifli

Publication Chair / Editor
Syed Abdul Mutalib Al Junid

Technical Program Chair
Hashimah Hashim

Local Arrangement Chair
Ahmad Sabirin Zoolfakar

International Advisory Board

Kamran Eshraghian	-	Australia
Amine Bermak	-	Hong Kong
Hassan Hamdan Al Alkim	-	UAE

IOP Conf. Series: Materials Science and Engineering **99** (2015) 011001 doi:10.1088/1757-899X/99/1/011001

WELCOME NOTE

On behalf of the Organizing Committee of ICEDSA2015, and with great pleasure, we would like to welcome all delegates to Kuala Lumpur, Malaysia. Being held from 14th to 15th September 2015 at Hotel Istana, Kuala Lumpur, the event is co-sponsored by the Faculty of Electrical Engineering, Universiti Teknologi MARA (UiTM), Malaysia, and the American University of Ras Al Khaimah, (AURAK), United Arab Emirates. This joint conference is an event held under the MoU signed between UiTM and AURAK, the aim being for this conference to be organized by both universities in the future.

The conference creates a forum for engineers, scientists and practitioners of varied disciplines and their associated applications. Throughout this event we hope to create an opportunity for researchers, academics and students to get together and, more importantly, to welcome new peers in diverse areas of expertise. On behalf of the organizing committee we would like to take the opportunity to express our gratitude to all reviewers who have been working hard to review the papers on time and share their expertise. Special mention goes to the distinguished Keynote Speakers for their willingness to spend their time and share their valuable knowledge. My deepest gratitude goes to the Organizing Committee who have been working hard behind the scenes for the success of this event.

To all participants, we would like to encourage you to fully utilize this platform for knowledge sharing and networking while enjoying your stay in Kuala Lumpur.

Sukreen Hana Herman
ICEDSA2015
General Chair

Peer review statement

All papers published in this volume of *IOP Conference Series: Materials Science and Engineering* have been peer reviewed through processes administered by the proceedings Editors. Reviews were conducted by expert referees to the professional and scientific standards expected of a proceedings journal published by IOP Publishing.

IOP Conf. Series: Materials Science and Engineering **99** (2015) 012001 doi:10.1088/1757-899X/99/1/012001

Slip detection with accelerometer and tactile sensors in a robotic hand model

Abdulrahman Abdulkareem S. Al-Shanoon, Siti Anom Ahmad and Mohd. Khair b. Hassan
Department of Electrical and Electronics Engineering,
University putra Malaysia, 43400 Selangor, Malaysia

E-mail: abdulrahman89eng@gmail.com

Abstract. Grasp planning is an interesting issue in studies that dedicated efforts to investigate tactile sensors. This study investigated the physical force interaction between a tactile pressure sensor and a particular object. It also characterized object slipping during gripping operations and presented secure regripping of an object. Acceleration force was analyzed using an accelerometer sensor to establish a completely autonomous robotic hand model. An automatic feedback control system was applied to regrip the particular object when it commences to slip. Empirical findings were presented in consideration of the detection and subsequent control of the slippage situation. These findings revealed the correlation between the distance of the object slipping and the required force to regrip the object safely. This approach is similar to Hooke's law formula.

1. Introduction

Human hands are an essential part of the human body. Tremendous structures have been created through human hands with precise capability. Human hands have an adequate ability to maneuver motions and accomplish complex actions (1,2). In addition, they consist of an array of hundreds receptive tactile sensors for sensory feedback mechanism. Thus, human hands can hold a particular object and automatically increase grasping force when the object begins to slip. Imperceptible changes can be easily detected through human skin sensibility. However, human hands still have some limitations in executing other tasks, such as dangerous or sophisticated operations in military and medical applications. In addition, high probability of errors can occur in repetitive and maintenance tasks because the capabilities of human hands have restrictions similar to limitations of other human abilities.

Thus, in recent years, significant efforts have been devoted to improve robotic manipulators, such as robotic hand or robotic claw. Currently, robotic hands begin as indispensable, especially because robotic hands have been broadly implemented in significant applications(3,4). In addition, robotic hand is widely used in automotive manufacturing industries for functions such as picking and placing, sorting, packaging, and palletizing, as well as in assembly and material handling production lines that require human hand substitution(5–7).

The most important issue in robotic hand applications is that robotic hands require distributed tactile pressure sensors that enable robotic hands to recognize the external environment. The primary emphasis placed on tactile sensors widely used in robotic hand applications is to measure the physical interaction that occurs through the contact points between the pressure sensor and a subjected object.

These tactile pressure sensors are designed and implemented based on different principle actions and materials, such as strain gauge, piezocapacitive, piezoelectric, pressure conductive rubber, and piezoresistive pressure sensors(8,9). Researchers have focused on the comprehension of tactile sensor characteristics, and different useful illustrative applications have been performed in various fields (10).

2. Background study

Robotic hand is considered a mechatronic instrument that can do some activities that are impossible for humans. Robotic hand is widely used in manufacturing and dangerous nuclear industries, as well as in precise applications such as military or medical implementations. In addition, repetitive and maintenance tasks could be achieved with high performance accuracy. Consequently, evolving robotic hand is required to cover a wide range of tasks and to provide robotic hand with special types of sensors that can measure the grasping force for a particular object. Grasping objects could be achieved using the dexterous robotic hand presented in Ref. (11) which is illustrated with the ability to grasp both soft and hard objects. In Ref (12), gripping operation has been implemented by robotic hands that use special types of tactile sensors, which employ physical properties and events through contact with objects. Many tactile sensors have been developed, and the sensor hardware has evolved to achieve certain gripping tasks. Moreover, to accomplish the gripping mechanism using robotic hand, some efforts have been expended in developing tactile pressure sensor structures, such as in Ref(13). In most recent studies, advanced robotic manipulations have used tactile pressure sensors implemented in different applications. The main interesting issue in advanced robotic manipulation tasks is that robotic hand is required to be equipped with distributed tactile pressure sensors that can continuously provide information about the magnitude and direction of forces at all contact points between the sensing area and a subjected object. Numerous studies have reported the proposed method that uses tactile sensor information through physical contact between the sensor and an object to detect both pressure force and hardness of the object(14). In addition, several studies have documented that tactile pressure sensors have been utilized successfully in different design concepts and action principles. These tactile sensors have presented the process of determining physical features with the environment (15), measuring applied forces exerted over an object and the art in tactile sensing and investigating trends (16).

The well-known piezoresistive tactile pressure sensor (force-sensing resistor)(FSR) has been broadly used in robotic gripping implementations, such as gripping an object with different weights and shapes (17,18). A pressure sensor indicates touch situation or continuous pressure force that occurs between the subjected object and the sensor. This indication depends on sensor resistance changes that correspond to the applied force. Thus, using a piezoresistive pressure sensor is appropriate in robotic hands for gripping operations.

However, FSR does not provide information on slipping of a particular object. Thus, the proposed robotic hand model that uses FSR remains to have some defects in gripping operation. In the other words, FSR does not provide information on slipping that may occur during grasping implementations. In this study, FSR was developed to realize the algorithm of regripping an object after a slip situation has been detected. Subsequently, object gripping operation without slip situation detection is not adequately reliable, and objects may not be gripped perfectly. The evidence presented thus far supports the significance of slip detection. Thus, slipping motions in different orientations can be detected with accelerometer sensors that can measure tilt, vibrations, and acceleration force under gravity. An accurate motion of this sensor generates an acceleration signal on the basis of the principle of action and concept of design of that accelerometer sensor. Accelerometer devices based on acceleration forces were used in robotic applications to detect slip information (19–21). These accelerometers were protected by a foam layer to lessen noise effect. The system design in (22) includes two accelerometers: one of which is located at the contact point, and the other is situated far from the peripheral frame. The control system of this method calculates the difference between these two accelerometer sensors. In this study, the main objectives of using an accelerometer sensor are to detect the acceleration signal caused by force gravity and to provide a control algorithm that deters slip continuity. Further data analysis was applied to investigate the features of the slipping signal, such as distance and velocity. Thus, the control feedback mechanism should accomplish the regripping algorithm that utilizes the slip distance signal to determine the required regripping force. This approach is presented with Hooke's law, which estimates the required force for object regripping on the basis of its slipping distance. In this study, a completely autonomous robotic hand model that can detect whether or not an object exists in between robotic hand claws was designed and implemented. To achieve this purpose, an infrared (IR) sensor was used to provide an automatic gripping operation.

3. EXPERIMENTAL SETUP

3.1 Proposed robotic hand model

This study focused on designing and implementing the proposed robotic hand model for gripping an object with different weights. This robotic hand can detect a slip situation and can control the consequences of this situation. As shown in Figure (1), the simplicity and uncomplicated traits of the robotic hand model are its key features. Furthermore, the researcher designed and implemented a printed circuit board (PCB) to fulfill the purposes of the empirical study. The crucial aim of this research was sufficiently achieved.

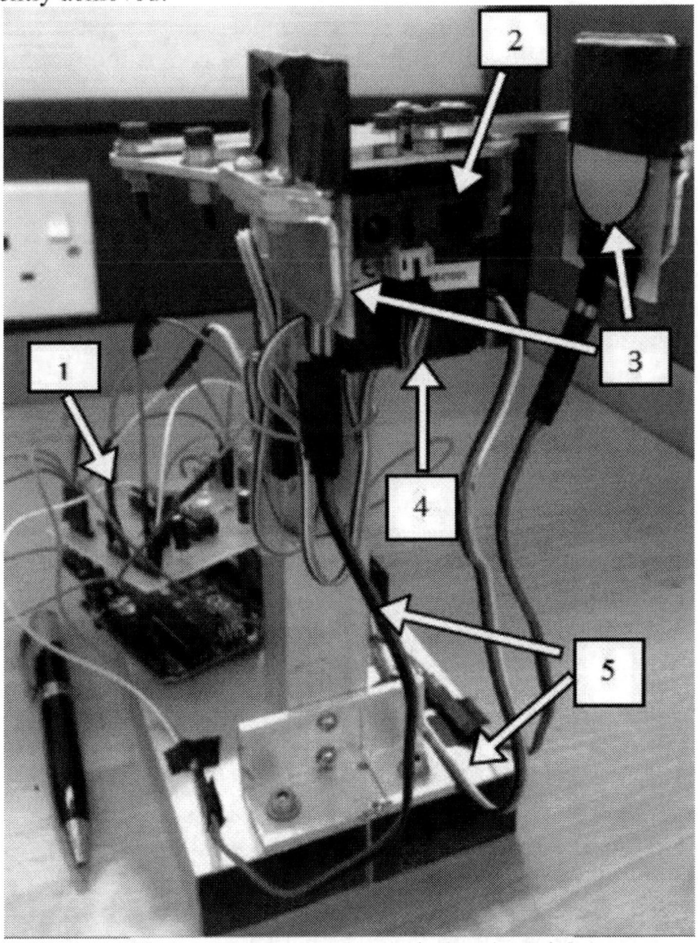

Figure (1) proposed robotic hand model

Where 1 denotes PCB, MCU, and power supply; 2 denotes IR sensor; 3 denotes pressure sensors; 4 denotes DC servo motor; and 5 denotes the proposed robotic hand structure.

Figure (2) illustrates that the accelerometer sensor is attached on the lateral part of the subjected object. The proposed robotic claws can be automatically closed after declaring that the object is

IOP Conf. Series: Materials Science and Engineering **99** (2015) 012001 doi:10.1088/1757-899X/99/1/012001

inserted between the robotic claws. The placed IR sensor indicates when the object is ready to be gripped. The subjected object was gripped with respect to the control feedback mechanism that was measured from the signals of acceleration response and tactile pressure sensors.

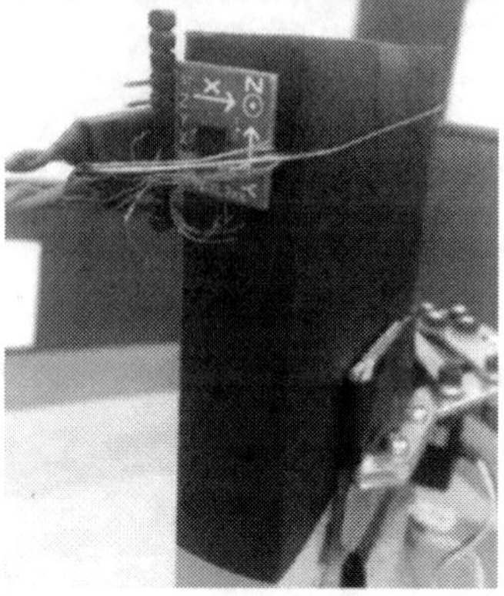

Figure (2) accelerometer sensor has been attached on object

The selected microcontroller unit (MCU) is Arduino Uno R3. This MCU has been chosen with respect to the following purposes: analog-to-digital converter, motor driver, pulse width modulation signals, digital in/out pins, analog input pins, low cost, and excellent compatibility with external components. The integrated circuit that has been implemented in MCU is an ATmega328. Moreover, the PCB in this study has been designed and implemented utilizing electronic components with high-quality performance and low power cost. As shown in Figure (1), these circuits are used to obtain experimental works that grip an object safely.

3.2 Schematic diagram of the experiment

In this study, the measured voltage of FSR can be specified through an analog readout circuit that was designed within PCB. Moreover, the output real-time data of the FSR and acceleration signal were displayed as graphs via LabVIEW application. Figure (3) illustrates the schematic of data extraction and signal processing. Further data analysis was conducted to process the data and to obtain signal features, such as velocity and distance. These features were provided after a double integral signal was applied to the acceleration signal, which was detected when the object slipped.

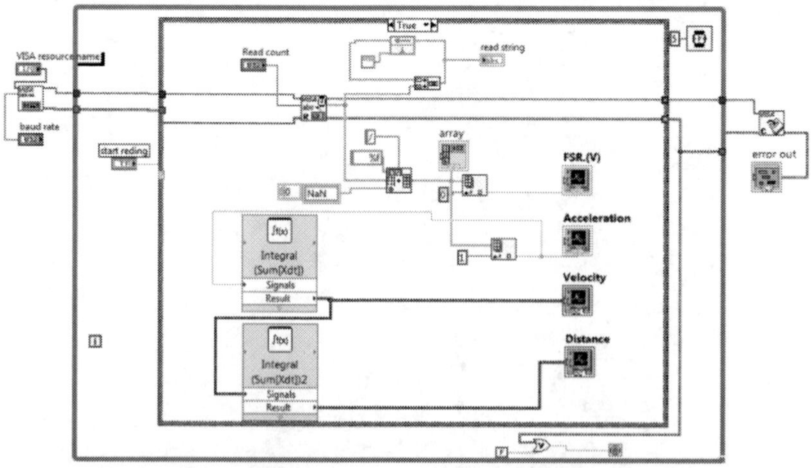

Figure (3) schematic diagram of the experiment

4. Results and discussion

Figure 4(b) illustrates that the gripping force dramatically decreased when the object started to slip. The voltage dropped to less than 0.2 V when the object slipped. This force denotes the output measured voltage of pressure force sensors. The regripping force voltage increased again at the first attempt of regripping the object. The acceleration signal can be clearly observed in Figure 4(a). The control system for regripping the object after slipping was stimulated when the system utilized the acceleration signal as a feedback control signal. When the object was held, the measured output voltage was approximately linear and the acceleration signal was quiet. However, when the weight of the object was increased, the acceleration signal triggered the system to prevent slip continuity and regrip the object safely.

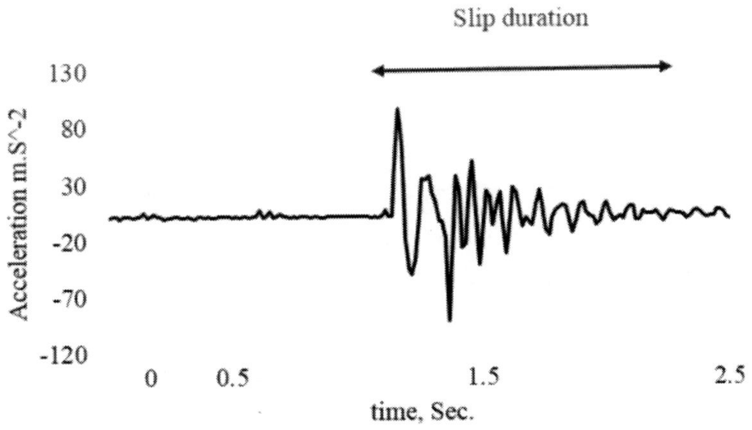

Figure 4 (a)

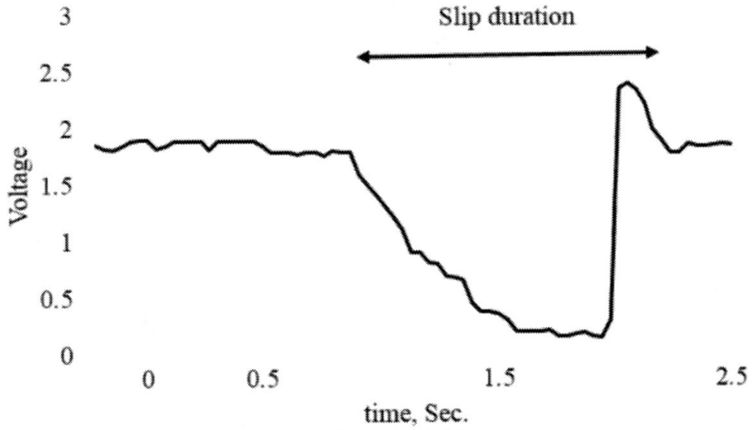

Figure 4 (b)

Figure (4) signal response from accelerometer and pressure sensor during an object slipping

After acceleration signal detection, the extracted data were analyzed to calculate slippage features, such as distance and velocity. These features were obtained by applying double integration to the acceleration signal. Figure (5) illustrates that an acceleration signal was generated when the object commenced to slip. At this point, the system was informed by the acceleration signal that triggers the system to regrip the object. This control feedback mechanism utilized the measured signal of the acceleration response after being analyzed.

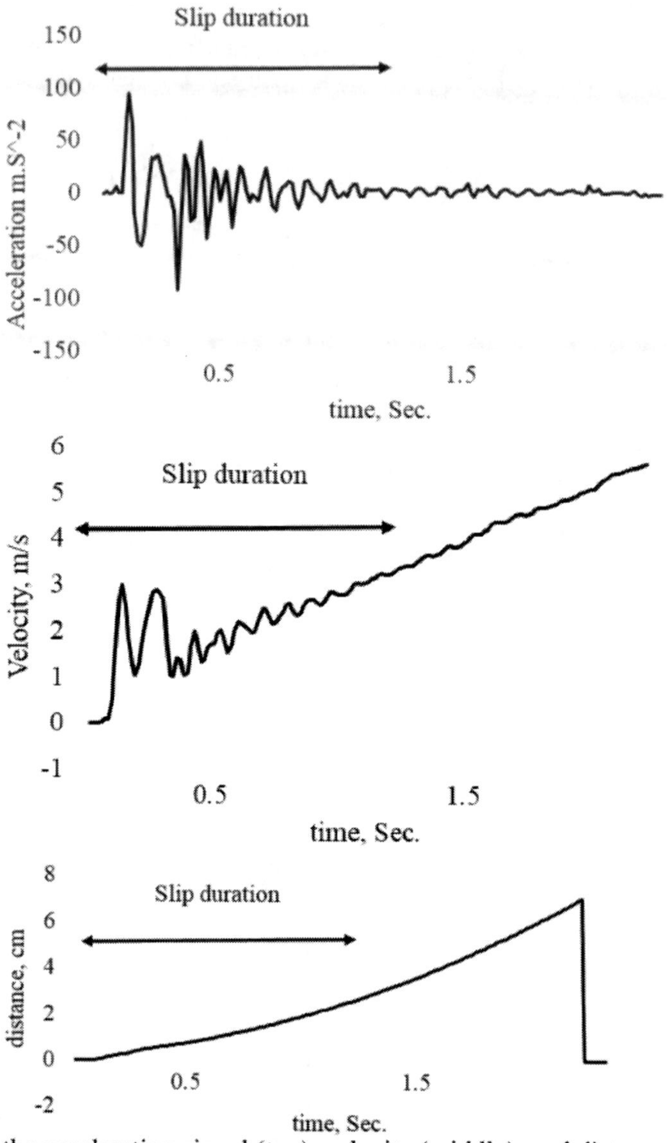

Figure (5) the acceleration signal (top), velocity (middle), and distance (bottom).

Figure (5) demonstrates the velocity of the slipping situation and distance. On the basis of the experimental results, the automatic control algorithm was developed by utilizing the feedback mechanism. This approach is presented in Hooke's law in Equation (1).

$$\boldsymbol{F = K \times X} \quad (1)$$

Where F is force, K is a constant, and X is the distance of an object slipping that is used to determine the amount of the required force to regrip the object after the slipping signal was measured. To calculate the required re-gripping force based on Hooke's law, the values of the distance of an object slipping and K are required to be calculated in this equation. Consequently, K in Equation (1) was calculated by repeating the experiment 10 times with different weights to determine an appropriate value of K.

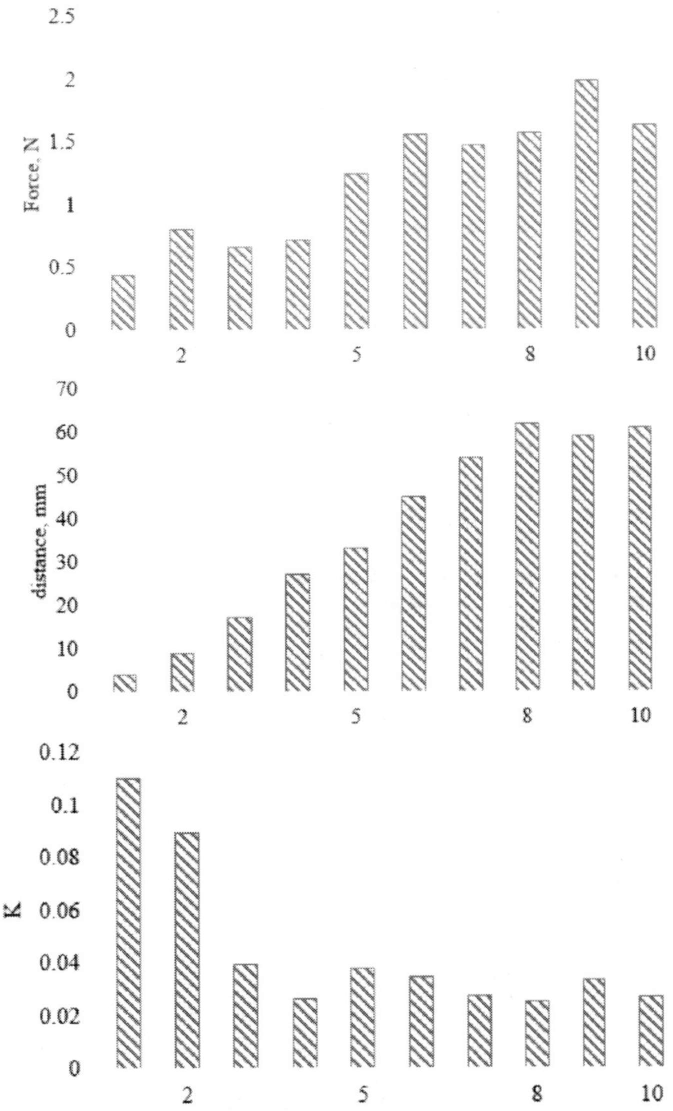

Figure (6) force (top), distance (middle) and *K* (bottom) values over ten attempts

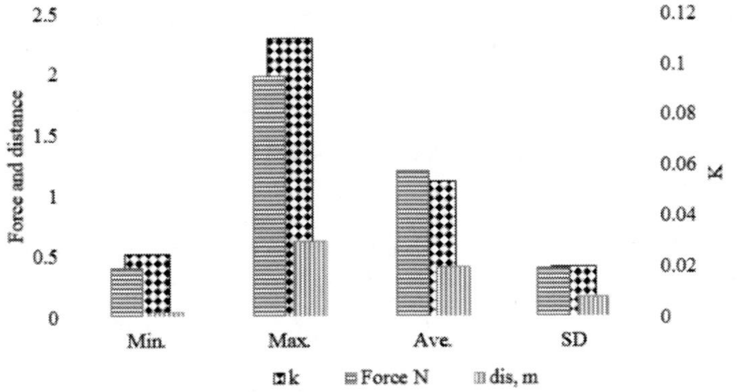

Figure (7) Statistics of force, distance, and K values

In data analysis, the required force when the robotic hand tried to regrip the object until the object being gripped was calculated, and the distance of object slipping was measured during the same regripping attempt. Figure (6) shows three plots that represent the force at the top, the distance at the middle, and the constant *K* at the bottom. Additionally, simple statistical analysis was conducted to calculate further parameter related to the force, distance, and K values. Figure (7) illustrates that the *K* value ranged from 0.1103 to 0.0142 and that the average of *K* was 0.0535. The standard deviation was calculated to be approximately 0.02, indicating the wide extent of the measured data.

5. Conclusion

Various types of tactile pressure sensors have been widely used for robotic grasping operations. In this study, a robotic hand was modeled by utilizing piezoresistive pressure sensors for gripping force purposes. Moreover, an effective accelerometer sensor was used to detect the acceleration signal under gravity. Further analysis was conducted to obtain slippage information, such as distance and velocity. The control feedback algorithm was achieved to provide an automatic regrip by applying Hooke's law, which evaluates the required regripping force from the distance where the object slipped. Tests and findings from the proposed robotic hand model illustrated three phases of gripping approach, namely, when the object was initially held, when the object started to slip, and when regripping attempt was secure.

6. References

1. Koike Y. *Human Hand Stiffness During Discrete Point-to-point Multi-joint Movement.* Proceedings of the Annual International Conference of the IEEE Engineering in Medicine and Biology Society. Ieee; 1992;**4**:1628–9.

2. Kunii J lee, L. T. *Model-based Analysis of Hand* Posture. IEEE Computer Graphics and Applications (1995). 1995;**15**(5):77–86.

3. Lin P, Bekey G, Abney K. *Autonomous military robotics*: Risk, ethics, and design. Cal Poly Stat Univer, San Luis. 2008;4–112.

4. Dharia SP, Falcone T. *Robotics in reproductive medicine.* Fertility and sterility. 2005 Jul;**84**(1):1–11.

5. Blomdell A, Dressler I. *Flexible application development and high-performance motion control based on external sensing and reconfiguration of ABB industrial robot controllers.* 2010 IEEE International Conference on Robotics and Automation. 2010;62–6.

6. Brogårdh T. *Present and future robot control development—An industrial perspective.* Annual Reviews in Control. 2007 Jan;**31**(1):69–79.

7. Pan Z, Polden J, Larkin N, Van Duin S, Norrish J. *Recent progress on programming methods for industrial robots.* Robotics and Computer-Integrated Manufacturing. Elsevier; 2012 Apr;**28**(2):87–94.

8. Lee M., Nicholls H. *Tactile sensing for mechatronics—a state of the art survey.* Mechatronics. 1999 Feb;**9**(1):1–31.

9. Dahiya RS, Metta G, Valle M, Sandini G. *Tactile Sensing—From Humans to Humanoids.* IEEE Transactions on Robotics. 2010 Feb;**26**(1):1–20.

10. Dargahi J, Najarian S. *Advances in tactile sensors design/manufacturing and its impact on robotics applications* – a review. Industrial Robot: An International Journal. 2005 Jun;**32**(3):268–81.

11. Alcazar J, Barajas L. *Dexterous robotic hand grasping method for automotive parts. Humanoid Robots (Humanoids)*, 2010 10th IEEE-RAS International Conference. 2010;282–7.

12. Goeger D, Ecker N, Woern H. *Tactile sensor and algorithm to detect slip in robot grasping processes*. 2008 IEEE International Conference on Robotics and Biomimetics. Ieee; 2009 Feb;1480–5.

13. Shimojo M, Namiki A, Ishikawa M, Makino R, Mabuchi K. *A Tactile Sensor Sheet Using Pressure Conductive Rubber With Electrical-Wires Stitched Method.* IEEE Sensors Journal, 2004;**4**(5):589–96.

14. Shikida M, Shimizu T, Sato K, Itoigawa K. *Active tactile sensor for detecting contact force and hardness of an object*. Sensors and Actuators A: Physical. 2003 Jan;**103**(1-2):213–8.

15. Souza R De, Wise K. *A very high density bulk micromachined capacitive tactile imager*. Proceedings of International Solid State Sensors and Actuators Conference (Transducers '97) (1997). 1997;**2**:1473–6.

16. Allen P, Miller A, Oh P, Leibowitz B. *Integration of vision, force and tactile sensing for grasping*. International Journal of Intelligent Machines. 1999;**4**:129–49.

17. Fiorillo AS. *A piezoresistive tactile sensor*. IEEE Transactions on Instrumentation and Measurement. 1997;**46**(1):15–7.

18. Wisitsoraat a., Patthanasetakul V, Lomas T, Tuantranont a. *Low cost thin film based piezoresistive MEMS tactile sensor*. Sensors and Actuators A: Physical. 2007 Sep;**139**(1-2):17–22.

19. Howe RD, Cutkosky MR. *Sensing skin acceleration for slip and texture perception*. Proceedings, 1989 International Conference on Robotics and Automation. IEEE Comput. Soc. Press; 1989;**1**:145–50.

20. Lowe RJ, Chappell PH, Ahmad S a. *Using accelerometers to analyse slip for prosthetic application*. Measurement Science and Technology. 2010 Mar 1;**21**(3):035203.

21. Tremblay M, Packard W, Cutkosky M. *Utilizing sensed incipient slip signals for grasp force control*. Technical Report NASA STI/Recon. 1992;N **93**(14434):1237–43.

22. Tremblay M, Cutkosky M. *Estimating friction using incipient slip sensing during a manipulation task. Robotics and Automation*, 1993 Proceedings, 1993 IEEE International Conference. 1993;**1**:429–34.

Fabrication of Flexible Au/ZnO/ITO/PET Memristor Using Dilute Electrodeposition Method

F B Fauzi[1], M H Ani[1], R Othman[1], A Z A Azhar[1], M A Mohamed[2], S H Herman[3]

[1] Department of Manufacturing and Materials Engineering, International Islamic University Malaysia (IIUM), Jalan Gombak, 53100 Kuala Lumpur, Malaysia
[2] Institute of Microengineering and Nanoelectronics, Universiti Kebangsaan Malaysia, 43600 UKM-Bangi, Selangor, Malaysia
[3] NANO-Electronic Centre (NET), Faculty of Electrical Engineering, Universiti Teknologi MARA, 40450 Shah Alam, Selangor, Malaysia

Email: mhanafi@iium.edu.my

Abstract. DRAM has been approaching its maximum physical limit due to the demand of smaller size and higher capacity memory resistor. The researchers have discovered the abilities of a memristor, a Non Volatile Memory (NVM) that could overcome the size and capacity obstacles. This paper discussed about the deposition of zinc oxide (ZnO) on indium tin oxide (ITO) coated polyethylene terephthalate (PET) substrate by electrodeposition. Metallic Zn film was deposited on substrates with varying deposition time from 15 to 120 seconds in very dilute zinc chloride ($ZnCl_2$) aqueous and subsequently oxidized at 150 °C to form ZnO/ITO coated PET junction. The deposited thin film was characterized via x-ray diffraction (XRD) and field emission scanning electron microscopy (FESEM). The results from *I-V* measurement show the deposited ZnO exhibits pinched hysteresis loop. The hysteresis loop becomes smaller with increasing deposition time. The 15 seconds electrodeposition gave the largest hysteresis loop and largest value of resistive switching ratio of 1.067. The result of the synthesized ZnO on the flexible substrate can be one of the alternatives to replace the current memory system as the flexible memory system.

1. Introduction

The usage of organic materials in electronic device has shown a rapid growth as the size demand of devices is increasingly smaller and faster. Future electronics industry depends on the development of organic base semiconductor devices due to their advantages. Flexible electronics possesses a lot of advantages of excellent portability, lightweight, low cost and ease of manufacture [1][2]. Many researches have been done on flexible electronic devices such as integrated circuits (ICs), organic light emitting diodes (OLEDs), and radio frequency identification (RFID) antennas. However, only few researches have been done on flexible memory resistor (memristor)[3][4]. The flexible substrate (PET) is favorable to be used in line with the growing interest from device manufacturers in thin films at thinner thicknesses. Memristor, a fourth fundamental element found by Chua in 1971 becomes the possible answer to replace the current charge based memories such as dynamic random access memory (DRAM) and flash memory [5].

Experiments have proven that a memristor can be made through a phenomenon known as bipolar resistive switching, and mostly found in transition metal oxides [6]. Hence, the titanium dioxide (TiO_2) film which is extensively used in most experimentation as the active layer to exhibit the memristive characteristics [7]. However, improvement on the use of other materials and different fabrication techniques as an alternative is crucial to overcome the drawbacks of using TiO_2 which is expensive and moreover, it's the fabrication process is complex, involving atomic layer deposition and other additional processes [8][9].

Owing to the similar resistive switching characteristics of the transition metal oxides (TMO), the TiO_2 thin film can be replaced by other TMO. For example, zinc oxide (ZnO) which has properties favorable for the electronics industry. ZnO is one of just several oxides that can be grown at low deposition temperatures as low as room temperature on various amorphous substrates [10][11]. Sauki et al. [12] has successfully deposited nanoparticle sized ZnO using RF sputtering method at low temperature. By optimizing the parameters, they are able to control the crystalline structure of ZnO. Yusoff et al. [11] has demonstrated that thin film ZnO is possible to be developed using simple electrodeposition method.

In this paper, thin film ZnO were fabricated on ITO coated PET substrate by using electrodeposition and thermal oxidation methods which replaces the expensive, complex and time consuming existing methods. Their memristive performance of Au/ZnO/ITO coated PET samples is evaluated.

2. Experimental procedures

Electrodeposition process consists of ITO coated PET substrate as working electrode, platinum foil as counter electrode and silver nitrate (Ag/AgCl) as the reference electrode. 0.005M zinc chloride solution ($ZnCl_2$) was choose as the electrolyte because of its high solubility in water thus important for producing low concentration of electrolyte solution while possessing good ionic conductivity. The flexible substrate was cut into 30 x 20 mm, cleaned in ultrasonic cleaner for 10 minutes followed by rinsing with ethanol and distilled water. 2.0 V DC voltage was set to generate voltage potential between anode and cathode using the AUTOLAB Potentiostat (PGSTAT302N). Figure 1 (i) show the schematic diagram of the electrodeposition process. Both electrodes were clamped in electrolytic cell holder made by acrylic with fixed distance of 20 mm and deposition area of 113 mm^2.

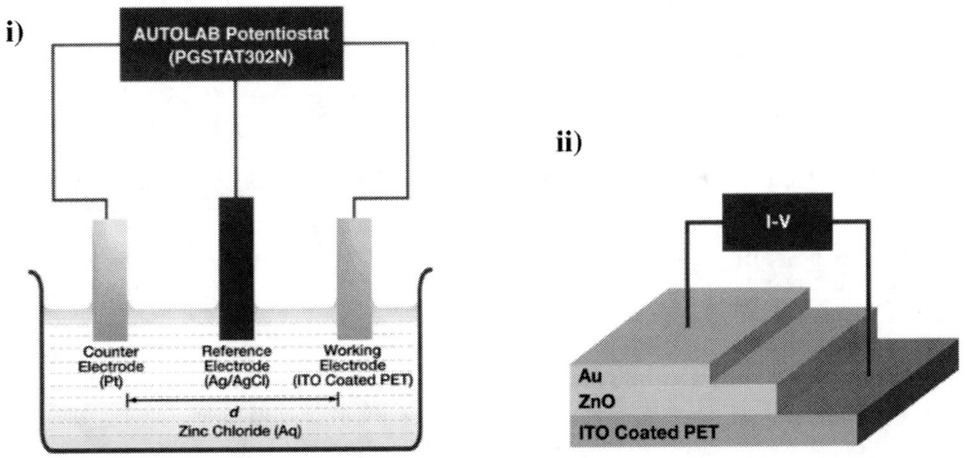

Figure 1 The Schematic Diagram of Electrodeposition Process (i) and Au/ZnO/ITO coated PET Sample in The *I-V* Measurement (ii)

After the electrodeposition process, the samples were heated for 60 minutes at temperature 150 °C on a hot plate subsequently cooled to room temperature (RT). Sputter coater was used to coat gold on top of the deposited ZnO thin film to create another metal contact between the ZnO insulators making an Au/ZnO/ITO coated PET sandwich structure of the memristor device. Before executing characterizations was performed using FESEM and *I-V* measurement. Figure 1 (ii) is a schematic diagram of *I-V* measurement of Au/ZnO/ITO junction. The measurement was conducted by applying DC voltage between -4 V and 4 V with scan rate of 0.1 V.

3. Results and Discussion

Figure 2 shows the X-ray diffraction (XRD) patterns of the 15 seconds deposited thin film before and after the thermal oxidation with the identification of Miller indices at each peak. The peak of Zn at 2-theta, 2θ of 29.1°(220), 31.8°(221), 32.9°(100) and 43.2°(101) were clearly seen right after the electrodeposition. After the thermal oxidation process, Zn was fully oxidized to ZnO as indicated by the formation of peaks, 2θ of 47.2°(102). There also the appearance of gold (Au) peak at 38.9° because of the gold coating on top of ZnO/ ITO sandwich. The corresponding peaks of the ITO coated PET substrate were found at 23.6° and 54.1°. The metal was slowly oxidized entirely in air with slow heating rate under the setting condition of thermal oxidation process. The result shows that the ZnO is well synthesized on flexible substrate.

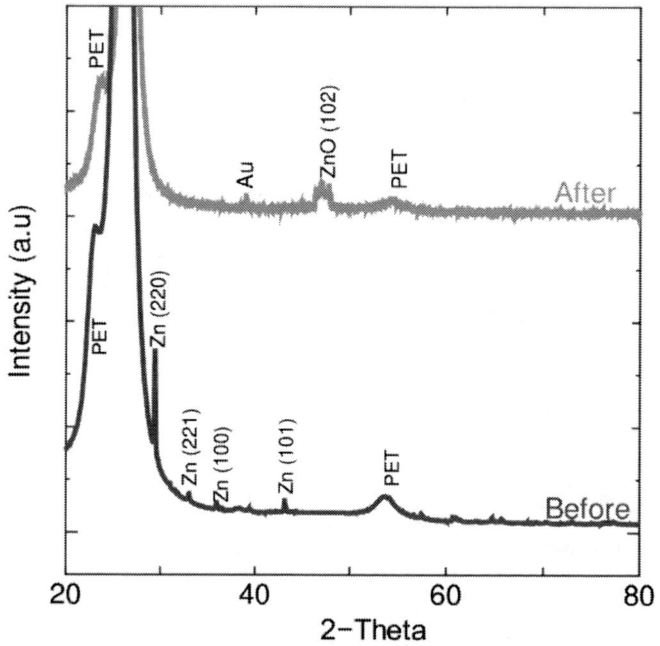

Figure 2 XRD Results Of 15 Seconds Deposited Zn Before And After Thermal Oxidation

Figure 3 (i) and (ii) show the morphology of surface for Zn, ZnO structures before and after thermal oxidation respectively at different deposition time. Fig 3 i) (a)-(d) exhibits a granular shape of Zn microstructures. Longer deposition time increased the number of small pallet structure of Zn. After the thermal oxidation process, the size of grain particles of ZnO become larger as shown in Fig 3 ii) (a)-(d). The increased of size in synthesized ZnO morphology are due to sintering of Zn during thermal oxidation. It is clearly shown that surface morphology has changed after the thermal oxidation. ZnO microstructure has different shape from the deposited metallic Zn.

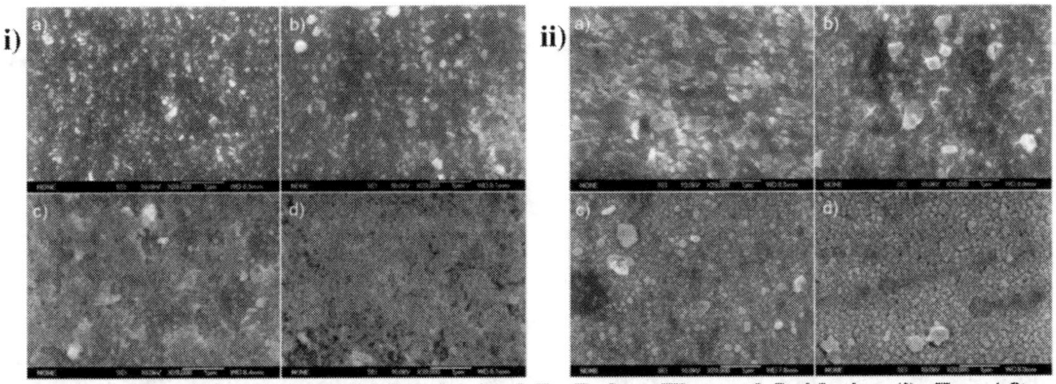

Figure 3 Surface Morphology Of Synthesized Zn Before Thermal Oxidation (i), Zno After Thermal Oxidation (ii) For Depositing Time at 15 (a), 30 (b), 60 (c) And 120 (d) Seconds at 20k Magnification

Figure 4 (i) show the *I-V* characteristics on the Au/ZnO/ITO coated PET memristor at deposition time from 15 to 120 seconds. The result show pinched hysteresis loops which describe the characteristic of a memristor. The size of pinched hysteresis loop gets smaller with the increase of the deposition time. 15 seconds of deposition time gave the biggest *I-V* curve with highest maximum current reading at 4 V. Figure 4 (ii) shows the reproducible of the memory switching characteristic of 100 cycles in semi log scale. The Au/ZnO/ITO coated PET memristor demonstrates good reproducibility memristic behaviour of the device because it can successfully performed 100 cycles without obvious deterioration. The pinched hysteresis loop is actually a bipolar resistive switching where the resistance depends on the polarity of the applied voltage [11].

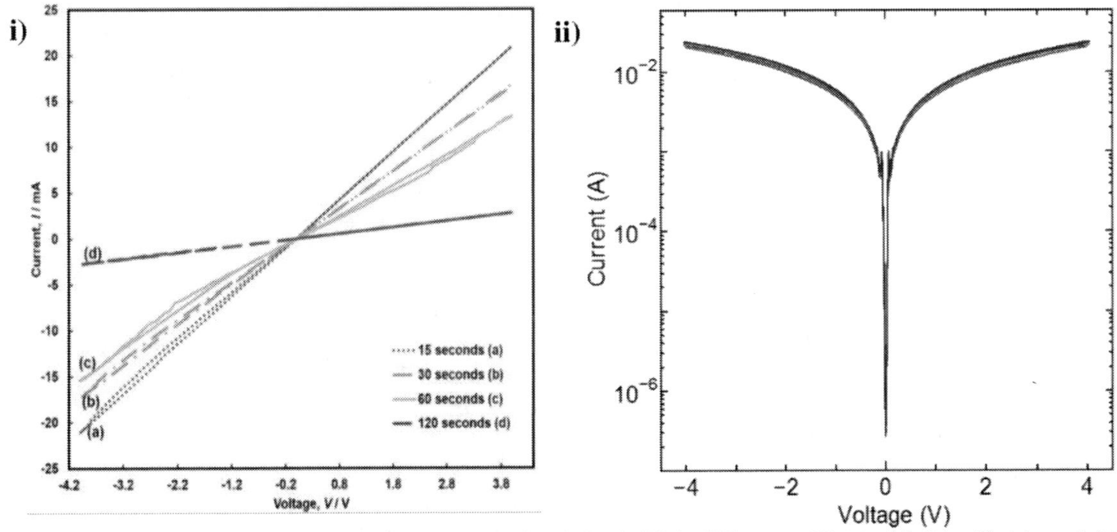

Figure 4 *I-V* Hysteresis Loops From Synthesized ZnO Thin Films at 15 (a), 30 (b), 60 (c) and 120 (d) Seconds (i), Semi Log Scale of *I-V* Curves in 100 Cycles (ii)

The existence of resistive switching characteristic could be seen from the changes of current due to changes of resistance from high resistance state (HRS) to low resistance state (LRS) from the curve. The metal vacancy defects inside the transition metal oxide of the active layer gives significant effect to the for memristive behaviour of the memristor. The ratios of R_{OFF}/R_{ON} (HRS/LRS) of Au/ZnO/ITO coated PET memristor are plotted in figure 5 (i). From the graph, it is observed that the R_{OFF}/R_{ON} ratio is decreases when the deposition time increased. This could easily understand because memristance is a function of thin film thickness. Memristance defined in the equation expressed by William et al. [13] below.

$$M(q) = R_{OFF}\left(1 - \frac{\mu_v R_{ON}}{D^2} q(t)\right) \quad (1)$$

Where, R_{ON} and R_{OFF} are the low resistance state (LRS) and high resistance state (HRS), μ_v is the dopant mobility, $q(t)$ is charge-dependent with time and D is the thin film thickness. The R_{OFF}/R_{ON} ratio of 15 seconds Au/ZnO/ITO coated PET is 1.067 and 120 seconds Au/ZnO/ITO coated PET is 1.051. Figure 5 (ii) shows compilation of HRS-LRS from previous works on ZnO thin film memristor and comparison with this work [14][15]. Suitability of the method reproduces the comparable results with the previous study. In order to obtain higher R_{OFF}/R_{ON} ratio, thinner scale of thin film is required which means in this technique it can be done by the controlling the deposition time and the concentration of electrolyte.

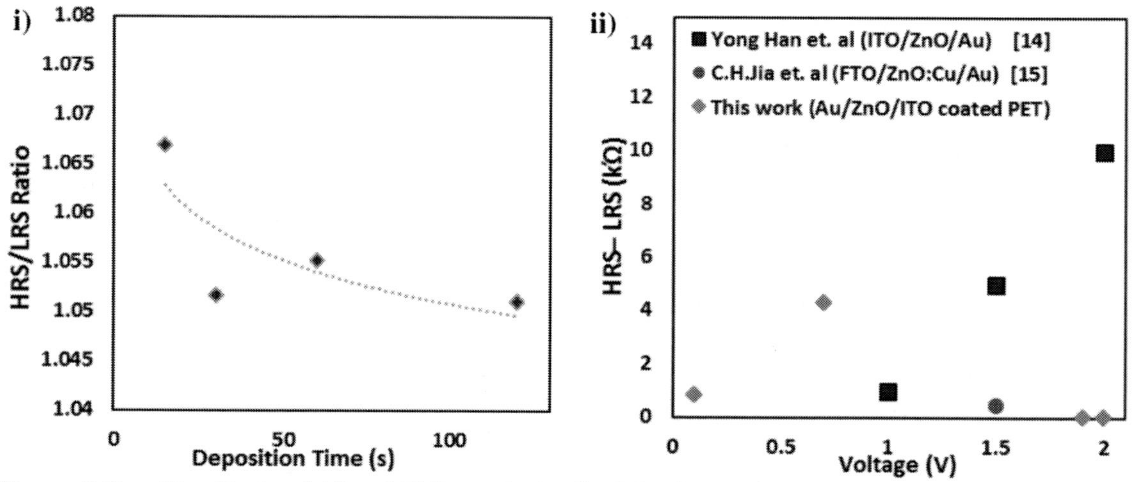

Figure 5 R_{OFF}/R_{ON} Ratio of 15 to 120 Seconds Au/ZnO/ITO coated PET Memristor (i) And Comparison of Difference Of R_{OFF}/R_{ON} With Previous Works (ii) [14][15]

4. Conclusion

An Au/ZnO/ITO coated PET thin film memristor was synthesized using electrodeposition and thermal oxidation method. XRD analysis shows that deposited metallic Zn was successfully oxidized into ZnO. The increment in particles size of the deposition surface structure proves that Zn was fully oxidized to ZnO after the thermal oxidation process. From the *I-V* measurement, a pinched hysteresis loop was observed proving the resistive switching behavior of Au/ZnO/ITO coated PET memristor. The measured *I-V* curve becomes smaller with the increase of deposition time. The largest size of hysteresis loop was obtained at 15 seconds deposition time with the largest HRS/LRS ratio, 1.067. It can be concluded that the results obtained from the synthesized ZnO of this work is comparable and demonstrated a good agreement with previous work using rapid, low cost and simpler ultra-diluted electrodeposition method. Furthermore, we believe this is the first reported attempt on fabrication of memristive device by electrodeposition method on a flexible substrate.

Acknowledgement

This work was financially supported by the Niche Research Grant Scheme, Ministry of Education Malaysia (600-RMI/NRGS 5/3 (7/2013)) and Research Acculturation Grant Scheme (RAGS13-002-0065), Universiti Teknologi MARA (UiTM) and International Islamic University Malaysia (IIUM).

References

[1] Reuss R H *et al* 2005 *Proc. IEEE* **93** 1239-1256
[2] MacDonald W A, Looney M K, Mackerron D, Eveson R, Adam R, Hashimato K and Rakos K 2007 *J. Soc. Inf. Disp* **15** 1075-1083
[3] Zidan M A , Hassan Fahmy H A, Hussain M M, Salama K N 2012 *Microelectronics Journal* **44** 1-9
[4] Gergel-Hackett N *et al* 2009 *IEEE Electron Device Lett.* **30** 706-708
[5] Chua L O 1971 *IEEE Transations Circuit Theory* **CT-18** 507-519
[6] Williams R S 2008 *Nature* **45** 28–35
[7] Gale E *et al* 2014 *Mater. Chem. Phys.* **143** 524–529
[8] Rosli A B, Marbie M M, Herman S H and Ani M H 2014 *J. Nanomater.* **2014** 1–7
[9] Kumar A, Rawal Y and Baghini M S 2012 *Proc. Int. Conf. Emerg. Electron (Mumbai)* 1–3
[10] Lincot D 2005 *Thin Solid Films* **487** 40–48
[11] Yusoff M M, Ani M H and Suryanto 2013 *Adv. Mater. Res.* **701** 172–175
[12] Sauki N S M, Herman S H, Ani M H and Mahmood M R 2013 *Adv. Mater. Res.* **795** 403–406
[13] Strukov D B, Snider G S, Stewart D R and Williams R S 2008 *Nature* **453** 7197
[14] Han Y, Cho K and Kim S 2011 *Microelectron. Eng.* **88** 2608-2610
[15] Jia C H, Dong Q C and Zhang W F 2012 *J. Alloys Compd.* **520** 250-254

IOP Conf. Series: Materials Science and Engineering **99** (2015) 012003 doi:10.1088/1757-899X/99/1/012003

Low Power Phase Change Memory using Silicon Carbide as a Heater Layer

M.S. Aziz, Y.Yin, S. Hosaka, Z. Mohammed, R.I. Alip

Faculty of Electrical Engineering,
Universiti Teknologi MARA 40450 Shah Alam, Selangor

shariman_1285@yahoo.com

Abstract. The amorphous to crystalline transition of germanium-antimony-tellurium (GST) using two types heating element was investigated. With separate heater structure, simulation was done using COMSOL Multiphysic 5.0. Silicon carbide (SiC) and Titanium Sitride ($TiSi_3$) has been selected as a heater and differences of them have been studied. The voltage boundary is 0.905V and temperature of the memory layer is 463K when using SIC as a heater. While the voltage boundary and temperature of memory layer when using $TiSi_3$ are 1.103V and 459K respectively. Based on the result of a simulation, the suitable material of heater layer for separate heater structure is Silicon carbide (SiC) compared with Titanium Sitride ($TiSi_3$).

1. Introduction

The phase change memory (PCM) is a unique behavior of chalcogenide glasses that changes from amorphous to crystalline with application of heat [1]. There are many advantages of the phase-change memory, such as high endurance, low programmable energy, fast switching speed, good data retention and excellent scalability [1] [2] [3]. The amorphous have a high resistance and can used as '0' state for data storage. The crystalline have a low resistance and can used as '1' state for data storage [3]. Now, the lower power consumption of PCM is a new creation of technology development.

Phase change material, germanium-antimony-tellurium (GST) is a chalcogenide glass which is widely used phase change memory industry. The advantages of GST are good reversibility between crystalline and amorphous and high thermal stability in room temperature [4]. The melting temperature of GST is about 900K and the crystallization temperature is about 450K [5]. The changing of resistance from amorphous to crystalline state can allow data to be stored. This is a unique characteristic of phase change material.

Silicon carbide (SiC) is one of the suitable semiconductor materials for operating at high temperature and high power [8]. The advantages of silicon carbide are low thermal expansion, high thermal conductivity and excellent thermal shock resistance [8] [9] [10]. Silicon carbide (SiC) is a wide band gap material [9]. That's mean; the resistance of SiC did not change significantly at higher temperature. Based on this fact, silicon carbide (SiC) is suitable material for heater in this separate heater structure.

In this paper, the COMSOL 5.0 Multiphysic software was used for simulating the separate heater structure of phase change memory. This structure consists 4 electrodes, the memory layer (GST), insulator layer, heater layer ($TiSi_3$ or SiC) and capping layer. The voltage 0.9V-1.1V with 100ns SET pulse were applied to the heater layer to investigate the changes of the memory layer (GST).

A. Lacita et al. (2006) proposed the PCM with 50ns-100ns but memory layer was heated directly by heater [2]. While the Irma R, Kobayashi R, Zhang YL et al (2013) was proposed the phase change memory structure with separate heater using Titanium Sitride ($TiSi_3$) as a heater layer [6]. The voltage 0.5V-3.0V with 100ns SET pulse were applied to phase change memory structure with separate heater to control the crystallization process.

2. Experimental

The structure that used in this simulation is the separate heater structure [6] [7]. Using the COMSOL 5.0 Multiphysics Software, the simulation was made with 2 types of heater, Silicon Carbide (SiC) and Titanium Sitride ($TiSi_3$). Figure 1 shows the simulation model and thickness of materials with GST as a memory layer and SiC and $TiSi_3$ as a heater. Table 1 shows the physical properties of material that used in this simulation.

The structure was designed as a figure 1 for the simulation process [7]. Titanium Nitride (TiN) was used as an electrode material and Zinc Sulfide-Silicon Dioxide (ZnS-SiO_2) was used as a capping and an insulator layer with thickness at 150nm and 20nm respectively. GST has 150nm thinned and was used as a memory layer in this simulation. The lastly material that used in the simulation is silicon carbide; SiC and titanium stride $TiSi_3$ as a heater. The heater has 50nm of thickness. The first model uses SiC as a heater and the second model uses $TiSi_3$ as a heater. From this simulation, the difference temperature can be obtained with using different material as a heater.

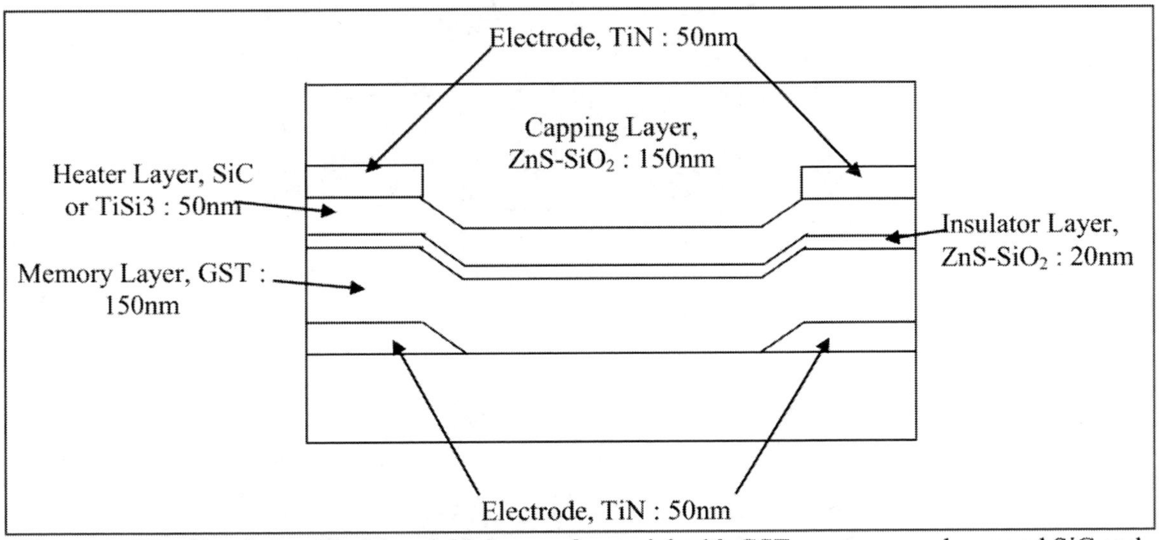

Figure 1 : The Simulation Model and thickness of material with GST as a memory layer and SiC and $TiSi_3$ as a Heater [7].

Table 1 : Physical properties of material used in simulation.

Material	Heat capacity (J/kg.K)	Thermal conductivity (W/m.K)	Density (kg/m^3)	Electrical conductivity (S/m)
Capping Layer,ZnS-SiO_2	263	0.657	3650	0.02
Insulator Layer,ZnS-SiO_2	263	0.657	3650	0.02
Electrode, TiN	784	22	5240	$5.0X10^6$
Heater, SiC	670	120	3200	$4.3X10^4$
Heater, $TiSi_3$	800	20	4043	$1.75X10^4$
Memory layer, a-GST	202	0.46	6200	$3.6X10^{-1}$
Memory Layer, c-GST	202	0.46	6200	$2.0X10^2$

From this simulation, the lower voltage will be measured when memory layer is changed from amorphous to crystalline state with different heater. The first voltage that makes changes from amorphous to crystalline state at the memory layer is called as voltage boundary. The temperature and voltage boundary can be obtained from this simulation.

3. Result and Discussion

From the simulation, the temperature can be obtained using the SiC and $TiSi_3$ as a heater to heat the memory layer, GST. The result of simulation can be seen in the figure 2 below. Figure 2 (a) show the temperature is 463K at memory layer using the SiC material as a heater and figure 2 (b) shows the temperature is 459K at memory layer using the $TiSi_3$ material as a heater. The SET voltage 0.91V with 100ns pulse width was applied at the SiC as a heater and SET voltage 1.11V with 100ns pulse width was applied at the $TiSi_3$ as a heater.

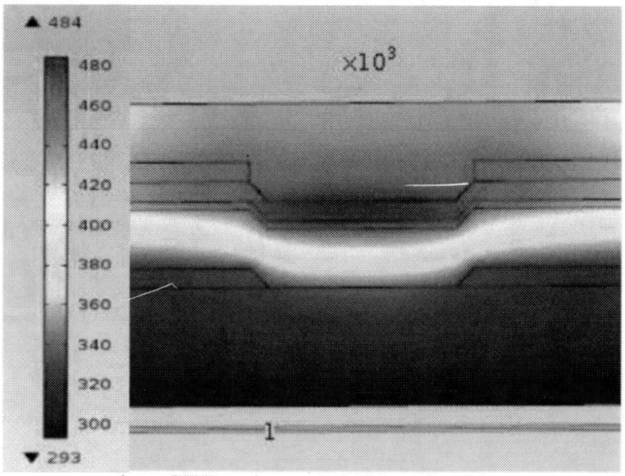

Figure 2 (a): The temperature when SET voltage 0.91V with 100ns pulse width was applied at SiC material.

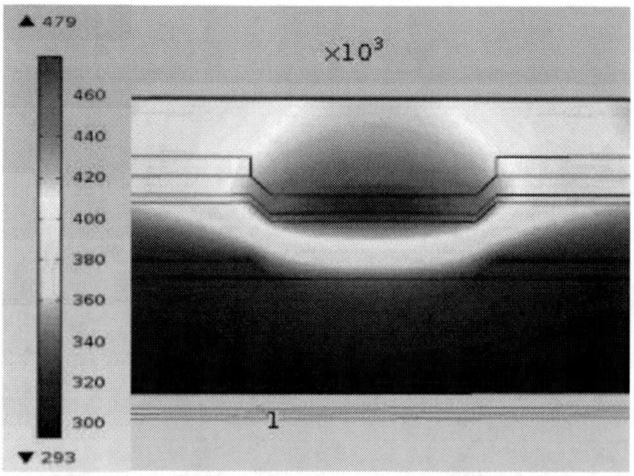

Figure 2 (b): The temperature when SET voltage 1.11V with 100ns pulse width was applied at $TiSi_3$ material.

From the figure 2 (a), the temperature of structure is widely and spread compared with the figure 2 (b) which is the temperature of the structure is more focused at the center of the structure. The one characteristic of silicon carbide is high thermal conductivity. The definition of thermal conductivity is the rate of heat flow through a material. So, the amount of heat transfer is controlled by the amount of dissipation, the amount of thermal energy present and the nature of the heat carrier in the material. From these factors, the silicon carbide has a number of high heat transfers and make the silicon carbide is a very attractive material for high temperature application.

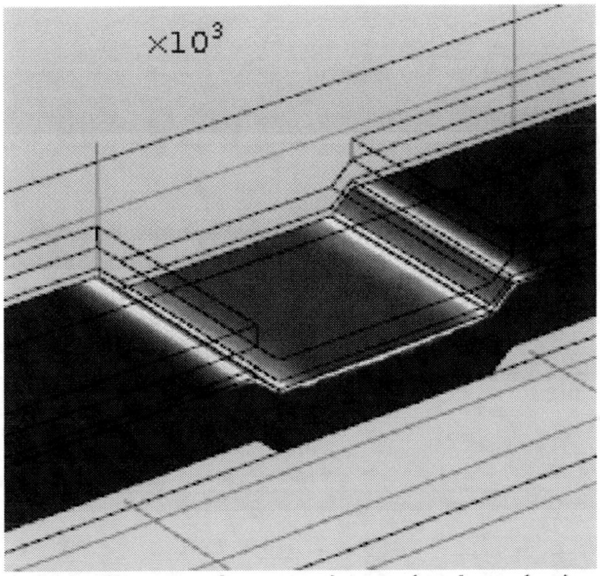

Figure 3(a): Changes of memory layer when heated using SiC

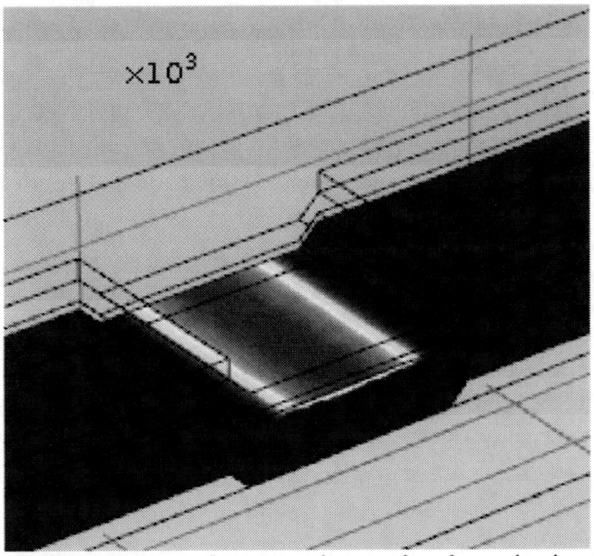

Figure 3(b): Changes of memory layer when heated using $TiSi_3$

The result of simulation can be seen in figure 3 below. The voltage boundary can be obtained when the memory layer will be changed from the amorphous to crystalline state for different heater. Figure 3 (a) shows the changes in memory layer when it is heated using SiC. Figure 3 (b) shows the changes in memory layer when it is heated using $TiSi_3$. From the figure 3(a), the changes of memory layer are uniformly on the surface when using SiC as a heater compared with the figure (b) using $TiSi_3$ as a heater. Based on the figure 3(a) and 3(b), the changes in memory layer are a difference because the difference characteristic of the heater. When using the silicon carbide as a heater, the changes of the memory layer is wider compared with the $TiSi_3$ as a heater. When there is a change in memory layer from amorphous to crystalline state, the data can be stored. That's time, the resistance of chalcogenide material is low and it is also known as SET state. These changes allow data to be stored and this process requires a voltage lower than RESET process.

The voltage boundary can be calculated using the following equation:

$$E = \frac{\Delta V}{r} \dots\dots\dots\dots\dots (1)$$

Where the E is Electrical Field for the memory layer, ΔV is different voltage at the 2 point and r is distance between different voltage points. From this equation, the different voltage can be obtained from the applied voltage at electrode and the voltage boundary at the memory layer. The distance can be obtained between the point of the applied voltage and point of the voltage boundary. The value of electrical field can get from the simulation. From this simulation, the result can be concluded as a Table 2 below

Table 2 : The result of simulation between two Heater, Silicon carbide and Titanium Sitride

Material of Heater	Applied Voltage (V)	Temperature (K)	Electrical Field (V/m)	Voltage Boundary (V)	Current Density (A/m^2)
SiC	0.91	463	16531	0.905	$3.31X10^6$
$TiSi_3$	1.11	459	20302	1.103	$4.06X10^6$

Based on result of simulation, the Silicon carbide (SiC) is a good heater than Titanium Sitride ($TiSi_3$). The voltage boundary when using SiC as a heater is 0.905V. That's mean, the voltage drop when using SiC as a heater is 0.005V equal 0.55% compared with the voltage drop using TiSi is 0.007V equal 0.63%. Using SiC, the voltage can be reduced 0.2V equal 18% voltages can be saved because Silicon Carbide has a high thermal conductivity and high temperature stability, making it suitable semiconductor material for operating at a high temperature. When the voltage can be saved, automatically the power can be reduced and make the structure more applicable.

4. Conclusion

In conclusion, the silicon carbide (SiC) can be categorized as a good heater compared with the titanium sitride ($TiSi_3$). The applied voltage when using SiC and TiSi3 as a heater are 0.91V and 1.11V respectively. With lower applied voltage, the memory layer will be changed from amorphous to crystalline state. The temperature of memory layer will be high when using SiC as a heater compared with the $TiSi_3$ as a heater. From the calculation, the voltage boundary can be obtained and the voltage boundary lower when using SiC as a heater.

References

[1] Wong H-SP, Raoux S, Kim S, Liang J, Reifenberg JP, Rajendran B, et al. 2010 *Phase ChangeMemory.* Proc IEEE **98(12)** 2201–27.

[2] Lacaita a. L. 2006 *Phase change memories: State-of-the-art, challenges and perspectives. Solid State Electron.* **50(1)** 24–31.

[3] Hu Y, Feng X, Zhai J, Wen T, Lai T, Song S, et al. 2014 *Superlattice-like Ge 8 Sb 92 / Ge thin films for high speed and low power consumption phase change memory application.* **93** 4–7.

[4] Liu J, Yu B, Anantram MP. 2011 *Scaling analysis of nanowire phase-change memory.* **32(10)** 1340–2.

[5] Crespi L, Ghetti A, Boniardi M, Lacaita AL. 2014 *Electrical conductivity discontinuity at melt in phase change memory.* **35(7)** 747–9.

[6] Yin Y, Alip RI, Zhang YL, Hosaka S. 2012 *Material Engineering for Low Power Consumption and Multi-Level Storage in Lateral Phase-Change Memory.* **490-495** 3286–90.

[7] Irma R, Kobayashi R, Zhang YL. 2013 *A Novel Phase-Change Memory with a Separate Heater Characterized by Constant Resistance for Multilevel Storage.* **534** 136–40.

[8] Daouahi M, Omri M, Ghani A, Kerm Y, Al-agel FA, Rekik N. *2015 Spectrochimica Acta Part A : Molecular and Biomolecular Spectroscopy An investigation on the effect of high partial pressure of hydrogen on the nanocrystalline structure of silicon carbide thin films prepared by radio-frequency magnetron sputtering.* **136** 1409–17.

[9] Amirthan G, Udaya A, Balasubramanian M. 2011 *Thermal conductivity studies on Si / SiC ceramic composites.* **09** 003.

[10] Wang D, Xue C, Bai H, Jiang N. 2015 *Silicon carbide nanowires grown on graphene sheets.* **12** 117

Predictions of Energy Content in Stretched Ribbons of Segmented Polymer Piezoelectrics

K. Prakash, S. Ramgopal[1*], Sanjiv Sambandan[1*]

1. Department of Instrumentation and Applied Physics, Indian Institute of Science, Bangalore 560012, India

*Email address of corresponding author: sanjiv@iap.iisc.ernet.in

Abstract

Polymer piezoelectrics such as poly vinyldene fluoride (PVDF) offer advantages such as low temperature fabrication into large area flexible sheets. A primary advantage of this is that large sheets of PVDF can be stretched into membranes to build sensors and energy harvesters that have low resonant frequencies. A key result reported recently is that when polymer piezoelectrics membranes are provided non-uniform stress, the use of segmented electrodes helps improve energy content by reducing surface currents. Using this key result in conjunction with experiments performed on stretched ribbons of PVDF, we predict energy densities obtained via electrode segmentation in this paper.

Keywords: energy harvesting, polymer electronics, poly-vinyldene fluoride

1. Introduction

Polymer piezoelectrics such as poly vinyldene fluoride (PVDF) are easily fabricated into large sheets at low temperatures (Kawai, 1965). Stretching these large sheets of PVDF into large membranes allows for the design of sensors and energy harvesters that have a low resonant frequency (Erika, 2005; Sodano, 2004). Freely available source in nature is mechanical vibration; using these natural vibrations energy harvesting systems are proposed. Many applications have been developed and are proposed to use piezo-polymers as sensors in the research (Ottman, 2002; Meninger, 2001; Kaya, 2007). Considering the low weight of these materials compared to electromagnetic based systems (also used for low frequency), there are distinct advantages in using PVDF based piezoelectrics for wide area sensor networks.

The study of stretched membranes of polymer piezoelectrics for energy harvesting has been considered recently and two aspects of the study are of interest in this paper (Kodali, 2013; Kodali, 2014). First, when polymer piezoelectric PVDF was stretched into ribbons of various lengths and subject to periodic vibration, it was found that the open circuit voltage of the piezoelectric reduced with length. When the frequency of the incoming vibration equaled the resonant frequency of the ribbon, the amplitude of oscillations of the ribbon maximized and the energy output maximized. With such a setup it was shown that a power of 2 mW/gm could be delivered at resonance. This power was shown to be sufficient to power a CMOS ring oscillator at 10 kHz (Kodali, 2013). Second, it was shown the energy content and power delivery of piezoelectric ribbons improved when the contact electrodes were segmented. This was due to the reduction in the loss of energy due to surface currents (Kodali, 2014).

In this paper we use the above two pieces of information to make predictions on the energy content and power delivery in 1D-stretched ribbons with segmented electrodes, subject to periodic vibrations at resonant frequency.

2. Method

2.1 Experimental Setup

Experiments were performed on ribbon like membrane of PVDF, 9 cm long and 3 mm wide, stretched across two supports as shown in Fig. 1(a). Studies were performed using periodic pulse inputs that were provided with a stepper motor and a computer controlled audio speaker coupled to the shaft. Fig. 1(b) illustrates the peak-peak voltage across the 9 cm ribbon as a function of frequency. The peak-peak voltage maximizes at around 700 Hz, which was resonance. The typical time domain response of the piezoelectric for 9 cm and 17 cm ribbon for a pulse like input is also shown figure 1(c), 1(d). In these experiments the typical amplitude of the vibrating ribbon was about 200μm.

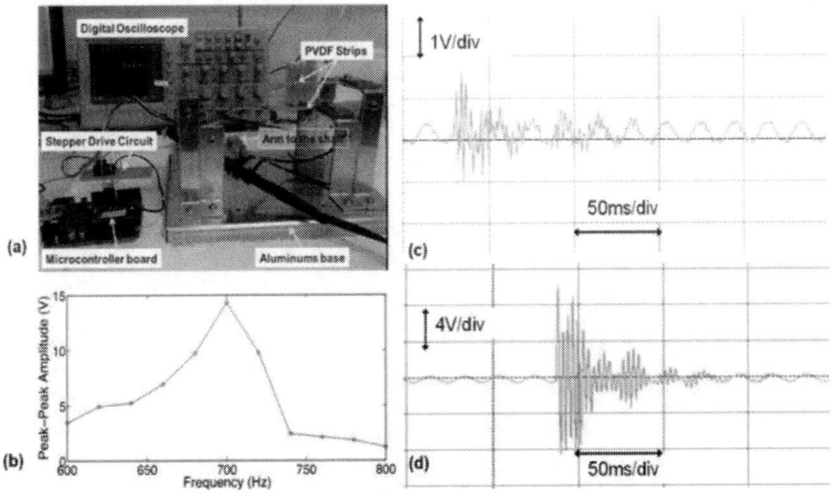

Figure 1: Experimental setup, time domain response for the 17cm ribbon, frequency domain response for the 9cm ribbon.

Under loading the typical current voltage characteristics for the 9cm ribbon under resonance was found to be typically as shown in Fig. 2. The peak power deliver is about 80μW which translates to about 2 mW/gm of power density.

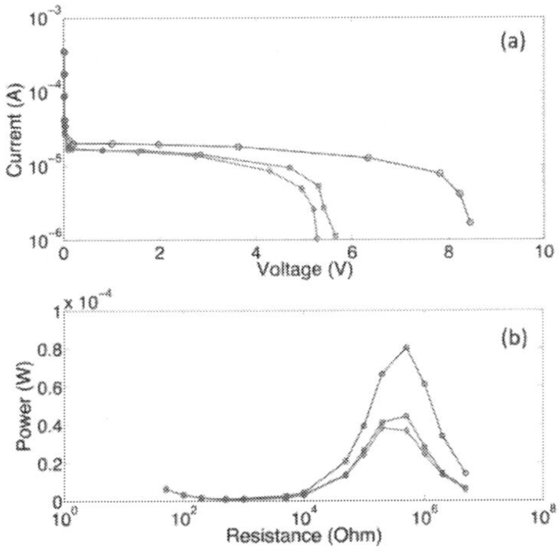

Figure 2: Current Voltage characteristics of the 9cm ribbon under loading and at resonance.

2.2 Theoretical Basis

Using this setup, the open circuit voltage for various lengths are obtained when the ribbon is provided periodic vibration at resonance. The theoretical variation of the peak open circuit voltage at is expected to be

IOP Conf. Series: Materials Science and Engineering **99** (2015) 012004 doi:10.1088/1757-899X/99/1/012004

$$V_{oc} \sim \frac{d_{31}Yh}{4\varepsilon}\left(\frac{a_0}{l}\right)^{\gamma} \tag{1}$$

Here d_{31} is the piezoelectric coefficient in the 31 direction, Y the modulus of elasticity, l the length, a_0 the amplitude of vibration, h the thickness of the ribbon, and ε the permittivity of PVDF. However in reality the actual open circuit voltage may vary as

$$V_{oc} \sim \frac{d_{31}Yh}{4\varepsilon}\left(\frac{a_0}{l}\right)^{\gamma} \tag{2}$$

Where a_0 is an exponent coefficient. This implies the energy content of the ribbon defined as the energy stored in the ribbon under open circuit condition varies as

$$E = \frac{CV_{oc}^2}{2} \sim \frac{d_{31}^2Y^2lwh}{32\varepsilon}\left(\frac{a_0}{l}\right)^{2\gamma} \tag{3}$$

Hence the energy density varies as

$$E_{\rho} \sim \frac{d_{31}^2Y^2}{32\varepsilon\rho}\left(\frac{a_0}{l}\right)^{2\gamma} \tag{4}$$

Where ρ is the mass density of PVDF.

From experiments on segmented piezoelectric ribbons, it was seen that the energy content of the ribbon varied as $1+A(n-1)/n$ where n is the number of segments, and 'A' a coefficient defining the properties of the piezoelectric (Kodali, 2014). We approximately implement this information with the above relation for energy density by stating that the energy density of the piezoelectric for n segmented electrodes is

$$E_{\rho}(n) \sim \frac{d_{31}^2Y^2}{32\varepsilon\rho}\left(\frac{a_0}{l}\right)^{2\gamma}\left(1+\frac{n-1}{n}\right) \tag{5}$$

When the piezoelectric ribbon is subject to non uniform stress, the charge stored in each region of the piezoelectric is different. This charge flows across a continuous electrode till the entire electrode surface reaches equipotential. This movement of charge (surface current) results in a loss in energy content due to Joule heating. As the electrode is segmented, the surface currents are reduced and the energy content improves.

We now perform experiments to determine γ so as to derive a complete relation for energy density and number of segments.

3. Results

Extracting the open circuit voltage for different lengths at their respective resonant frequencies, it was found that the open circuit voltage decreases as $l^{-0.83}$ as described in Figure 3, with l being the length of the ribbon.

IOP Conf. Series: Materials Science and Engineering **99** (2015) 012004 doi:10.1088/1757-899X/99/1/012004

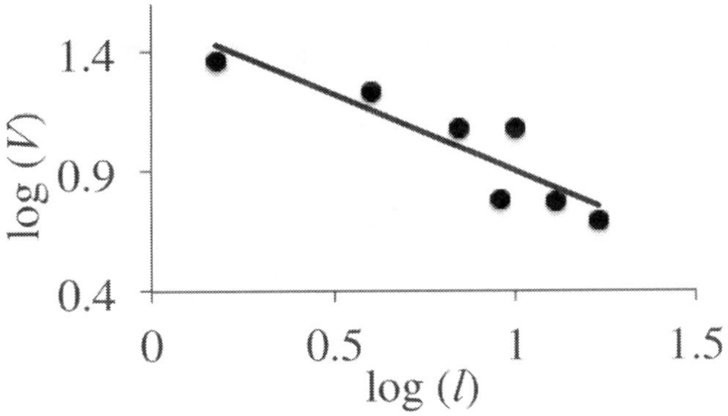

Figure 3: The open circuit voltage varies as $l^{-0.83}$

The energy content improvement for segmented electrodes is shown in Fig. 4. The saturative behavior in energy content fits the definition of the dependence on n as shown in Eq. 5 reasonably well.

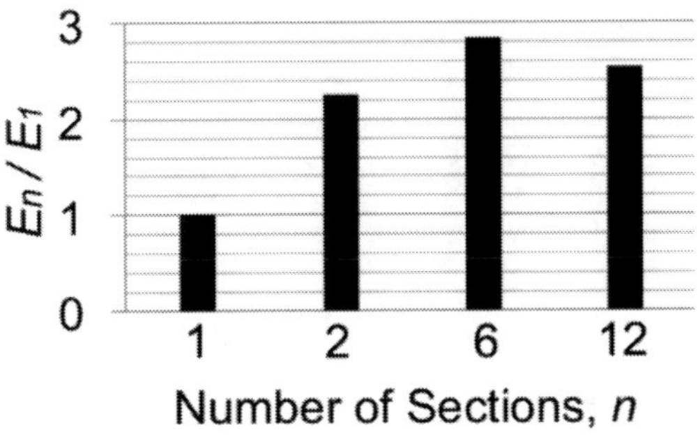

Figure 4: Energy content improvement with electrode segmentation.

4. Discussions

Based on the results observed in Fig. 3 and Fig. 4, and using the value of γ=0.83 in Eq. 5 we obtain the complete model for predicted energy content of the piezoelectric ribbon under tension and non uniform stress with n electrode segments as,

$$E(n) \sim \frac{d_{31}^2 Y^2}{32\varepsilon\rho}\left(\frac{a_0}{l}\right)^{1.66}\left(1+\frac{n-1}{n}\right) \quad (6)$$

For poled PVDF d_{31}~10 pC/N, Y~2 GPa, ε~8.854 10^-12 F/m and ρ~1780 kg/m^3. Using these parameters in Eq. 6, Fig. 5 shows the predicted variation in energy content in with a_0/l and n.

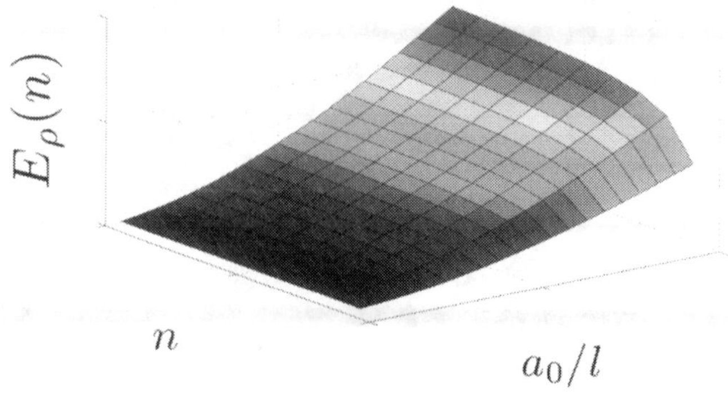

Figure 5. Prediction of energy content based on Eq. 6.

Conclusion

The study of segmented PVDF piezo effect on the efficiency of energy harvesting was carried out. The experimental setup to arrange energy scavengers of different lengths with different segments were studied to realize the energy efficiency. Responses of desired segments of selected strips were simulated by means of the pulsed tapping with stepper motor and its driving unit. Thus it concludes that energy content is improved and maximized with optimum number of segments.

Acknowledgements

The authors acknowledge the Department of Science and Technology for the funding received Grant No. DST/TSG/PT/2012/26.

References

Erika, M. (2005). Energy harvesting from vibration using a piezoelectric membrane, *J. Physique*, 128, 187-193.

Kawai, H. (1965). The piezoelectricity of Poly (vinylidene Flouride). *Jpn. J. of Appl. Phys.*, 8, 975-976.

Ottman, G K. (2002). Adaptive piezoelectric energy harvesting circuit for wireless remote power supply. *Power Electronics, IEEE Transactions on*, *17*(5), 669-676.

Meninger, S. (2001). Vibration-to-electric energy conversion. *Very Large Scale Integration (VLSI) Systems, IEEE Transactions on*, *9*(1), 64-76.

Kaya, T. (2007, September). A new batteryless active RFID system: smart RFID. In *RFID Eurasia, 2007 1st Annual* (pp. 1-4). IEEE.

Kodali, P. (2013). Harvesting energy at resonance from standing waves on polymer piezoelectric ribbon-like membrane. *Electronic Letters*, 49 (20), 1294-1296.

Kodali, P. (2014). Segmented Electrodes for Piezoelectric Energy Harvesters. *IEEE Electron Device Letters*, 35 (4), 485,487.

Sodano, H A. (2004). A review of power harvesting from vibration using piezoelectric materials. *Shock and Vibration Digest 36*, 197-205.

Capacitance Voltage of P3HT:Graphene Nanocomposites Based Bulk-Heterojunction Organic Solar Cells

Nur Shakina Mohd Shariff, Dr. Puteri Sarah Mohamad Saad, Prof Dr. Mohamad Rusop Mahmood
NANO-Electronic Centre, Faculty of Electrical Engineering, Universiti Teknologi MARA, Shah Alam, 40450, Malaysia

e-mail: kina_jb@yahoo.com

Abstract. After the discovery of conjugated polymer and bulk-heterojunction concept, organic solar cell has gain many interest in the photovoltaic world. The main problem for organic solar cells is that the power conversion efficiency (PCE) is still considered low even though it is much more low cost compared to inorganic solar cell such as Silicon (Si). Therefore, the objective of this research is to investigate the effect of Poly(3-hexylthiophene) (P3HT) thickness and concentration towards the capacitance voltage of the P3HT:Graphene solar cells. A simulation software called SCAPS is used in this research to simulate the effect on the solar cells. SCAPS is specialized for photovoltaic simulation studies. The solar cell's structure will be drawn inside the simulation and the parameters for each layers is inserted. The voltage range will be fixed and the capacitance voltage will be calculated by the software and all the results will be put into one graph. For thickness results, P3HT's layer at a thickness of 100nm has the lowest value of capacitance and clearly shows a peak at 0.86V. Where for the concentration, $1\text{x}10^{16}$ cm^{-3} is the only value that clearly shows there is the built-in voltage (V_{bi}) in the solar cells. Therefore, P3HT's thickness of 100 nm and concentration of $1\text{x}10^{16}$ cm^{-3} has the best overall results.

1. Introduction

Recent research has shown by adding Graphene into polymer solar cells can improve the solar cell's efficiency [1], [2], [3]. More understanding needed to be done on the physics of charge transport in order to improve the solar cells performance [4],[5]. To analyse the charge transport in organic semiconductor devices, capacitance (*C*) is being used. The capacitance-voltage characteristics can be give a better insights into these devices when combined with current density-voltage (*J-V*) data [6]. *C-V* measurements can also provide wealth info of the device characteristic and can determine the semiconductor's parameters. Recent research has showed that the built- in voltage (V_{bi}) of the device can be obtained at the peak in the *C-V* characteristic [7]. In *C-V* characteristic, it is usually observe that only one peak is visible when the polarity is changed from reverse to forward when the bias is applied through the organic semiconductor [8],[9],[10]. Deep traps also affect the capacitance through dynamic response to the small-signal frequency. This forms the capacitance – frequency (C-f) measurements, also termed as admittance spectroscopy, which relies on a frequency differential (where the DC bias is constant) and the AC small signal frequency is varied. Many research has been done on P3HT:Graphene composite solar cells but the power current efficiency is still low due to low photocurrent. The photocurrent is low because P3HT electron mobility is low ($\mu_{electron} = 10^{-14} - 10^{-14}$ cm^2/Vs) since P3HT is a p-type semiconductor [4],[11]. The incomplete absorption of the incident light power (P_{in}) in P3HT is also one of the factors that limits the solar cell's efficiency. Both parameters are affected by the thickness of the active layer [11],[12].

Poly(3-hexylthiophene) (P3HT), which has many excellent properties such as high electrical conductivity, high solubility in various solvent and can be deposit over a large area, under low deposition temperature and using low-cost deposition technique. The main problem for P3HT is it possesses a high electrical mobility ($\mu_{hole} = 0.2$ cm^2/Vs) [2],[11] in conjugated polymers but it is still low compared to the electrical mobility of inorganic solar cells. P3HT is p-type semiconductor and also a donor material, that is why it is high in hole mobility while the electron mobility is low.

Graphene will be used as the carbon polymorph composites and will be mixed with P3HT to form an active layer in the organic solar cells. Since the discovery of Graphene in 2004, many research on Graphene's properties has been done and it has been proven that Graphene has an excellent electrical characteristic such as an electron mobility of 10000 cm^2/Vs and has a higher mobility & conductivity than carbon nanotubes (CNTs). Graphene is also said to be an excellent electron acceptor [2]. Therefore Graphene will be an excellent combination with P3HT in order to achieve a higher power conversion efficiency. With the aid of Graphene, the mobility can be increase and will effect positively on the photocurrent.

A simulation software that is used in this research is call SCAPS. SCAPS was developed by Marc Burgelman from University of Gent, Belgium. This simulation program is to simulate the electrical characteristic of a thin film heterojunction solar cell. This simulation program will be used in this research to calculate the *C-V* characteristic of P3HT:Graphene solar cells.

The main propose of this research is to investigate the *C-V* characteristic of the P3HT:Graphene nano composite and see whether the presences of Graphene in P3HT effect the active layer's performance. The *C-V* measurements will be simulated using SCAPS and its characteristic will be measured.

2. Methodology

The structure of the solar cell is drawn in the software definition panel. The parameters such as the bandgap are fixed at 2.1 eV for P3HT layer and 0.25 eV for Graphene layer while electron affinity are fixed at 3.40 eV for P3HT layer and 4.5 eV for Graphene layer. For Graphene, the thickness will be fixed at 20 nm while the thickness of P3HT layer will be evaluated from 60nm to 100nm. For each thickness of P3HT, the carrier concentration will be evaluated from $1x10^{13}$ till $1x10^{16}$ cm^{-3}. The capacitance voltage will be calculated by the simulation software and all the result will be plotted into a graph.

3. Results & Discussion

The effect of the P3HT's concentration towards the device's capacitance is shown in Figure 1a). Here, the thickness of the active layer is fixed at 100nm and the concentration is evaluated from 1.0×10^{13} cm^{-3} to 1.0×10^{16} cm^{-3}. From Figure 1, only (A) shows a drastic increase at 0.7 V while (B),(C) and (D) slowly decrease as the voltage increase. (A) also shows there is a built-in voltage inside solar cells. The capacitance increases in the forward direction and reach a peak at 0.86V. It has been reported that the peak came from mobile holes [8].This might occur due to the concentration of P3HT which increase from 1.0×10^{13} cm^{-3} to 1.0×10^{16} cm^{-3}. P3HT is p-type semiconductor, therefore it is more rich in hole concentration. When the hole concentration increases, the capacitance value also increase.

Figure 1b) shows a closer look the capacitance at a carrier concentration of 1.0×10^{16} cm^{-3}. It can be seen that the capacitance decrease at first until it reach a voltage of 0.58V. Starting form 0.58V the capacitance increased rapidly till reach its peak at around 0.86V. The capacitance then decrease till 0.94V due to the limitation of the minority carriers concentration in the neutral regions of the device [7]

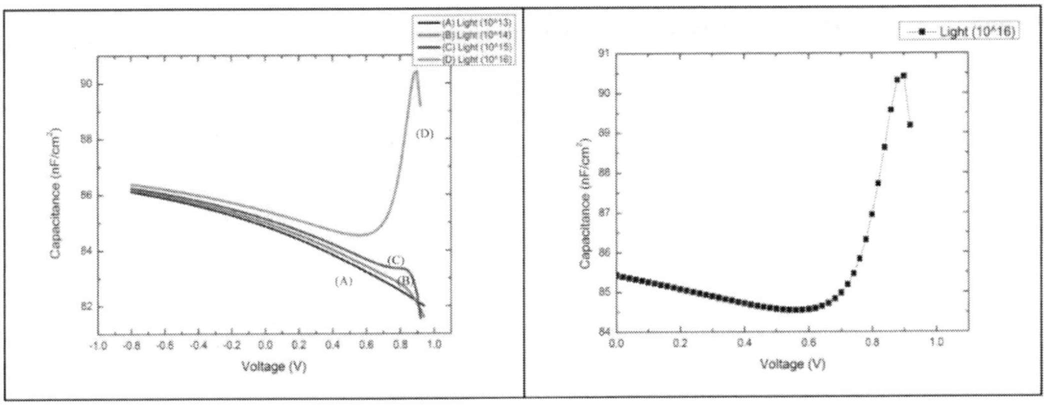

(a) (b)

Figure 1: a) shows the C-V characteristic for every concentration. b) C-V characteristic for concentration of 1.0 x 10^{16} cm^{-3}

Figure 2 show the effect of thickness towards the device's capacitance voltage. The voltage is fixed from -0.8V to 1.0V and the concentration is fixed at 1.0 x 10^{16} cm^{-3}. The thickness of P3HT is measured from 60 nm till 100 nm. As seen from Figure 2, capacitance is high when the thickness of the active layer is at 60 nm and the capacitance is at its lowest when the thickness is 100 nm. Therefore, the capacitance increases as the thickness decreases. There is a slight rise at 0.84V for very thickness but as the thickness decrease, the peak also decreases. As explained, the peak decreases due to the limitation of minority carriers concentration [7] in P3HT which is the hole concentration. When the thickness of the P3HT layer decreases, the hole concentration also decreases. This will lead to decrease in the peak. The peak for all thickness is also the same since it has been reported that the peak voltage is related to the bias voltage V_{bi} [6]. The capacitance is inversely proportional to the thickness of the active layer, therefore the capacitance is lower for thicker active layers.

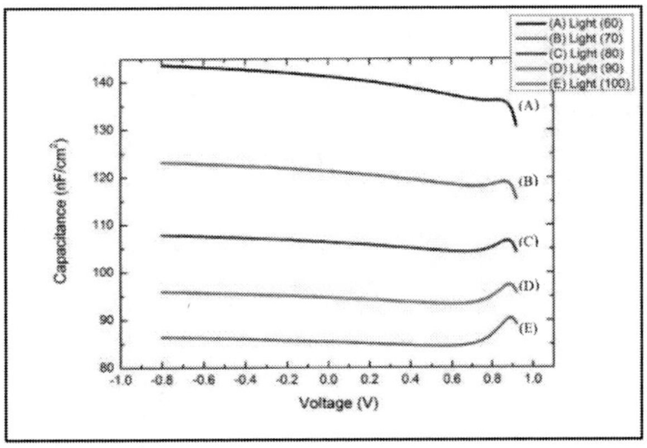

Figure 2: C-V characteristic for different active layer's thickness

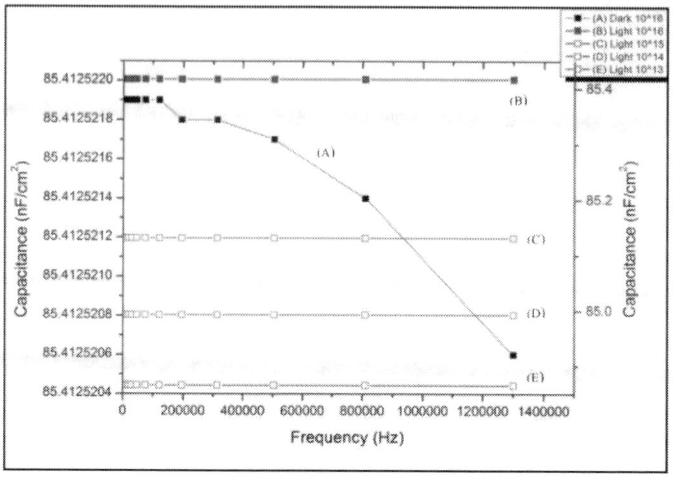

Figure 3: C-f characteristic for different P3HT concentration

From the capacitance – frequency (C-f) result shown in Figure 3, it can be seen that the highest capacitance value is when the concentration is at 1.0 x 10^{16} cm^{-3} and under illumination. The capacitance is not really affected by carrier concentration as the frequency increases and it shows that even the concentration increases from 1.0 x 10^{13} cm^{-3} to 1.0 x 10^{16} cm^{-3}, the capacitance range did not change a lot. (B) shows the highest capacitance value around 85.4125 nF/cm^2 while (E) has the lowest capacitance value at 84.832 nF/cm^2. The capacitance value remains constant when the active layer is under illumination occurs due to fixed temperature in the simulation program.

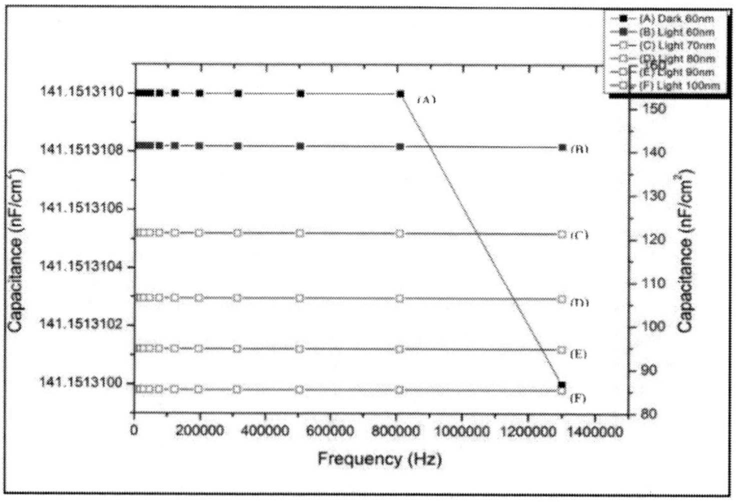

Figure 4: C-f characteristic at different thickness

Figure 4 shows the effect of the active layer's thickness towards the capacitance of P3HT:Graphene solar cell. The capacitance decrease as the active layer's thickness increases. The thickness of 60nm has the highest capacitance value while a thickness of 100nm has the lowest. As seen in Figure 4, the capacitance of P3HT:Graphene stays static as the frequency increases but the capacitance value between (B) and (F) has a larger gap compared to the one in Figure 3. This shows that the thickness has a greater effect towards the solar cell's capacitance value than the carrier's concentration. As shown in Figure 4, (A) shows a decrease in capacitance when the frequency increases. For all thickness under illumination, the graph is actually quite the same as (A) but because the decrease of capacitance is really small, it seen as a straight line. From the graph, it can be concluded that the thickness of the thin film has a small effect on the capacitance as the frequency increases because capacitance is mainly affected by the temperature [13]. The reason why the capacitance decrease when the active layer is thicker is because the distance between two electrodes of the solar cells increase, which cause the capacitance to slowly decrease.

4. Conclusion

In summary, P3HT at 100nm and the concentration at 1.0×10^{16} cm^{-3} shows the best results out of all the thickness and concentration of P3HT layer. This is because the peak at C-V characteristic appears really clearly at both results. Besides that, at forward bias, the capacitance value shows a strong sign that the minority charge carriers plays an important role. As the thickness increase, the capacitance decreases.

For C-f measurements, it can be concluded that the thin film's thickness & concentration have a little effect on the capacitance as the frequency increases. Capacitance is greatly affected by the temperature and in this simulation, the temperature is fixed. The capacitance decrease as the thickness of the active layer increases due to the increase of distance from the two electrodes. The thickness of the active layer will separate the electrodes even more, which leads to the decrease in capacitance.

5. Acknowledgement

Deepest gratitude to my Supervisor for all the help and support in this research. Also to Kementerian Pelajaran Malaysia and UiTM for the Tenaga Pengajar Muda (TPM) scholar.

References

[1] R. Bkakri, A. Sayari, E. Shalaan, S. Wageh, A. A. Al-Ghamdi, and A. Bouazizi, *Superlattices Microstruct.*, vol. 76, no. 0, pp. 461–471, 2014.
[2] V. Saini, O. Abdulrazzaq, S. Bourdo, E. Dervishi, A. Petre, V. G. Bairi, T. Mustafa, L. Schnackenberg, T. Viswanathan, and A. S. Biris, *J. Appl. Phys.*, vol. 112, no. 5, pp. 54327–54329, 2012.
[3] Q. Liu, Z. Liu, X. Zhang, L. Yang, N. Zhang, G. Pan, S. Yin, Y. Chen, and J. Wei, *Adv. Funct. Mater.*, vol. 19, no. 6, pp. 894–904, 2009.
[4] G. Garcia-Belmonte, A. Munar, E. M. Barea, J. Bisquert, I. Ugarte, and R. Pacios, *Org. Electron.*, vol. 9, no. 5, pp. 847–851, 2008.
[5] S. W. Tsang, S. K. So, and J. B. Xu, *J. Appl. Phys.*, vol. 99, no. 1, p. 13706, 2006.
[6] A. T. Mallajosyula, S. Sundar Kumar Iyer, and B. Mazhari, *Org. Electron.*, vol. 13, no. 7, pp. 1158–1165, 2012.
[7] S. L. M. van Mensfoort and R. Coehoorn, *Phys. Rev. Lett.*, vol. 100, no. 8, p. 86802, 2008.
[8] J. W. Zhang, Y. He, X. Q. Chen, Y. Pei, H. M. Yu, J. J. Qin, and X. Y. Hou, *Org. Electron.*, vol. 21, no. 0, pp. 73–77, 2015.
[9] D. C. Tripathi and Y. N. Mohapatra, *Appl. Phys. Lett.*, vol. 102, no. 25, p. 253303, 2013.
[10] P. Pahner, H. Kleemann, L. Burtone, M. L. Tietze, J. Fischer, K. Leo, and B. Lüssem, *Phys. Rev. B*, vol. 88, no. 19, p. 195205, 2013.
[11] A. J. Moulé, J. B. Bonekamp, and K. Meerholz, *J. Appl. Phys.*, vol. 100, no. 9, p. 94503, 2006.

[12] M. Lenes, L. J. A. Koster, V. D. Mihailetchi, and P. W. M. Blom, *Appl. Phys. Lett.*, vol. 88, no. 24, p. 243502, 2006.
[13] Y. Zhang, L. Li, S. Yuan, G. Li, and W. Zhang, *Electrochim. Acta*, vol. 109, pp. 221–225, 2013.

IOP Conf. Series: Materials Science and Engineering **99** (2015) 012006 doi:10.1088/1757-899X/99/1/012006

TiO_2 Thin Film via Sol-Gel Method: Investigation on Molarity Effect

Puteri Sarah Mohamad Saad, Hanis Binti Sutan, Shafinaz Sobihana Shariffudin, Hashimah Hashim, Uzer Mohd Noor

NANO-ElecTronic Centre, Faculty of Electrical Engineering, Universiti Teknologi MARA, 40450 Shah Alam, Selangor, Malaysia

puterisarah@salam.uitm.edu.my

Abstract. We have systematically investigated the current-voltage (*I-V*), absorbance and optical band gap of TiO_2 thin film prepared through varying the molarity of the TiO_2 precursor by sol-gel spin coating technique. In addition to the electrical and optical characteristics, the surface morphology was examined by using Atomic Force Microscope (AFM). From the image of the AFM, we were able to observe the uniformity of the TiO_2 thin film. From the experimental results, we found that the uniformity of the TiO_2 thin film is optimized at 0.2M sample. It is also found that, as the molarity increased, there is tendency of the resistivity to decrease. Not only that, the absorbance measurement and optical band gap also gave its best value for 0.2M sample. Therefore, in this work it is concluded that 0.20M of TiO_2 gave the best characteristics for all measurements.

1. Introduction

A lot of works were done on TiO_2 due to its interesting properties such as chemical stability, non-toxic, highly oxidative photogenerated holes, high energy conversion, high transparency and high refractive index [1]. Nanoparticles of TiO_2 are widely-known in many applications such as antireflection coatings for photovoltaic cells and passive solar collectors [2], photocatalytic refinement of polluted air or wastewater and excellent degradation for organic pollutants [3]. Moreover, the TiO_2 thin film deposited on conducting glass can be used in new types of solar cells which are liquid and solid dye-sensitized photo-electrochemical solar cells [4, 5].

There are several methods to prepare TiO_2 thin film. TiO_2 films can be synthesized by several deposition procedures including sol-gel [6, 7], chemical vapour deposition [8], electrophoretic [9] [10], screen printing [11] and sputtering method [12]. Among the methods available to prepare TiO_2 thin film, sol-gel is one of the most favorable methods. The advantages of using this method are able to vary the properties of the thin film, cheaper and convenient to coat on large areas [7]. Researchers found that the molarity effect and the preparation conditions such as solution preparation arrangements, ageing temperature and time taken [13] have effect on the TiO_2 properties.

In this work, the thin films were fabricated by a sol–gel method combined with spin coating technique on glass substrate. The concentration amounts of TiO_2 in the solution were kept at 0.01M, 0.05M, 0.10M, 0.15M, and 0.20M. The annealing procedures applied to achieve the higher electrical conductivity of the TiO_2 thin films. The effects of different molarity of the films are discussed.

2. Methodology

2.1. Glass Substrate Preparation

The glass was cut with the size of 2 cm x 2 cm. Then, it was cleaned with acetone in ultrasonic bath, followed by ethanol and lastly with di-ionized (DI) water for 10 minutes. The last step is blowing the glass substrate with N_2 gas for drying purpose.

2.2 TiO_2 Solution By Sol-Gel Method Preparation

The starting material used was Titanium (IV) Isopropoxide (TTIP). The molarity of the precursor was varied from 0.01M, 0.05M, 0.1M, 0.15M and 0.2 M. Glacial acetic acid and Triton-X were added to the solution as stabilizer and surfactant respectively. Small amounts of DI water were also added to increase the hydrolysis reaction rate.

2.3 Deposition of TiO_2

The TiO_2 solution was deposited on the glass substrate using spin coating technique with 6000 rpm for 1 minute and 36 seconds. The deposition was done for 5 times with 10 drops of TiO_2 solution. The deposited thin film was then dried at 150° C for 10 minutes to vaporize the solvent. The drying process was repeated for 5 times. It was then undergoes the annealing process at 450° C for 1 hour to restructure the crystallinity of the thin film.

2.4 Metal contact deposition

A metal contact of gold (Au) 99.9% with 60nm thickness was deposited on TiO_2 thin film. The deposition was conducted using metal sputter coater. The sputter coater was set with current 50mA and time taken was 4 minutes. Argon gas was purged into the chamber with pressure $1x10^{-1}$ mbar. Au was chosen rather than other metal because of its properties; the work function difference between thin film and metal contact materials.

2.5 Sample Characterization

The samples were characterized using atomic force microscope (AFM, Park System XE-100) for structural images. Meanwhile, the thicknesses have been measured using Dektak 150 Stylus Veeco Dektak 750 Surface Profiler. The optical properties were studied using UV-Visible (Vis)-near infrared. Current-voltage measurement has also been carried out for all the samples using two point probes Bukoh Keiki (CEP-2000) Spectral Sensitivity Analyzing System Solar Simulator.

3. Results and discussions

3.1 Surface morphology and thickness

Table 1 shows the thickness of TiO_2 thin film with different molarity from 0.01M, 0.05M, 0.10M, 0.15M and 0.20M.The thickness are increasing from 9.24 nm, 19.9 nm, 29.0 nm, 60.6 nm and 79.7 nm respectively. It is found that as the thickness increase it will affect the electrostatic contact between solute particles [14]. The particles become larger and resulted in increasing the probability of more solute particles to be grouped creating agglomerate particles. This result was supported by structural analysis.

Table 1: Thickness of TiO_2 Thin Film

Molarity (M)	Thickness (nm)
0.01	9.24
0.05	19.9
0.1	29.0
0.15	60.6
0.2	79.7

Figure 1 shows the AFM images of the TiO_2 thin films. The images shown are for samples at 0.01M and 0.2M. The lowest molarity give the less roughness with 2.2nm while the higher molarity 0.20M gives the best uniformity and highest roughness about 7.66nm. This was probably due to scattering mechanism in which influence the particle size. Uniformity of the particle size across the thin film is important because the nanoparticle size significantly affects the electrical conductivity of the film [15]. With the highest uniformity of the thin film, there are providing higher surface contact of the TiO_2 thin film and resulted in higher density and in return, producing high conductivity of the thin film.

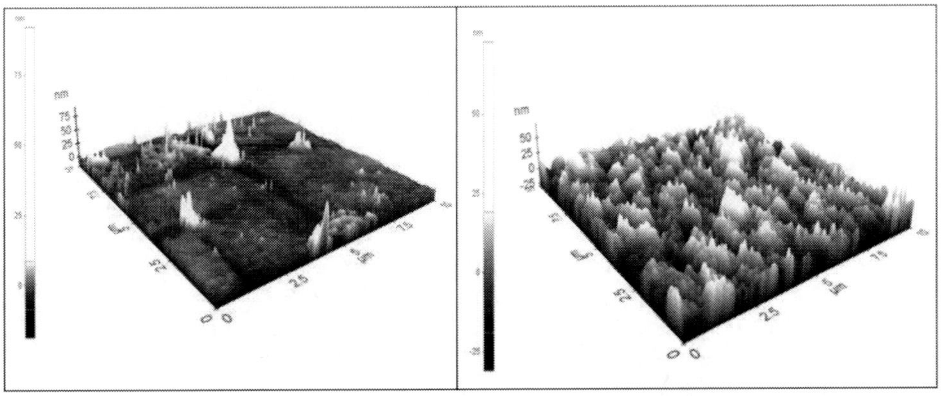

Figure 1: AFM images for a) 0.01 M, b) 0.2 M

3.2 Electrical Properties

Figure 2 shows the current-voltage (*I-V*) characteristics under illumination (100 mWcm^{-1}, AM 1.5) of the TiO_2 thin film with different molarity. The TiO_2 samples exhibit a linear and symmetrical curve, indicating ohmic contact characteristics between the thin films and the Au metal contact [16]. From the figure, we could see that the molarity strongly influences the *I-V* characteristics. When the molarity of TiO_2 is 0.2 M, the thin film exhibited the highest current (4.5×10^{-8} A). On the other hand, the current at 0.1 M is approximately 0.5×10^{-8} A. The electrical resistivity of the thin film was calculated from the *I-V* curve and was plotted and shown in Figure 3. The resistivity decreases with the increment of molarity and reaches a minimum value at 0.2M. The high density of TiO_2 causing an increase of electron carrier concentration. This will contribute to the high electrical conductivity.

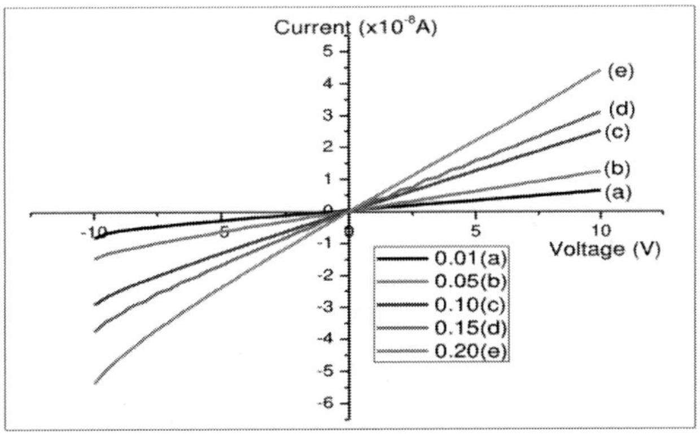

Figure 2: *I-V* measurement plots of TiO_2 thin film

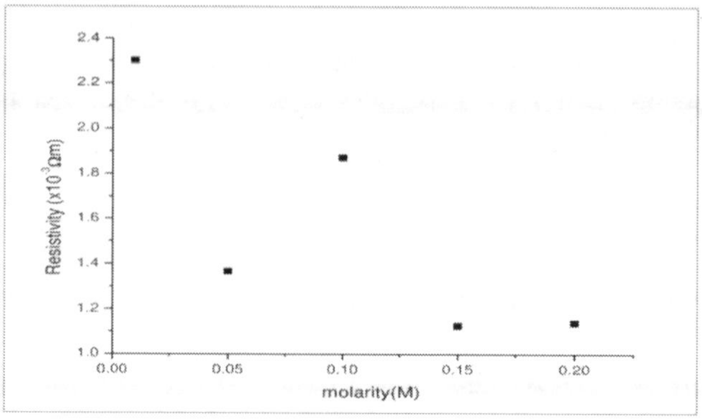

Figure 3: Resistivity of TiO_2 thin film with different molarity

3.3 Optical Properties

Figure 4 represents the absorbance spectrum of the prepared samples. High molarity and low molarity concentration samples show comparable absorbance spectrum behavior. When the molarity concentration reaches 0.2M, the absorbance shows a significant increment in the Vis region. It is a known fact the molarity increment can improve the absorbance of TiO_2 by relating it with high density of TiO_2 particles. High molarity concentration of TiO_2 can produce a much thicker thin film as in Table 1 [17]. Therefore, the photon energy will be absorbed more, thus increasing the absorbance value in the Vis region. Minimum absorbance value implies that electron did not absorb the photon energy. This means that the electrons are transmitted or reflected [18]. Meanwhile, Figure 5 shows the optical band gap for 0.1M and 0.2M samples. The optical band gap energy (Eg) is derived assuming direct transition of electron between the edges of the valence band and the conduction band. From the extrapolation of the graph in Figure 5, the band gap energy for 0.01M and 0.2 M are 3.9eVand 3.7eV respectively. The value for optical band gap that has been reported elsewhere is 3.2 eV [19]. With the molarity concentration being analyzed in parallel with the optical band gap properties, the significant relationship between this parameter can be concluded; the results showed that high molarity exhibited lower optical band gap.

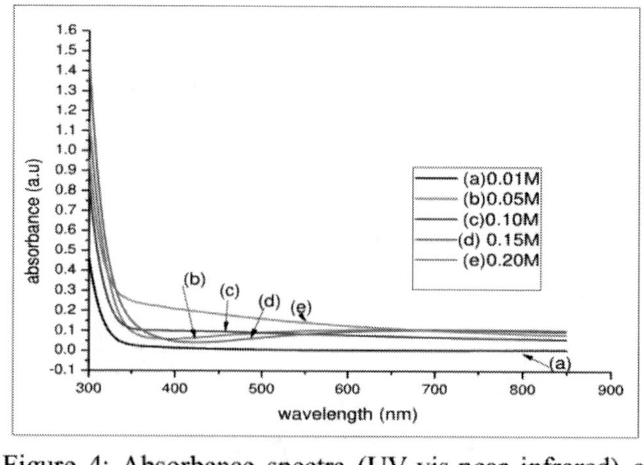

Figure 4: Absorbance spectra (UV-vis-near infrared) of prepared samples

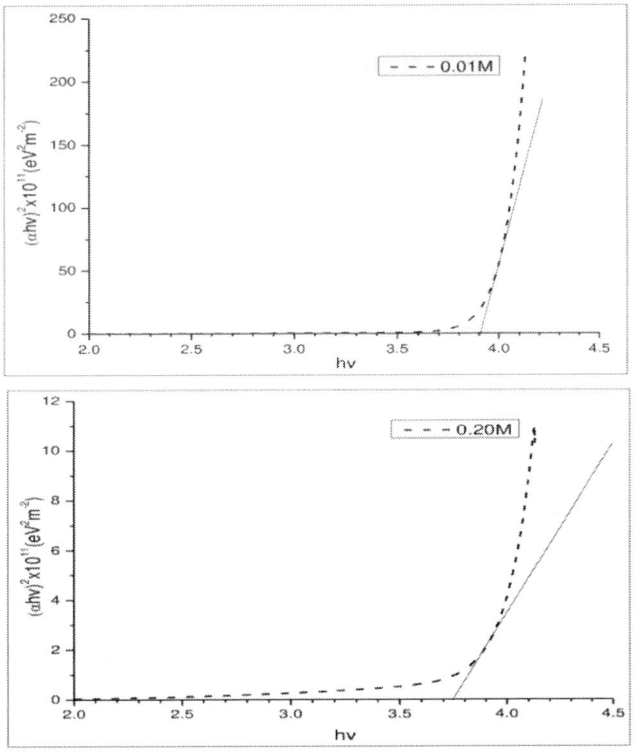

Figure 5: Optical band gap for 0.01M and 0.2M samples

Conclusion

As a conclusion, TiO_2 thin film with different molarity has been successfully prepared using the solution based method. 0.2M shows the roughest surface as compared to 0.01M. Not only that, *I-V* results also shows the highest current for the same sample. In conjunction to that, the results for the rest of the characterizations also show that 0.2M gives the same trends.

Acknowledgement

The authors would like express their deepest appreciation to all the staffs of NANO-ElecTronic Centre (NET) and NANO-SciTech Centre (NST) for their assistance in completing this work. Special thanks also go to the Faculty of Electrical Engineering, UiTM Shah Alam for the funding.

References

[1] A. Fujishima, T. N. Rao, and D. A. Tryk 2000 *Journal of Photochemistry and Photobiology C: Photochemistry Reviews* **1** 1-21.
[2] A. S. R. A. Shanaghi, T. Shahrabi, and M. Aliofkhazraei 2008 *Journal of Materials Science* 233-247.
[3] X. X. Wei Wu, Shaofeng Zhang, Feng Ren, and C. Jiang 2011 *Nanoscale Research Letters* **6** 1-15
[4] M. G. B. O'Regan 1991 *Nature* **353** 737.
[5] D. L. U. Bach, P. Comte, E. Moser, F. Weissortel, Salbeck,H. Spreitzer, M. Gratzel 1998 *Nature* **395** 583.
[6] M. J. Alam and D. C. Cameron, *Surface and Coatings Technology* 2001 **142-144** 776-780.

[7] S. Sung, S. Park, W.-J. Lee, J. Son, C.-H. Kim, Y. Kim, *et al,* 2015, *ACS applied materials & interfaces,* **7**, 7456-7461.
[8] M. L. Hitchman, Tian 2002 *J. Electroanal. Chem* F. **538 – 539** 165 – 172.
[9] J. A. Bryne, Davidson, A., Dunlop, P. S. M., Eggins, B. R. J. 2002, *Photochem. Photobio. A: Chem,* **148** 365 – 374.
[10]A. Dakka, Lafait, J., Lefdil, M. A., Sella, C. 1999 *M.J. Condensed Matter* **2** 2 – 4.
[11]T. Ma, Kida, T., Akiyama, M., Inoue, K., Tsunematsu, S., Yao, K., Noma, H., Abe 2003 *E. Electrochemistry Communications* **5** 369 – 372.
[12]M. J. Alam, Cameron, D. C. 2001 **142 – 144** 776 – 780.
[13]J. Y. L. C. M. Shin, J. H. Heo, J. H. Park, C. R. Kim, H. Ryu, J. H. Chang, C. S. Son, W. J. Lee, S. T. Tan, J. L. Zhao, and X. W. Sun 2009 *Applied Surface Science* **255** 8501–8505.
[14]C. Firdaus, M. S. Rizam, M. Rusop, and S. R. Hidayah 2012 *Procedia Engineering* **41** 1367-1373.
[15]C. Lin, C. Hung, M. Venkateswarlu, and B. Hwang 2005 *Journal of power sources* **146** 397-401.
[16]L. M. Porter and R. F. Davis 1995 *Materials Science and Engineering: B* **34**, 83-105.
[17]L. N. Lau, N. B. Ibrahim, and H. Baqiah 2015 *Applied Surface Science* **345** 355-359.
[18]A. W. Nuayi, H. Alatas, I. S. Husein, and M. Rahmat 2014 *International Journal of Optics* 8.
[19]K. Madhusudan Reddy, S. V. Manorama, and A. Ramachandra Reddy 2003 *Materials Chemistry and Physics* **78** 239-240.

Preparation and Characterization of Nanostructured CuO Thin Films using Sol-gel Dip Coating

S.S. Shariffudin, S.S. Khalid, N.M. Sahat, M.S.P. Sarah, H. Hashim

Faculty of Electrical Engineering, Universiti Teknologi MARA, 40450 Shah Alam, Selangor, Malaysia

sobihana@salam.uitm.edu.my

Abstract. Nanostructured CuO thin films were deposited onto quartz substrates by sol-gel dip coating technique. The precursor solution was prepared by dissolving copper acetate powder into isopropanol with molarity of 0.25M. Preheating and annealing temperature were fixed at 250°C and 600°C respectively. This study focused on various film thicknesses by varying the frequent number of deposited layers. The effect of thickness on electrical, surface morphology and optical properties of CuO thin film were studied. The surface morphology was examined using field emission scanning electron microscopy (FE-SEM), surface profiler for thickness measurement, optical properties of CuO thin film were characterized by using ultraviolet-visible spectroscopy (UV-VIS) for transmittance and absorbance, and the electrical property was examined by using two point probes method. The films were found to be denser at higher film thickness due to lesser porous observed on the surface. The thickness of these CuO thin films varied from 87.14 – 253.58 nm and the direct band gap energy was observed in between 1.9 to 2.35 eV. Lowest resistivity was found for sample with a thickness of 253.58 nm.

1. Introduction

Copper oxide is known to exist in two semiconducting phases, namely cupric oxide or copper (II) oxide (CuO); and cuprous oxide (Cu_2O) [1]. An intermediate compound between the previous two, a metastable copper oxide, (Cu_4O_3) has been also reported [2]. CuO, which has a band gap between 1.4 – 2.1eV [3], exhibits versatile range of applications [4]. CuO crystallizes in a monoclinic structure with the lattice parameters of a=4.684 Å, b=3.425 Å, c=5.129 Å, and β=99.28°. CuO thin film has transmittance of 20% in the visible spectrum and it could reach 90% for the high wavelength of the visible region with 2 to 2.5 refractive index value [5]. CuO thin films were reported as p-type conductivity due to copper vacancies in the structure [6].

CuO is a promising semiconductor for solar cell fabrication, such as a solar absorber [7] due to high solar absorbency and a low thermal emmitance. In addition, this compound has been employed for the fabrication of glucose sensors [8], as anodes in lithium-ion batteries [9], gas sensor [10], and catalysts [11]. An important advantage of using CuO is that it is non-toxic and its constituents are available in abundance.

In preparation of CuO thin films, various techniques have been used such as spray pyrolysis, chemical vapor deposition, electro deposition, thermal oxidation, sputtering process and also sol-gel [12] and so forth. Among these techniques, the sol-gel technique has been utilized as a potentially useful technique for preparation of even nanoparticles CuO because this technique very simple and low cost technique, which requires no sophisticated specialized setup [13]. Other than that, dip coating represents an efficient route to coat large surfaces [14].

This study focused on preparation of CuO thin film by sol-gel dip coating technique by varying the number of layers deposited to examine the thin film thickness. The effect of various film thicknesses to physical, optical and electrical properties of CuO thin film was studied and discussed in this paper.

2. Experimental

Several steps were involved in the preparation of the CuO thin films such as preparation of solution and substrates, deposition process, preheating, annealing process and characterization. The steps are explained in details as below.

2.1 Quartz Substrate Preparation & Solution Preparation

Quartz substrate was immersed in the ultrasonic bath machine for 10 minutes using certain chemicals that was used to remove the contaminants. Firstly, quartz substrate was cleaned using acetone, followed by ethanol and lastly with DI water. Then, the quartz substrates were blown with argon gas to dry it.

The starting precursor, copper acetate powder was considered based on molar concentration of 0.25M and dissolved into (di) ethanolamine (DEA) and isopropanol (IPA). Then, (poly) ethylene glycol was added as a stabilizer. The solution was stirred on a hot plate stirrer for 10 minutes to dissolve all the ingredients. The final solution was dark blue and clear without any suspension of particles.

2.2 CuO Thin Film Deposition and Characterization

Quartz substrates were dipped into the solution with a constant withdrawal speed 2mm/s. This deposition process was repeated for one to five times to achieve various thicknesses of the thin films. The samples were named as S1 for 1 layer, S2 for two layers, S3 for three layers, S4 for four layers and S5 for five layer films. After the deposition process, the CuO thin film was dried in the furnace at 250°C for 10 minutes immediately. Lastly, the CuO thin film was annealed at 600°C for 60 minutes for purpose of recrystallization.

For the measurement of surface morphology, field emission scanning electron microscopy (FE-SEM; model: JEOL JSM-7600F) with accelerating voltage of 5 kV was used. The film thickness was measured using surface profiler (model: Veeco Dektak 750) while to observed transmittance and absorbance, ultraviolet-visible spectroscopy (UV-VIS; model: JASCO V-670) was used. Gold metal contact with a thickness of 60 nm was sputtered onto the surface of the films. For *I-V* characteristic, two point probes system (solar simulator; model: CEP 2000) was used.

3. Result and Discussion

3.1 Physical Properties

FE-SEM micrographs were taken to investigate the surface morphology of the CuO films and are embedded in Fig. 1. It is clearly seen that the surfaces of the films are free of cracks. A strong influence of the film thickness to the morphology of the CuO is clearly observed. The films morphologies showed a homogeneous distribution of grains on the substrates with high porosity at lower film thickness. It is also observed that the films became less porous with the increased of the film thickness; therefore the surface of the films became denser. Changes in the film thickness also affect to the CuO grain size, where the average grain size was found to increase with the thickness and is shown in Table 1.

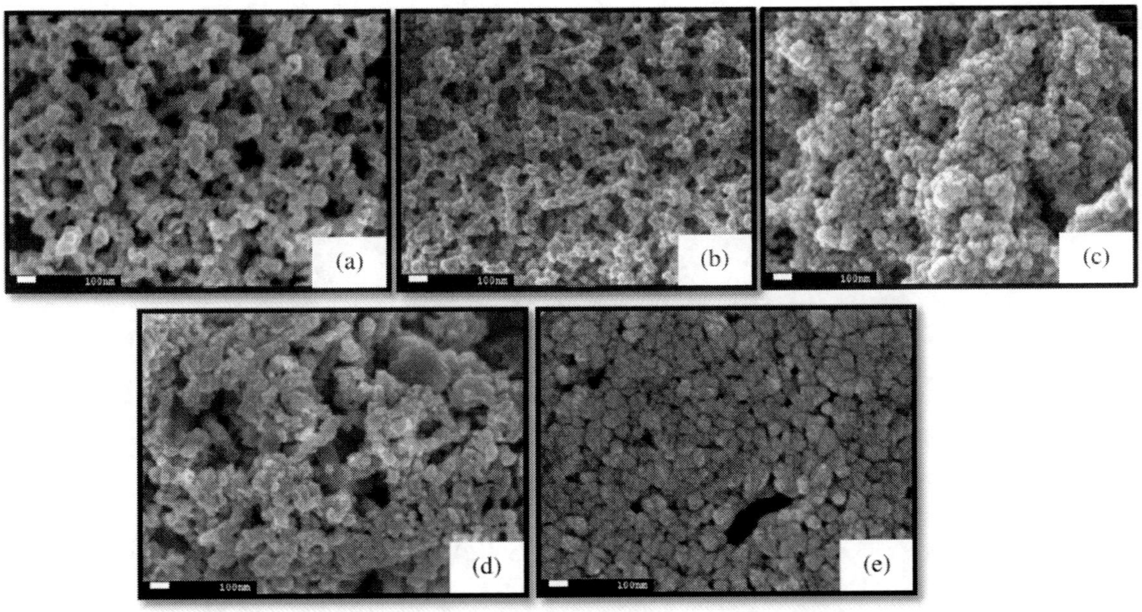

Fig.1: FE-SEM micrographs of CuO thin film at withdrawal speed of (a) S1, (b) S2, 9c) S3, (d) S4, and (e) S5

The thickness of the CuO thin films was measured and presented in Table 1. It is observed that the thickness of the CuO thin films increased when withdrawal speed decreased. The highest thickness was achieved at sample S5 with a thickness of 253.58 nm.

Table 1: Thickness of CuO thin film at different withdrawal speed

Samples	S1	S2	S3	S4	S5
Grain size (nm)	25.5	26.7	26.5	38.6	40.5
Thickness (nm)	87.14	95.22	105.06	209.29	253.58

3.2 Optical Transmittance and Absorbance

Transmittance of CuO thin film was observed by using UV-VIS spectrometer as shown in Fig. 2. Transmittances of the thin films were sensitive to the film thickness, where the highest average transmittance in the visible wavelength was shown by sample S1. The increase of film thickness induces the reduction of the transparent percentage from 64.7% to 25.1% in the visible range. This result was expected as more photons are adsorbed in a material when the thickness is increased. Besides, this might due to the the structures of crystalline grain became denser as the film thickness increased, which is already proven by the FE-SEM images. This makes the light more difficult to get through the surface of thin films. From this figure, all samples show sharp absorption edges in the visible region and these absorption edges slightly shifted to longer wavelengths (red shift) when the thickness of the films increased from 87.14 nm of S1 to 253.58 nm of S5.

IOP Conf. Series: Materials Science and Engineering **99** (2015) 012007 doi:10.1088/1757-899X/99/1/012007

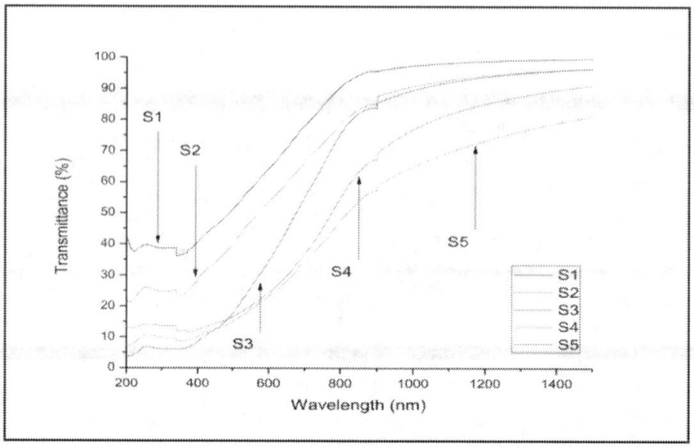

Fig. 2: Transmittance of CuO thin films

The optical band gap can be measured by extrapolation technique. In order to determine the direct optical band gap, was using the Taucs relationship as follows [15]:

$$(\alpha hv) = A\,(hv\text{-}Eg)^2 \qquad (1)$$

Where α is the absorption coefficient, h is Planck's constant, v is the photon frequency, and Eg is the optical band gap. An extrapolation of the linear region of a plot of the graph of $(\alpha hv)^2$ on the y- axis versus photon energy (hv) on the x – axis, gives the value of the optical band gap, Eg [15]. Optical band gap is calculated and showed in Fig. 3.

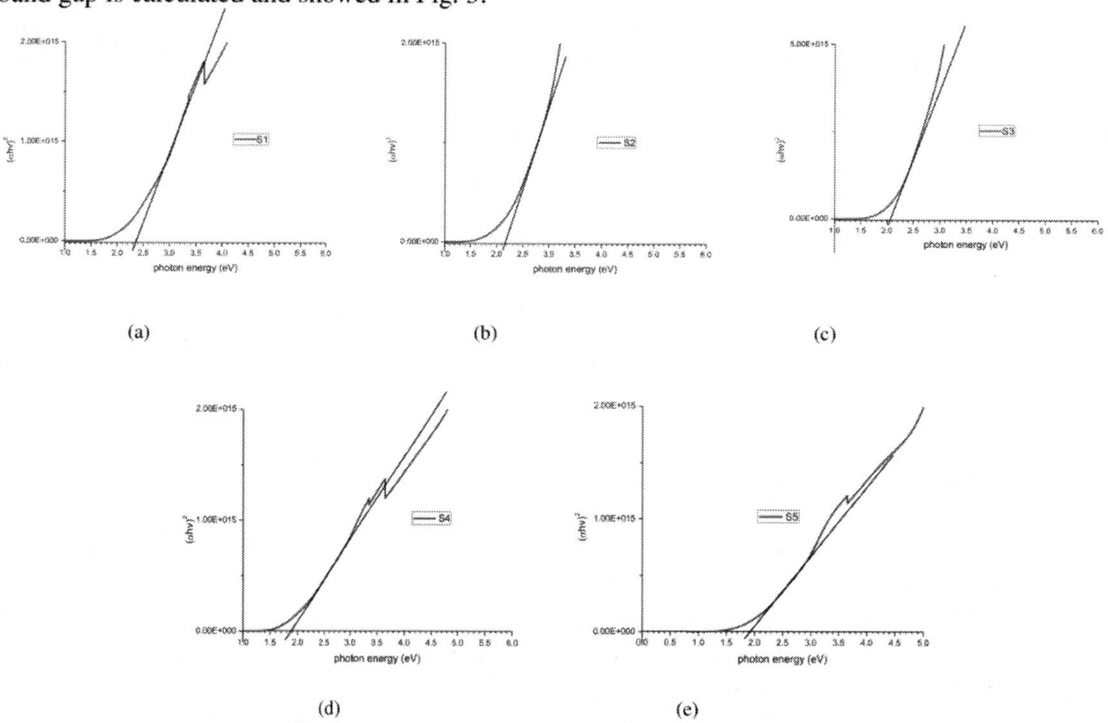

Fig. 3: Graph of optical band gap of CuO thin film for (a) S1 (b) S2, (c) S3, (d) S4, and (e) S5

From Fig. 3, the values of the optical band gap of CuO thin films were tabulated in Table 2. The direct optical band gap energy of CuO was reported in the range of 1.46 - 3.0 eV [3, 16], which shows that our findings are comparable with others. It is observed that the optical band gap decreased from 2.35 eV for sample S1 to 1.9 eV for sample S5. The decrease of band gap indicates the improvement of the films with the increased of the film thickness. The shifts of the optical band gap might due to the decreased of defects in the thin films as reported by Sengupta et. al. [17]. Besides, it is also reported that the optical band gap could decreased with the increased of the grain size [18].

Table 2: Optical band gap of CuO thin films values

Samples	S1	S2	S3	S4	S5
Optical band gap (eV)	2.35	2.15	2.1	1.9	1.9

3.3 Electrical Properties

The current–voltage (*I–V*) characteristics of CuO thin films are presented in Fig. 4a. The *I–V* dependence is linear within the studied voltage range of -5 to 5V, proving that the gold contact is ohmic with the thin films. The *I-V* characteristic of the CuO thin films were highly affected by the film thickness in the deposition process. The graph shows that the current increased with the film thickness.

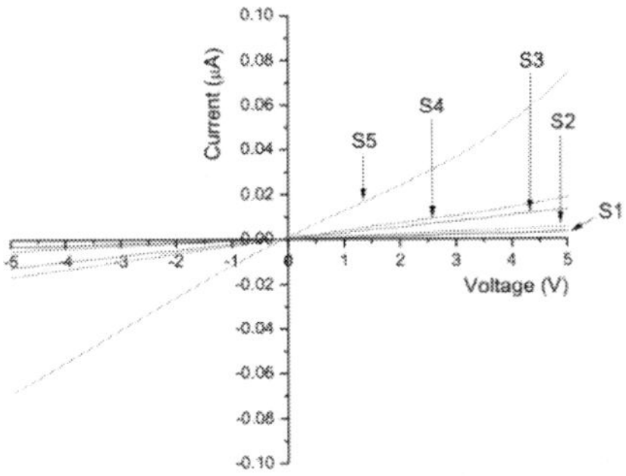

(a)

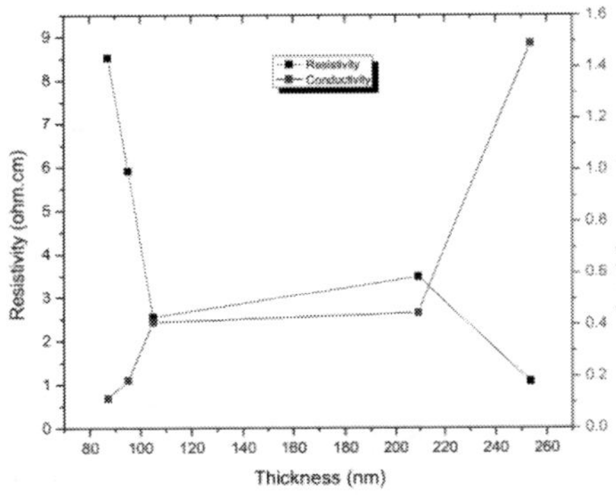

(b)

Fig. 4: (a) *I-V* characteristic of CuO thin film with different withdrawal speed (b) Resistivity and Conductivity of CuO thin films

Resistivity and conductivity of the CuO thin films were calculated and showed in Fig.4b. Highest resistivity was 8.53 Ω.cm for sample S1 and decreased to 1.01 Ω.cm for sample S5. It was found that the values of resistivity decreased with the increased of the film thickness. This might due to the improvement of the optical band gap. Besides, with the increased of the grain size, the grain boundary would decrease, which subsequently would decrease the scattering of carriers at the grain boundaries [19]. Therefore it would improve the conductivity of the films hence decrease the value of the resistivity.

4. Conclusion

In summary, thin film of CuO was deposited using sol-gel method on quartz substrate with different thickness by varying the frequent withdrawal times using dip coating method. The optical, electrical and physical properties of thin films measured by FE-SEM, UV-Vis spectrometer, surface profiler and two probe solar simulator. Film with lowest thickness showed the highest transmittance percentage. The decrease of the optical band gap with the increased of the thickness might attributed by the increased of the grain sizes. For *I-V* characteristics, the best output current was shown by film S5 with thickness of 253.58 nm, which showed lowest resistivity value of 1.01 Ω.cm.

Acknowledgement
The authors would like to thank staff of NANO-Electronic Centre (NET) and NANO-Scitech Centre (NST), Universiti Teknologi MARA for the laboratory facilities. Also special thanks to Faculty of Electrical Engineering, Universiti Teknologi MARA for funding this project.

References

[1] P. Vikas, J. Datta, P. Shailesh, C. Manik, G. Prsad, P. Sanjay, *et al.*, 2011 *Journal of Sensor Technology,* vol. 2011, pp 36-46.
[2] J. Li, G. Vizkelethy, P. Revesz, J. Mayer, and K. Tu, 1991 *Journal of applied physics,* vol. 69, pp. 1020-1029.
[3] R. A. Hammoodi, A. P. D. A. K. Abbas, and A. K. Elttayef 2014 *International Journal of Application or Innovation in Engineering & Management (IJAIEM),* vol. 3, pp 1-7.
[4] A. Umar, M. Rahman, A. Al-Hajry, and Y.-B. Hahn 2009 *Electrochemistry Communications,* vol. 11, pp. 278-281.
[5] M. Alhamed and W. Abdullah, 2013 *Journal of Electron Devices,* vol. 18, pp. 1563-1567.
[6] Y. Ohya, S. Ito, T. Ban, and Y. Takahashi, 2000 *Key Engineering Materials,* vol. 181, pp. 113-116.
[7] S. Karthick Kumar, S. Suresh, S. Murugesan, and S. P. Raj, 2013 *Solar Energy,* vol. 94, pp. 299-304.
[8] Y. Zhang, Y. Liu, L. Su, Z. Zhang, D. Huo, C. Hou, *et al.*, 2014 *Sensors and Actuators B: Chemical,* vol. 191, pp. 86-93.
[9] W. Zhang, G. Ma, H. Gu, Z. Yang, and H. Cheng, 2015 *Journal of Power Sources,* vol. 273, pp. 561-565.
[10] R. Mohammadpour, H. Ahmadvand, and A. Iraji zad, 2014 *Sensors and Actuators A: Physical,* vol. 216, pp. 202-206.
[11] R. Poreddy, C. Engelbrekt, and A. Riisager, 2015 *Catalysis Science & Technology,* vol. 5, pp. 2467-2477.
[12] Y.-F. Lim, C. S. Chua, C. J. J. Lee, and D. Chi, 2014 *Physical Chemistry Chemical Physics,* vol. 16, pp. 25928-25934.
[13] S. C. Ray, 2001 *Solar energy materials and solar cells,* vol. 68, pp. 307-312.
[14] P. H. G.G. Valle, S.H. Pulcinelli, C.V. Santilli, 2004 *Journal of the European Ceramic Society,* vol. 24, pp. 1009-1013.
[15] A. Ogwu, T. Darma, and E. Bouquerel, 2007 *Journal of achievements in materials and manufacturing engineering,* vol. 24, pp. 172-177.
[16] N. Mukherjee, B. Show, S. K. Maji, U. Madhu, S. K. Bhar, B. C. Mitra, *et al.*, 2011 *Materials Letters,* vol. 65, pp. 3248-3250.
[17] J. Sengupta, R. K. Sahoo, K. K. Bardhan, and C. D. Mukherjee, 2011 *Materials Letters,* vol. 65, pp. 2572-2574.
[18] R. Hong, J. Huang, H. He, Z. Fan, and J. Shao, 2005 *Applied Surface Science,* vol. 242, pp. 346-352.
[19] D. Jundale, P. Joshi, S. Sen, and V. Patil, 2012 *Journal of Materials Science: Materials in Electronics,* vol. 23, pp. 1492-1499.

IOP Conf. Series: Materials Science and Engineering **99** (2015) 012008 doi:10.1088/1757-899X/99/1/012008

Energy Consumption Analysis Procedure for Robotic Applications in different task motion

Iman Ahmed[1] , Ishak b. Aris[2], Mohammad Hamiruce Marhaban[3], Asnor Juraiza Ishak[4]
[1,2,3,4.] Electrical and Electronics Engineering, University Putra Malaysia UPM, 43400, Serdang, Malaysia

E-mail: eng.iman79@yahoo.com[1]

Abstract. This work proposes energy analysis method for humanoid robot, seen from simple motion task to complex one in energy chain. The research developed a procedure suitable for analysis, saving and modelling of energy consumption not only in this type of robot but also in most robots that based on electrical power as an energy source. This method has validated by an accurate integration using Matlab software for the power consumption curve to calculate the energy of individual and multiple servo motors. Therefore, this study can be considered as a procedure for energy analysis by utilizing the laboratory instruments capabilities to measure the energy parameters. We performed a various task motions with different angular speed to find out the speed limits in terms of robot stability and control strategy. A battery capacity investigation have been searched for several types of batteries to extract the power modelling equation and energy density parameter for each battery type, Matlab software have been built to design the algorithm and to evaluate experimental amount of the energy which is represented by area under the curve of the power curves. This will provide a robust estimation for the required energy in different task motions to be considered in energy saving (i.e., motion planning and real time scheduling).

1. Introduction

For the biped robot the energy conservation is one of the important factors that should be taken into account when design or develop any robot movement activity. In general, there are several methods to reduce the energy consumption in robot such as; optimizing the robot path planning, optimizing the robot parameters (speeds, payloads), scheduling the robot operation time, or throughout using integrated control strategy based on data of friction, speed and gravity of robot axes. Finally the on board process frequency. Many potential works have been devoted in order to reduce the consumed energy in robotic application to prolong the operation time. Wigström et al. 2013, Presents a dynamic programming method to achieve a decrease in energy consumption for any given cycle time. By generate new energy optimal trajectories, which follow the same path but in different execution time[1]. Brateman et al. 2006, is introduced an approach for energy saving for mobile robots through controlling both processor's frequency and motors speed[2]. Chemnitz & Schreck 2011, have showed that the slow motions are not the most energy- efficient and, the best strategy for energy conservation depends on a robot model[3]. Mei et al. 2005, Builds a power model for motion to analysis the energy consumer using two techniques dynamic power management(DPM) and real time scheduling(RTS) which provide greater opportunities to achieve better energy- efficiency[4]. Costa & Moreira n.d. have developed a simulator for humanoid robot which can be used to find the important parameters to optimize energy consumption [5]. Siregar 2005, Presents a technique for numerical optimization for power consumption for five-link planners by taken into account the transients in the circuit of electric motors[6]. Kormushev et al. 2011, Reduces the consumed energy by applying the reinforcement learning algorithm that will affect on the center of mass and the height of the humanoid robot[7].

In all the above mentioned literature related to the energy consumption there is no certain procedure that other researchers can follow to calculate the consumed energy.
The aims of this research are to analyze the energy consumption of the bioloid premium type A robot in different task motion using (single, double and multi servos), also to evaluate the energy consumption for the robot in different speed by utilizing Matlab software integration and bar chart explanation. Finally this integration can be used as a reference for any task motion of robot to calculate the consumed energy. Accordingly we can optimize the parameters that can affect on the reduction of the energy consumption.
To investigate the power consumption analysis practically in robot application, Humanoid Bioloid Robot Type A developed by the Korean company Robotis was used due to its flexibility. The kit with 18 Degrees of freedom (DOF) consists of a CM-530 main Controller, 18 AX-12A DC servo motor . Powered by 11.1v 3S LiPo battery.. An overview of the proposed biped robot is shown in figure 1.

2. Experimental procedure measurement for the energy consumption:

The devices used in the experimental measurement to calculate the energy consumption during motion task are shown in figure 2.

Figure 1. Bioloid premium robot type A

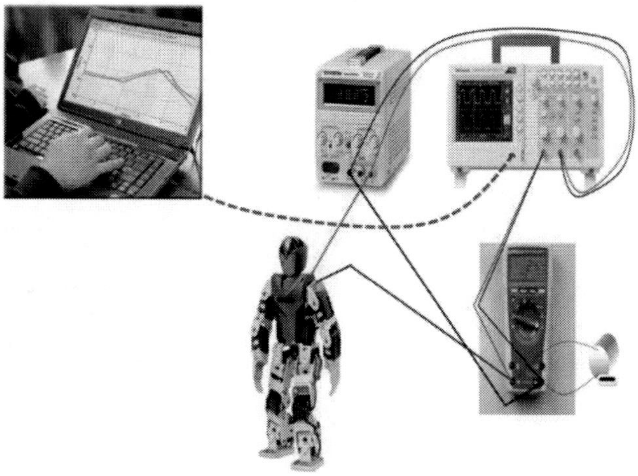

Figure 2 . Lab. Instruments used for power analysis.

In this research a measurement instruments have been adopted as an acquisition device to collect and store energy data by utilize the facilities that are available in these smart components. For example, to get the power consumption rate curve, a Fluke current multi-meter, with suitable scale is used as a current sensing element, one oscilloscope channel to measure the drop voltage over that multi-meter terminals to get the instantaneous value of the consumed current as voltage signal, while other channel is used to measure and store the variation in the voltage level. The products of the two channels would generate the power consumption rate curve. Parameters setting up are important to get high resolution and clear data. Finally the data of different task motions were collected using different values of speed or payloads and recorded data transferred for more analysis when USB memory is used to transfer the data points to laptop.

3. Simulation analysis

The Matlab software is utilized to create a program for calculating the area under the power curve which represents the consumed energy, this can be used as a reference for the energy calculation for different robot types. Integration software flow chart is represented in figure 3.

3.1. Power chart

The energy consumption of the biped robot affected directly by power dissipation of the driving and sensors components that accumulated to construct its body, therefore a need of study the power of these components is quite important at project planning aspects and conservation of energy. The estimation of the robot energy of a particular task of motion depends on servo motor power consumption and the number of those servos that participate to implement the task; therefore we studied the power variation range individually for each servo and similarly for multiple servos as well as the limits of stability in each case.

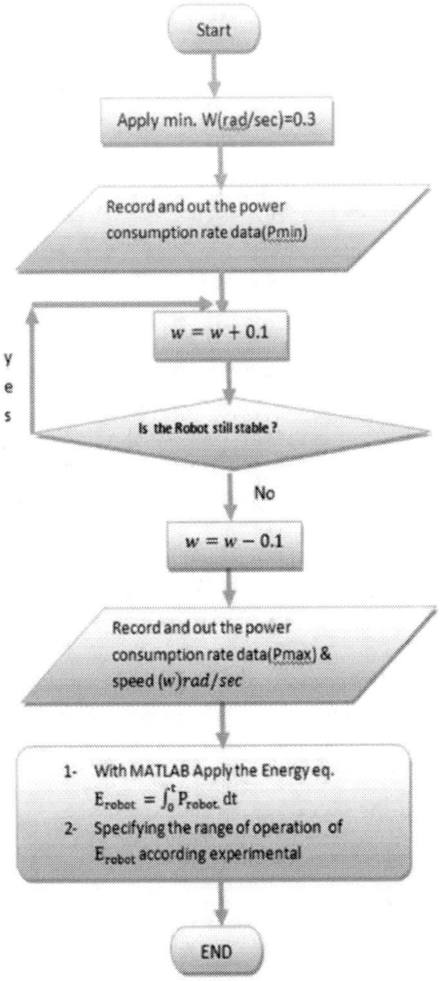

Figure 3. Integration Software Flow chart

3.1.1. Single Servo Motor Power chart

Three cases study of power measurements have been taken in this research as shown in figure 4
There is a similarity at the minimum value limits for both Servo 1, and Servo 5, this result is reasonable because that, each of them has consumed power due to controller torque-on magnitude, which is essential to maintain position control of the servo and the slight difference results from the servo location in the structure of humanoid body.
The power can be given by the equation:

$$P_{k\,servo}(\omega) = P_{T\,on} + P_k(\omega) \quad (1)$$

$P_{T\,on}$ = torque-on consumed power which is essential to maintain the stability, it is constant whenever the humanoid and the servo were settled on (no-motion state).
$P_k(\omega)$ = consumed power at a specific angular speed ω.

3.1.2. Double Servo Motor Power chart

The minimum and maximum limits of the 2 servo column chart in figure 4, represents another study of power consumption limits.
referring to equation (1), it's obviously that the minimum bar represents the consumed power of the two servos with torque-on power results from not only their own consumption but also from all other 16th servos that constitute this type of humanoid, so equation (1) is still valid in this case but with attention to construction, or the type of the motion task, and the number of servos.
The suitable expression to this type of power can be as follow:

$$P_{nservo}(\omega) = P_{Tall\,on} + \sum_{k=1}^{n} P_k(\omega) \quad (2)$$

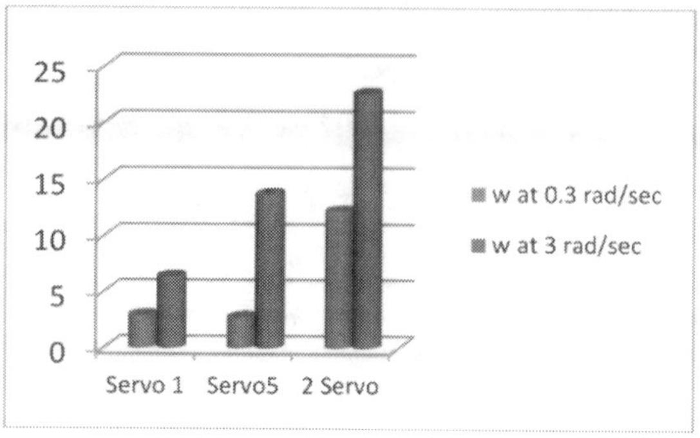

Figure 4. Single and Double Servo Motor Power chart

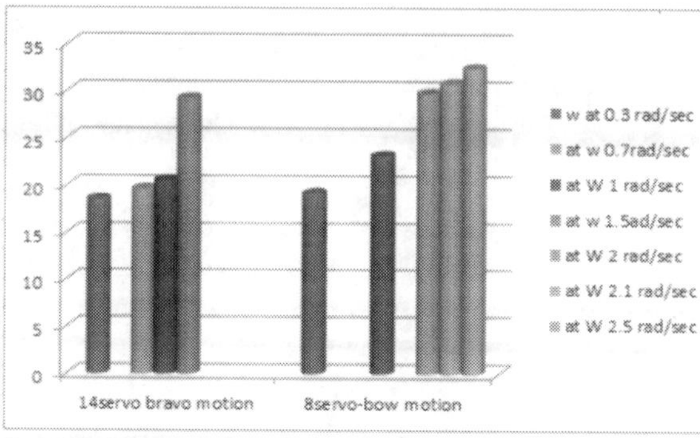

Figure 5. Multi Servo Motor Power chart

3.1.3. Multi servo Motion Power Chart

For motion planning design the energy consumption and the stability are the most essential parameters that should be taken to maintain the task specifications. Figure 5, shows two types of task motion, they are: 14 servo (Bravo motion), and 8servo (Bow motion), the main objectives of these experimental measurements are to specify the consumed power and variation range at different angular speed values, and assign the angular speed value at which the critical stability could be affected for each task. As seen in figure 5, both blue colour bars are relatively equal in values, the reasonable interpolation based on the torque-on power of all servos, similarly to the previous case, the slight difference between these two bars results from(due to) robot orientation(position) or the mass distribution of humanoid position. The peaks values of the bars in each motion indicate the critical angular speed on which the robot tends to be unstable, it is 1.5 rad /sec at 29.24 watt consumed power in (bravo motion), while it is 2.5 rad /sec at 32.46 watt power(in bow motion).

3.2. Energy Chart

3.2.1. Single Servo Motor Energy Chart

Servo1, end location servomotor in the robot hand , with lower load and less power consumption servo as shown in figure 6. This figure shows the photo, power rate curve, and the energy difference in both angular speed rate (1and 3)rad/sec.

Servo5, shoulder location servomotor, more power consumption within the same hand, which means more payload than previous servomotor. Figure 7, shows the difference in the power consumption rate through the two value of speed (1 and 3) rad/sec.

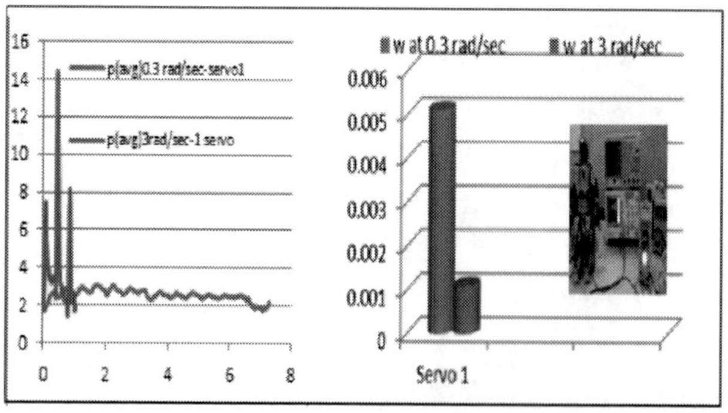

Figure 6. power rate & energy of servo1

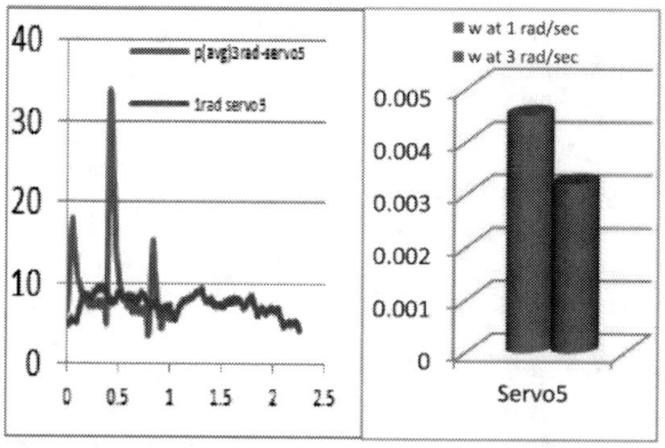

Figure 7. power consumption rate & energy of servo 5.

IOP Conf. Series: Materials Science and Engineering **99** (2015) 012008 doi:10.1088/1757-899X/99/1/012008

3.2.2. Double Servo Motor Energy chart(2 servos up moving)

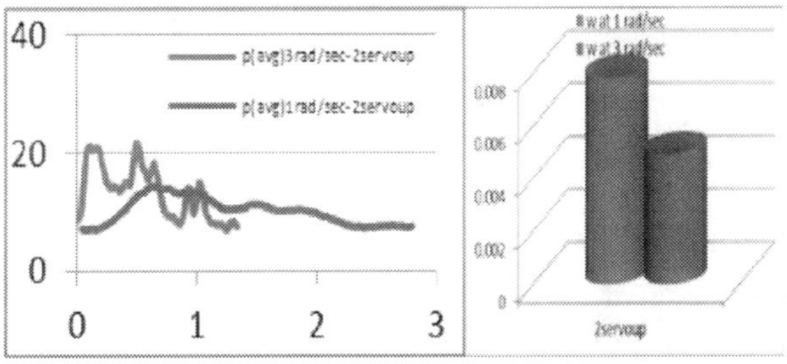

Figure 8. Power consumption rate & energy of 2 servos

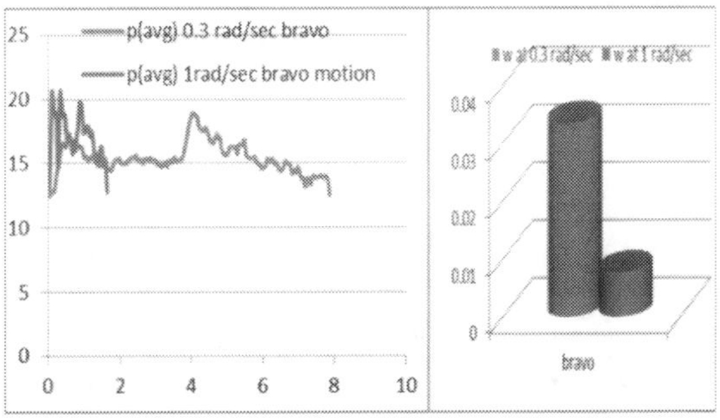

Figure 10. power consumption rate & energy of 12 servos *Bravo motion (12 Servos)*

3.2.3. Multi servo Motion Energy Chart

As depicted in the above figures (6-10), for all task motions the energy rate is larger in slow speed than in high speed rate for example in (2servo up) motion task the amount of the consumed energy at (1rad/sec) equal to 0.0078 Wh which is more than at (3rad/sec) 0.0049 Wh . And this is due to two reasons; first, the longer execution time for the slow motion and second, is the fast task motion has high power consumption for short pin as shown in the above figure. Put this is not common for all application since in certain task motion the slow motion consume energy for less than higher motion.

Several batteries have been selected based on market availability to evaluate the most important parameter in mobile robot battery which is the energy density as shown in table(1), a model equation have been conducted for each type of battery or hybrid battery.

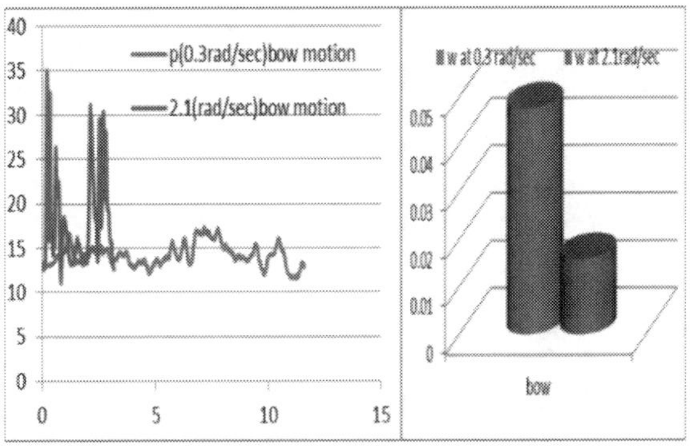

Figure 9. power consumption rate & energy of 8 servos. *Bow motion (8 servos)*

3.3. Total Battery Capacity

Energy consumption of the battery is tested under high power consumption motion task which includes 8 servomotors.

We have used the general energy equation, Eq.(3) to find out the experimental energy capacity in specific type of task motion which is shown in figure 11, this represents the energy consumption of 8 servos motion task which is elapsed 10 times and finished within 200sec.

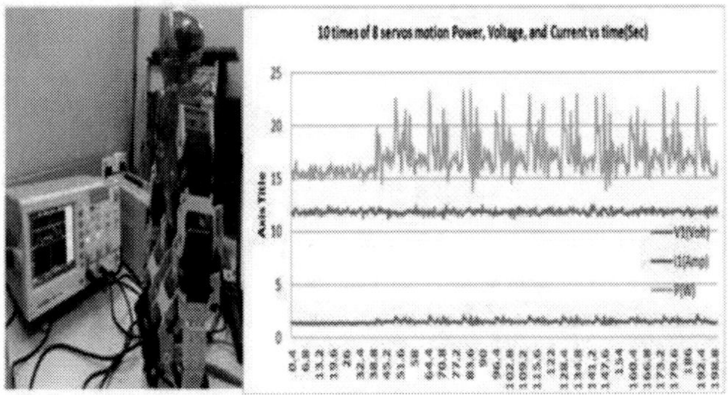

Figure 11 . Energy consumption of 8 servos motion task

$$W_{robot} = \int_0^t P_{robot.}\,dt \qquad (3)$$

The above case study has been used as a base to find out the modelling power equations for different power sources.
The difference between the manufacturer and experimental energy capacity can be shown in table 1, for several battery types, the differences are due to some battery parameters such as; the state of charge(SOC), discharge rate, and the battery life cycle.

3.4. Energy Density

The energy density is a measure of the amount of energy per unit weight or per unit volume which can be stored in a battery. Thus for a given weight or volume a higher energy density cell chemistry will store more energy or alternatively for a given storage capacity a higher energy density cell will be smaller and lighter. The table below shows some typical examples.
For example; for Lithium ion Polymer battery, Energy Density is 211 Wh/Kg.
However, there is a significant different between the experimental and the manufacturer value of the battery energy capacity, whereas the manufacturer value of energy can be calculated directly from the following Eqs.(4,5):

$$E_{th} = P * t \qquad (4)$$

$$\text{Where} \quad P_{th} = I * v \qquad (5)$$

E_{th} = Theoritical energy.; P_{th} = Theoretical Power.

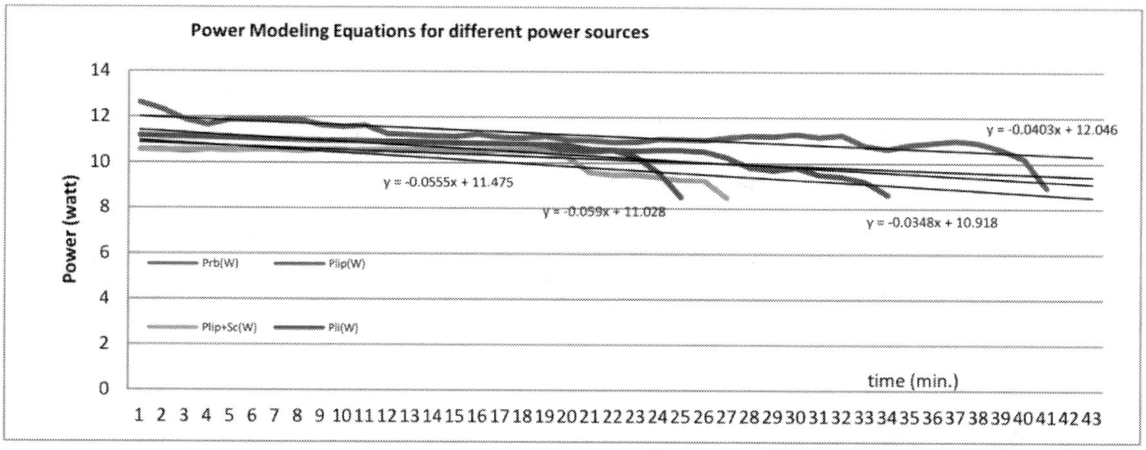

Figure 12. Power Modelling Equations for different power sources.

Table 1. The energy parameters for different sources

Battery Type	Weight (gm)	Voltage (Volt)	Specific energy (Wh/g) theoretical	Modelling Equation	time (min)	practical Energy (W.h)	Practical Specific energy (Wh/g)	Capacity (mA.h)
Robot on-board	79	11.1	(1*11.1)/79=0.14	y= - 0.0488x +12.168	41	7.631	0.0966	1000
Li-Polymer	84	11.1	(1.6*11.1)/84=0.211	y= - 0.0348x +10.918	34	5.851	0.0696	800
Li	134	11.1	(3.6*11.1)/134=0.298	y= - 0.059 x+11.028	27	4.604	0.0343	3600
SuperCap +LiP	116	11.1	0.211+2.1=2.311	y= - 0.0555 x + 11.475	25	4.492	0.0387	
SuperCap	7*4=28	11.1	60.5/28=2.1					60.5

3.5. Energy Modeling Equations for different power sources

It is obviously that the power tends to be linearly reduces down with a constant slope which is easily found by the equation:

$$P = 0.0488 * t + 12.168 \tag{6}$$

P = the instantaneous power given by (Watt); T = is the time in (minutes) in this measurements. The Total Energy can be found by integrating the power line over the time (0-39)mints
From Eq.(6), The energy consumption which is the area under the curve is given by

$$E = {}_0\int^{39} 0.0488 * t + 12.168 = -0.0488/2 * (39)\text{^}2 + 12.168 * 39 = \ 7.29066\ Watt.h$$

E = the experimental energy.

On the same concept of Eq.(6), we have found out the modelling equation and calculated the total capacity of each energy source, as described in figure 12, for the energy equivalent sources; the existence robot battery, lithium ion polymer battery, Lithium ion polymer+ super capacitor, and lithium ion respectively .

However, there is a significant different between the experimental and the manufacturer value of the battery energy capacity, whereas the manufacturer value of energy can be calculated directly from the following Eqs.(7,8):

$$\mathrm{E} = \mathrm{P} * \mathrm{t} \tag{7}$$

$$,\ \text{Where}\quad P = I * v \tag{8}$$

4. CONCLUSION.

Energy consumption in robot application is one of the most important issue in research area , this paper focuses on best, reliable, and simple method as well as an algorithm to be utilized in energy analysis and conservation, it is found that the main two energy parameters (current, voltage) can be measured based on the capabilities of the laboratory instruments that are considered as trusted and more reliable devices, thus an accurate measurements are obtained and recorded, this procedure can be considered as a reference simple method to any researcher in this field.

The percentage of power consumption rate is some times more less than the total power rate which is consumed thought task execution and consequently, we need to take into account this value which is called (on-torque energy) of the robot. The (on-torque) is the energy required to maintain the stability of the robot orientation and its servos position angles, this amount of energy can be reduced by energy saving motion planning design tasks.

It is found that the consumed energy is not always inversely proportional with the task execution speed, but it may have a direct proportional according to the task type.

The power source type and brand of the humanoid robot is most effective parameters on the performance. Practically, it is found that the capacity of the on-board battery is about 65.7% with respect to that value mentioned in its datasheet, while this percentage is not the same with other types and even the brand over the same type as described in figure 12, and table 1.

Thanks of Matlab software facilities; it's easy to integrate the area under the power rate curves of all types of batteries that subjected to energy measurements test.

This analysis can open a horizon to search a more effective, long life and motion duration with high energy density especially with mobile robot. In the future work, the researcher should tackle firefly in more various types of problem such as find optimum method that will effect on the energy saving , or try to find alternative power source .

Acknowledgments

Thank you for UPM for providing Financial support under the research Grand (GP-IPS)/(2014-9438720).

REFERENCE

[1] Wigström, O. et al., 2013. High-Level Scheduling of Energy Optimal Trajectories. , 10(1), pp.57–64.
[2] Brateman, J., Xian, C. & Lu, Y., 2006. Energy-Efficient Scheduling for Autonomous Mobile Robots. , pp.361–366.
[3] Chemnitz, M. & Schreck, G., 2011. Analyzing energy consumption of industrial robots. , pp.0–3.
[4] Mei, Y. et al., 2005. A Case Study of Mobile Robot ' s Energy Consumption and Conservation Techniques. , pp.492–497.
[5] Costa, P.G. & Moreira, A.P., HUMANOID REALISTIC SIMULATOR : The servomotor joint modelling.
[7] Siregar, H.P., 2005. Simulation of power consumption for walking robot. Proceedings of the Fifth International Workshop on Robot Motion and Control, 2005. RoMoCo '05., pp.27–32.

IOP Conf. Series: Materials Science and Engineering **99** (2015) 012009 doi:10.1088/1757-899X/99/1/012009

Mobile Carbon Monoxide Monitoring System Based on Arduino-Matlab for Environmental Monitoring Application

Nur Azieda Mohd Bakri, Syed Abdul Mutalib Al Junid*, Abdul Hadi Abdul Razak, Mohd Faizul Md Idros, Abdul Karimi Halim
Centre for Electronics Engineering Studies,
Faculty of Electrical Engineering, Universiti Teknologi MARA,
Shah Alam, Selangor, Malaysia

E-mail: samaljunid@salam.uitm.edu.my

Abstract. Nowadays, the increasing level of carbon monoxide globally has become a serious environmental issue which has been highlighted in most of the country globally. The monitoring of carbon monoxide content is one of the approaches to identify the level of carbon monoxide pollution towards providing the solution for control the level of carbon monoxide produced. Thus, this paper proposed a mobile carbon monoxide monitoring system for measuring the carbon monoxide content based on Arduino-Matlab General User Interface (GUI). The objective of this project is to design, develop and implement the real-time mobile carbon monoxide sensor system and interfacing for measuring the level of carbon monoxide contamination in real environment. Four phases or stages of work have been carried out for the accomplishment of the project, which classified as sensor development, controlling and integrating sensor, data collection and data analysis. As a result, a complete design and developed system has been verified with the handheld industrial standard carbon monoxide sensor for calibrating the sensor sensitivity and measurement in the laboratory. Moreover, the system has been tested in real environments by measuring the level of carbon monoxide in three different lands used location; industrial area; residential area and main road (commercial area). In this real environment test, the industrial area recorded the highest reading with 71.23 ppm and 82.59 ppm for sensor 1 and sensor 2 respectively. As a conclusion, the mobile real-time carbon monoxide system based on the Arduino-Matlab is the best approach to measure the carbon monoxide concentration in different land-used since it does not require a manual data collection and reduce the complexity of the existing carbon monoxide level concentration measurement practise at the same time with a complete data analysis facilities.

1. Introduction

Nowadays, the level of carbon monoxide concentration has become global issues since it produced a bad effect and harms the environment globally. A series of conference and symposium are organized all over the world to discuss and find the solution for the increasing of the carbon monoxide issue. New policy and convention are introduced as an outcome from the meeting, but it cannot solve the issue as the whole since the implementation required an agreement on the country law by itself. But, their findings on the main contributory factor of carbon monoxide contamination have produced a good guideline for each of the countries on the source of the pollution towards design and implement their own law and guidelines control the pollution. The industrial area and transportation have been identified as a major contributor for the carbon monoxide released which contaminate the air globally and it has been confirmed by the inmost of the report and publications [1]-[2].

In Malaysia, transportation has been identified as one of the major contributing factors for air pollution with corresponding to the increasing numbers of vehicles in major cities in Malaysia from year to year [2]. In 2011, the total numbers of registered vehicles on Malaysia road have passed 21.25 million units as reported in [1]. The data was supported by the estimation of air pollutant emission load in 2008 with 1.45 million metric tonnes of carbon monoxide concentration have been released [3]. On the other hand, the carbon monoxide is a poisonous gas that is colourless, odourless and tasteless, which is produced by a product of combustion, which is impossible to detect by the exposed person [4]-[5].

In health perspective, carbon monoxide reduces the flow of oxygen in the bloodstream and is particularly dangerous to persons with heart disease [2]. At the high level of carbon monoxide concentrations, it can impair vision and coordination, such as headaches, dizziness, confusion, nausea and also can cause flu-like symptoms. While at low level carbon monoxide concentrations, it can fatigue in healthy people and chest pain in people with heart disease and other lung related disease [7]. Over 500 people killed and more 20,000 people are sent to the hospital each year due to the effect of carbon monoxide gas in environment as reported in [7].

The situation will remain until the continuous monitoring for observing this scenario has been enforced and the high penalty rules implemented. The Real-Time carbon monoxide monitoring system is one of the best solutions to determine the level carbon monoxide concentration in a large scale area which capable to plot the area which contribute to the high level of the carbon monoxide content [8]. This will help in plotting the region or location of the carbon monoxide on Geographical Information System (GIS) which are currently used for controlling and observing the environmental issues globally.

Although several solutions have been proposed to monitor the environmental issue previously in [9]-[10] but, it focuses more on the measuring the thermal comfort issue which resulted from the carbon monoxide effect and improper land used development. Considering the global solution for the proposed design, a low cost controller based on the open source electronics prototyping platform will be used as the controller in the project, which has been reported used widely in environmental and prototyping projects in [10]-[13].

The report is organized in five important sections with the overview of the project is highlighted in section 1. Section 2 will be brief the method or process of accomplishment of the design, development and test of the proposed project. The details result of the implementation will be reported in Section 3 while the discussion of the finding will be detailed elaborated in the next section. The last section or section5 will conclude the all the finding of the project.

2. Methodology of design

Four phases of work have been used to accomplish the project until the implementation of the system in real environment. It starts with the development of sensor system for sensing the carbon monoxide gas concentration and followed by controlling and integrating the sensor with an Arduino UNO microcontroller. The third phase focussed on the process of data collection and the final phase of work emphasis on the data analysis measured and recorded using the proposed system in the real environment. The details of each phase are explained in the subsequent sections.

2.1. Sensor development

There are three stages have been carried out in the sensor development stage, which is the calibration of carbon monoxide sensor, module starter kit for Global Positioning System (GPS) and development of sensor circuit

2.1.1. Calibration of carbon monoxide sensor

Figure 1. Carbon Monoxide sensor (MQ-7)

Figure 1 shows the MQ-7 carbon monoxide sensor that has been used in the project. The calibration of sensor reading was done by referring to the MQ-7 carbon monoxide sensor datasheet. Moreover, the calibration of the sensor also required a complete reference to the level of carbon monoxide in Atmospehere (in parts per million) as shown in Table I for relate the sensor readings obtained precisely. However, the exposure to the carbon monoxide required high precaution and need be avoided because it gives various unhealthy effects and the detailed effects is summarized in Table 2 [15].

Table 1. Carbon Monoxide source concentration

Level of Carbon Monoxide	**Source**
0.1 ppm	Natural atmosphere level
0.5 to 5 ppm	Average level in homes
5 to 15 ppm	Near properly adjusted gas stove in homes
100 to 200 ppm	Exhaust from automobiles in the city
5000 ppm	Exhaust from a home wood fire
7000 ppm	Undiluted warm car exhaust without a catalytic converter

Table 2. Carbon Monoxide exposure symptoms

Level of Carbon Monoxide	**Source**
35 ppm (0.0035%)	Headache and dizziness within six to eight hours of constant exposure
100 ppm (0.01%)	Slight headache in two to three hour
200 ppm (0.02%)	Slight headache within two to

	three hours; loss of judgement
400 ppm (0.04%)	Frontal headache within one to two hours
800 ppm (0.08%)	Dizziness, nausea, and convulsions within 45 min; insensible within 2 hours
1600 ppm (0.16%)	Headache, tachycardia, dizziness, and nausea within 20 min; death in less than 2 hours.
3200 ppm (0.32%)	Headache, dizziness and nausea in five to ten minutes. Death within 30 minutes
6400 ppm(0.64%)	Headache and dizziness in one to two minutes. Convulsions, respiratory arrest, and death in less than 20 minutes.

2.1.2. SKM53 GPS Module starter kit

Figure 2 shows the GPS module starter kit used in this project. The SKM53 series with the embedded GPS antenna enables high performance navigation in the most stringent applications and solid fix even in harsh GPS visibility environments. It is specially designed starter kit which offer convenient yet safer GPS module for user. Moreover, the supply is from 5V of UART pin with operating temperature range from -40°C to 85°C. The SKM53 GPS module has ultra-high tracking sensitivity within -165dBm extends positioning coverage into place like urban areas and dense foliage environment where the GPS was not possible before. The SKM53 also used NMEA protocols which is a simple device generating NMEA sequence for testing embedded GPS reception firmware and hardware is described [16]. The device can work in standalone mode and also in conjunction with control software. Configuration program can be used to generate test strings without tester hardware as well as stated in [16].

2.1.3. Development of sensor circuit

The MQ-7 carbon monoxide sensor with simple drive circuit is used to detect the presence of carbon monoxide (CO) gas in this project. The sensor has been chosen because it has high sensitivity to carbon monoxide and is widely used in industrial for a CO detector. The circuit construction and position of the sensor is very important for obtaining the reading of carbon monoxide level at this stage.

2.2. Controlling and Integrating Sensor

The second stage involved in this project is controlling and integrating the carbon monoxide sensor with the microcontroller. This phase is divided into two parts which is hardware development and software development.

2.2.1. Hardware development

In this part, two carbon monoxide sensors and GPS are combining with the embedded system at the first step. The purpose of the two carbon monoxide sensors were used is to make sure the accurate reading obtained since the operation of the sensors required pre-heat process for every reading or measurement. Next, both of the sensors and GPS is configured for reading the latitude and longitude

IOP Conf. Series: Materials Science and Engineering **99** (2015) 012009 doi:10.1088/1757-899X/99/1/012009

of the GPS and level concentration CO sensor via the serial monitor of the microcontroller. The position of the reading obtained from the serial monitor is important since the data is very important and required when interface with the software part. The carbon monoxide sensors and GPS are supplied with 5V Direct Current (DC) which source from the Arduino UNO. The complete explained system architecture is illustrated in Figure 3.

Figure 2. GPS Module starter kit

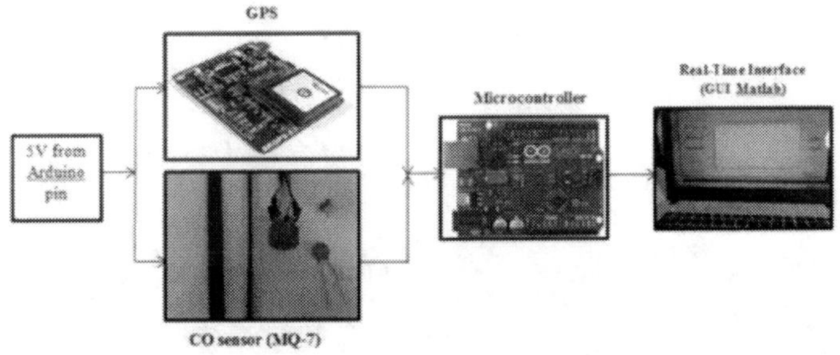

Figure 3. System Architecture of Mobile Monitoring System.

2.2.2. Software development

General User Interface (GUI) MATLAB is used in the project for development of the real time monitoring activity since it has a graphical display in more than one window and the user can use it to perform more interactive tasks. In addition, MATLAB version used in the project is 2011b with the Arduino interface package or tool box. Figure 4 shows the illustration of GUI MATLAB with the graphical representation for plotting the data.

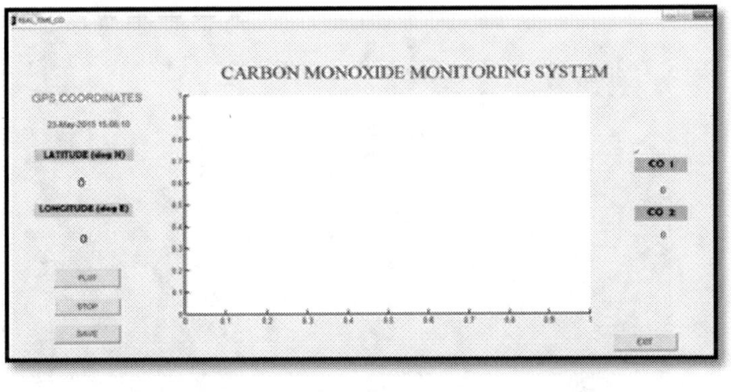

Figure 4. Front panel of Monitoring System

The reading obtained from the carbon monoxide sensors were viewed in the graphical representation. Moreover, the plot handles are used where the system started to illustrate the reading from the microcontroller. Besides that, the results from sensors obtained are displayed in the numeric value for ease the user to read the data. On top of that, the real-time graph with the interval changes are embed together with the sensor reading to ease the analysis of the measurement. The signal detected from GPS plotted the coordinates of latitude and longitude of the in the GUI. Besides that, there is a save handle that link to the Microsoft Excel as shown in Figure 5. The function is to save all data obtained during the process of collecting data to the Microsoft Excel format for further analysis.

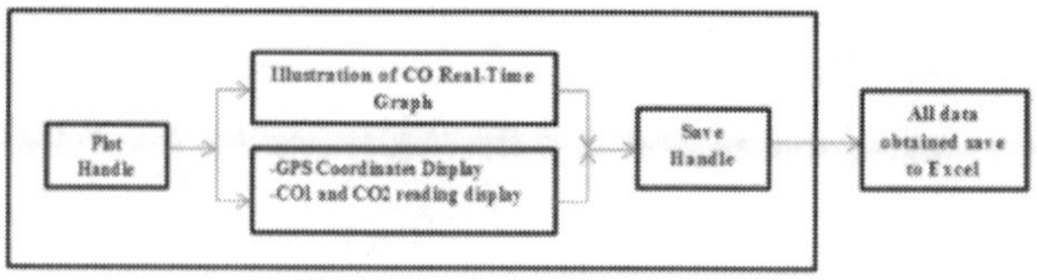

Figure 5. Data logging

2.3. Data collection

In the third stage, the data collection was conducted to verify and validate the system. A complete monitoring activities have been conducted at this stage. The monitored area is Industrial Area, Residential Area and main road (commercial area). Every area is being monitored within three minutes and every second data will be recorded and saved in data logger.

Figure 6. Integration of Carbon Monoxide sensors

Figure 7. Development of mobile monitoring facility.

Figure 8. Real-Time host equipped in car

Figure 6 shows the integration of carbon monoxide Sensor 1 and Sensor 2 to the monitoring facilities and it was positioned on top of the monitoring vehicle. The carbon monoxide sensor 1 and sensor 2 are placing about 40cm apart from each other as shown in the Figure 6. The purpose of placing two sensors is to observed data in more accurate and precise ways. The GPS module also has been positioned at the monitoring facilities. It is due to open space and clear sight which required by the GPS module in order to find the satellite connection and get the coordinates of the monitored area without interruption. Figure 7 shows the equipped mobile monitoring facility on top of the monitoring vehicle. Figure 8 shows the Real-Time Host (GUI MATLAB) placed inside the monitoring vehicle. Every recorded data and the changes will be displayed in real-time graphical representation.

2.4. Data analysis
Data analysis will be the final stage of work which involved analysis of sensor readings. The level of carbon monoxide in each monitored area will be carefully recorded and systematically analysed.

3. Result
The results of the monitoring process consist of three parts which is the maps (from MATLAB) that locate the observed area, the graphical representation of the carbon monoxide concentration level in each monitored area and analysis of tabulated data of carbon monoxide profiles. Each of the components will be detailed discussed in its respective sub-sections.

3.1. Maps Locate the monitored area
The monitoring activities has been done to observe the level of carbon monoxide in three different areas. Figure 9 shows the example coordinates locate by the GPS module in real-time during the monitoring activities. The mapping location data were recorded from the starting point up to the ending point of monitoring areas.

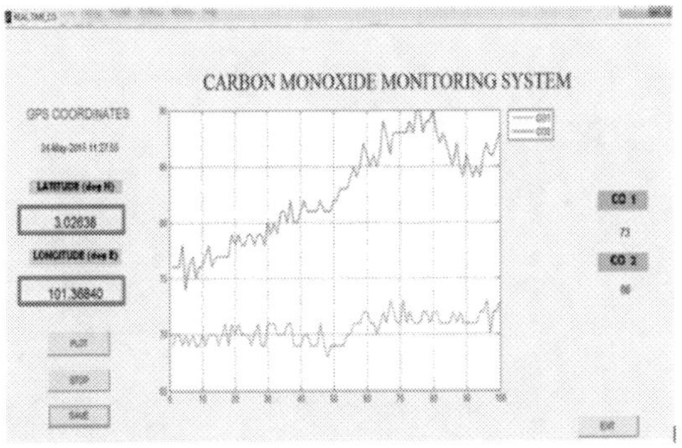

Figure 9. Latitude and longitude coordinates are located by the GPS.

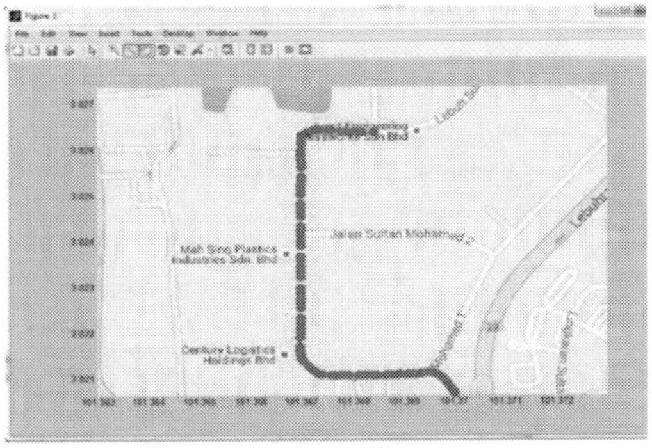

Figure 10. Industrial area mapping point

Figure 10 shows the mapping point of the monitored Industrial area. The location is at Kawasan Perusahaan Bandar Sultan Suleiman, Klang, Selangor. It starts from 166, Solok Sultan Mohamed 1 and ending at 19, Lebuh Sultan Mohamed 2. The duration of monitoring activities was set at 3 minutes. This area has been chosen due to the numbers of factories and the routes for trucks and trailers. The level carbon monoxide concentration is expected to be higher due to the reasons.

Figure 11 shows the mapping point of the monitored residential area. The location is at at Jalan Platinum, Seksyen 7, Shah Alam, Selangor. The point starts from Jalan Platinum 7/59, Seksyen 7 and ends at Jalan platinum 7/60, Seksyen 7. This location has been chosen due to the populated area.

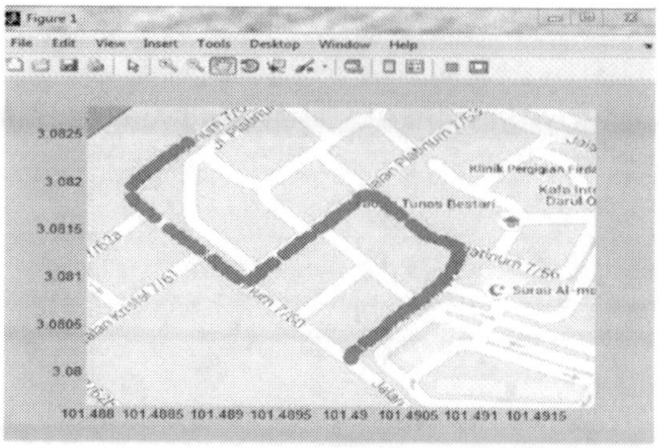

Figure 11. Residential area mapping point.

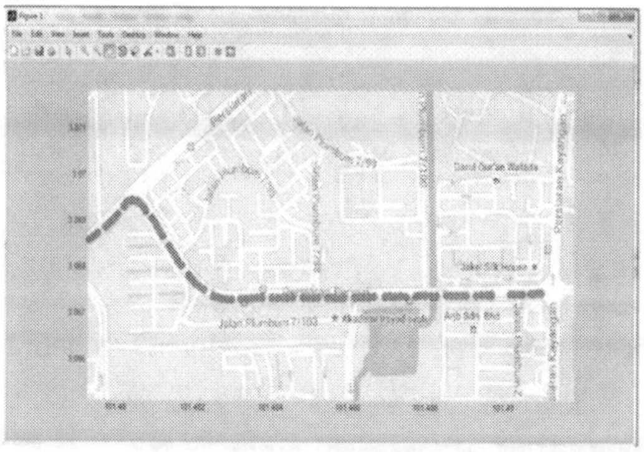

Figure 12. Main road (commercial area) mapping point

Figure 12 shows the main road (commercial area) mapping point which has been chosen to be observed in this monitoring activity. The location is at Seksyen 7, Shah Alam which is the main road for residents and community. The observation point starts from Persiaran Permai Seksyen 7 and ends at Lebuh Keluli, Seksyen 7.

3.2. Graph of Carbon Monoxide concentration in each monitored area.

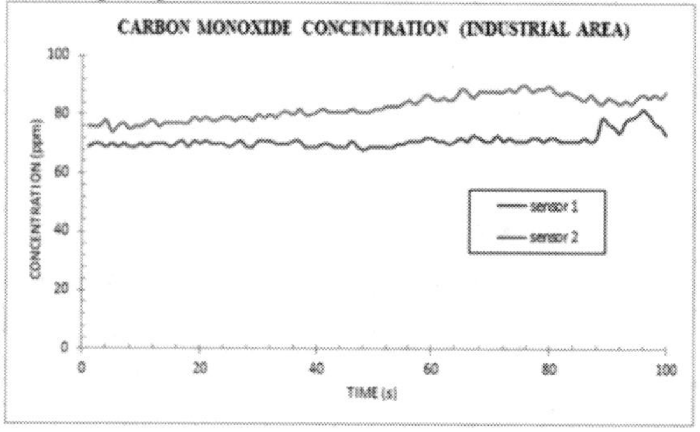

Figure 13. Carbon monoxide concentration in industrial area.

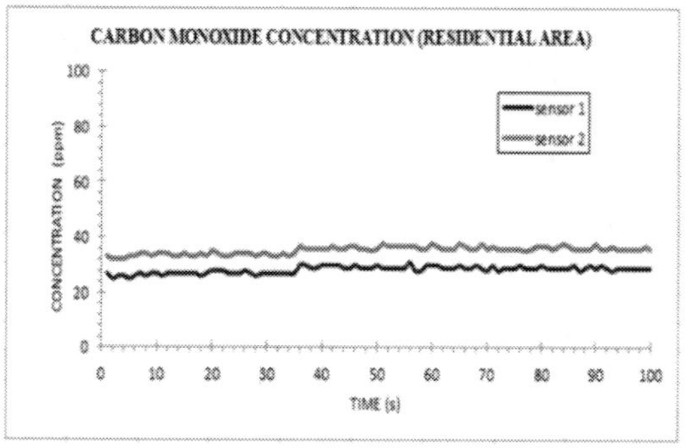

Figure 14. Carbon monoxide concentration in residential area.

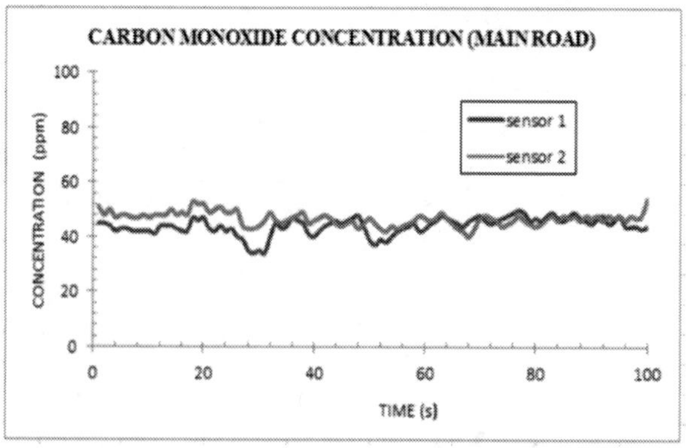

Figure 15. Carbon monoxide concentration in main road (commercial area)

Figure 13, Figure 14 and Figure 15 shows the graph of carbon monoxide concentration for each of the monitored areas during monitoring activities. The graph is plotted based on data recorded by both of the carbon monoxide sensors.

3.3. Carbon monoxide data analysis

Table 3. Average of carbon monoxide reading in monitoring areas

Monitored Area	**Sensor 1 (ppm)**	**Sensor 2 (ppm)**
Industrial	71.23	82.69
Residential	28.38	35.40
Main Road	44.10	35.40

Table 3 shows the average reading of carbon monoxide concentration recorded by both of the sensors in the monitored area. The highest average reading recorded in the Industrial Area with 71.23 ppm and 82.67 ppm for sensor 1 and sensor 2 respectively. The main road (commercial area) recorded the average of 44.10 ppm for sensor 1 and 35.40 ppm for sensor 2. The residential area recorded the lowest reading with 28.38 ppm and 35.40 ppm for sensor 1 and sensor 2 respectively.

4. Discussion

Carbon monoxide concentrations in three different areas have been measured and recorded during the monitoring activities. Industrial area recorded the highest average of carbon monoxide concentrations reading compared to others monitoring area. Smoke from factories and vehicle exhaust are the two major factors has been identified influence the reading. Besides that, the residential area recorded the lowest average of carbon monoxide concentration. The effect of tree planting around the housing estate may lead to lower the level of carbon monoxide emission in air while less source of air pollution compared to the industrial area can be considered as the main factor as well. The main road (commercial area) that has been observed in monitoring activities also recorded higher average of carbon monoxide concentration, it may due to the numbers of motor vehicles and traffic flow conditions passing through this area.

5. Conclusion

As a conclusion, the real-time mobile carbon monoxide monitoring system based on low cost microcontroller and interface to General User Interface (MATLAB) provides and offers a real-time and accurate carbon monoxide concentration monitoring for application in environmental monitoring. Moreover, the systems record accurate real-time data based on the industrial requirement and capable to work in a real environment. Besides that, the application of Global Positioning System (GPS) provides the accurate location of the monitored area complete with the measurement of carbon monoxide concentration. This will help to provide the accurate location of carbon monoxide concentration clearly for enforcement purposes.

Acknowledgments

Authors wishing to acknowledge Faculty of Electrical Engineering, Universiti Teknologi MARA for providing the financial support and facilities until completion of the project.

6. References

[1] Azmi, M., Saman, M. Z. M., Sharif, S., Zakuan, N., & Mahmood, S. 2006 Proposed Framework for End-Of-Life Vehicle Recycling System Implementation in Malaysia.

[2] Shuhaili, A., Fadzil, A., Ihsan, S. I., & Faris, W. F. 2013. Air Pollution Study of Vehicles Emission In High Volume Traffic: Selangor, Malaysia As A Case Study. WSEAS Transactions on Systems, 12(2), 67-84

[3] Choong, J. S. S. 2010. Conceptualization of Perceived Utilization of Public Bus Transport Service: Empirical Evidence for Malaysia Transport Sustainability (Doctoral dissertation, Universiti Sains Malaysia).

[4] Raub, J. A., Mathieu-Nolf, M., Hampson, N. B., & Thom, S. R. (2000). Carbon monoxide poisoning—a public health perspective. Toxicology, 145(1), 1-14.

[5] Nelson, V. C. (2011). Introduction to renewable energy. CRC press.

[6] Hamzah, N., Othman, K. A., & Zabidi, A. N. Z. M. (2013). Monitoring Carbon Monoxide Emission in the Air Using Wireless Application. In Proceedings of the World Congress on Engineering (Vol. 2).

[7] Voogt, J. (2007). How researchers measure urban heat islands. In United States Environmental Protection Agency (EPA), State and Local Climate and Energy Program, Heat Island Effect, Urban Heat Island Webcasts and Conference Calls.

[8] The Malaysian Times, "Global warming, a warning to all: The Malaysian Times," 30 July 2012. [Online]. Available: http://www.themalaysiantimes.com.my/. [Accessed December 2013].

[9] Rashid, Z.A., Al Junid, S.A.M. 2014. "Trees' cooling effect on surrounding air temperature monitoring system: Implementation and observation." *International Journal of Simulation: Systems, Science and Technology* 15 (2): 70-77.

[10] Amiruddin, A.I.A., Al Junid, S.A.M., Osman, F.N., Thani, S.K.S.O. 2014 "Real-time mobile microclimate monitoring system using labview RT." *International Journal of Simulation: Systems, Science and Technology* 15 (2): 91-97.

[11] Othman, N.; Manan, M.I.A.; Othman, Z.; Al Junid, S.A.M. 2013. "Performance analysis of dual-axis solar tracking system," in Control System, Computing and Engineering (ICCSCE), 2013 IEEE International Conference on , vol., no., pp.370-375

[12] Tahir, N. A. M., S. A. M. Al Junid, Z. Othman, Z. Abd Majid, and S. K. S. O. Thani. 2012. "Design Automatic Meter Reading (AMR) Data Logger with Xbee." *International Journal of Simulation: Systems, Science and Technology* 13 (1): 67-73.

[13] Ussdek, M. E. M., S. A. M. Al Junid, Z. A. Majid, F. N. Osman, and Z. Othman. 2013. "High-Sensitivity Gas Detection and Monitoring System for High-Risk Welding Activity.". doi:10.1109/SPC.2013.6735143.

[14] Husin, R., S. A. M. Al Junid, Z. Abd Majid, Z. Othman, K. K. Md Shariff, H. Hashim, and M. F. Saari. 2012. "Automatic Street Lighting System for Energy Efficiency Based on Low Cost Microcontroller." *International Journal of Simulation: Systems, Science and Technology* 13 (1): 29-34.

[15] Milton, R., & Steed, A. (2007). Mapping carbon monoxide using GPS tracked sensors. Environmental monitoring and Assessment, 124(1-3), 1-19.

[16] Sinivee, V. (2010). Simple yet efficient NMEA sentence generator for testing GPS reception firmware and hardware. In Novel Algorithms and Techniques in Telecommunications and Networking (pp. 207-210). Springer Netherlands.

4th International Conference on Electronic Devices, Systems and Applications 2015 (ICEDSA) IOP Publishing
IOP Conf. Series: Materials Science and Engineering **99** (2015) 012010 doi:10.1088/1757-899X/99/1/012010

Wi-Fi and GSM Based Motion Sensor for Home Security System Apllication

Mohamad Huzaimy Jusoh[1,2], Muhammad Firdaus Bin Jamali [2], Ahmad Faizal bin Zainal Abidin[1,2], Ahmad Asari Sulaiman[1,2] and Mohamad Fahmi Hussin[1,2]

[1]Applied Electromagnetic Research Group,Advance Computing and Communication Communities of Research,
[2]Faculty of Electrical Engineering, Universiti Teknologi MARA Shah Alam, Selangor

E-mail: huzaimy@salam.uitm.edu.my

Abstract. The Wi-Fi and GSM based home security system is a system designed to reduce the high rates of crimes in most personal housing. The overall project consists of three major parts; the input part that consists of sensors, the software part that operates the entire hardware structure, and the output part, which consists of camera, alarm system, and micro secure digital (SD) data storage card. It is based on the principle of infrared radiation generated by a human body heat which trigger the passive infrared (PIR) sensor. The microcontroller processes the received signal, then trigger the buzzer alarm, camera and alerts the home owner through an SMS. Once triggered, the camera will capture the image of the intruder and the image will be saved in SD card. As alert to the user (away), the Global System for Mobile Communication (GSM) will send the Short Message Service (SMS) from the device to the user's mobile phone. The image will be sent to Dropbox data cloud storage via Wi-Fi for further clarification. The prototype was successfully developed, tested and has been installed at residential area in Taman Cahaya Alam, Section U12, Shah Alam, Malaysia.

1. Introduction

In this modern day of digital technology, home security system is becoming one of the fastest developing application based technologies in the world because human needs for their living and property crime protection. The concept of this project is to integrate wireless technologies on the system and powered with clean renewable energy. Wi-Fi is a technology that allows electronic device to exchange data wirelessly via radio waves. Network access point (or hotspot) has 20 m (65 ft) range indoors and a greater range outdoors. However, most of security systems available in current market, needed wiring, expensive, bond with a contract, complex system, bulky size, and fixed. Hence, a comprehensive portable and simple security system should be develops to fill these loophole gaps. The implementation of this project is to provide the user monitoring ability of their premises from anywhere in real time, alert the user through SMS and photo capture based on plug and play system. From the paper [1], the authors propose a system which is works when there is a breach attempt at the house area, the owner will be alerted through e-mail and message is sent to home owner instantly. However, there is a lack of function camera capturing application. In paper [2], the system used the camera to capture the image when the sensor is triggered without real time notification to the owner. From the paper [3], the proposed system contains three kinds of security sensors. When any of these three sensors triggered, the alarm will triggered. In paper [4], the system is developed by using wireless sensor network based on the Zigbee technology. Besides, the system used the GPRS network for monitoring application. In paper [5], the security system consist of sensors, buzzers, relays and the owner will be alerted through SMS, alarm and e-mail. This developed Wi-Fi and GSM based motion sensor for home security application project aim is to comprehend the existing security system.

2. System Design

The initial system was designed by using block diagram as shown in Figure 1. The system consists of two parts which are transmitting and receiving part. For transmitting part, the motion sensor, camera, RF transmitter, SD card, solar panel and battery are connected to the microcontroller. RF link Transmitter and Receiver act as the medium of the data communication between the sensor and the microcontroller. At the receiving part, the system consists of RF receiver, alarm and GSM module.

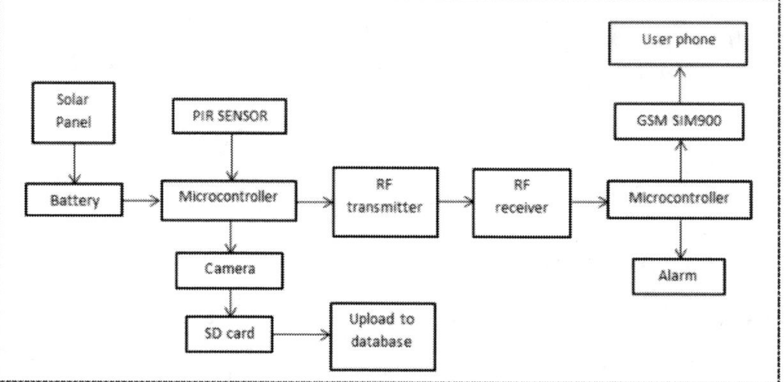

Figure 1. System Block Diagram

2.1 Arduino Yun Prototype Board

The Arduino Yun is a prototype board consists of one microcontroller; Atmel Atmega32u4) and one Wi-Fi System-on-Chip (WISOC); Atheros AR9331. It features built-in Ethernet (IEEE 802.3 10/100Mbit/s) and Wi-Fi (IEEE 802.11b/g/n) support, 2.0 USB-A port, micro-SD card slot, 20 digital input/output pins, 16 MHz crystal oscillator, and micro USB connection. This board is used as a processor unit for transmitter system.

2.2 Atmel Atmega328P Microcontroller Development Board

Arduino Uno Board functioning as microcontroller processor unit for receiver system. The microcontroller board operates at 5V. It has 32 KB flash memory in which 0.5 KB of it used for boot loader, 2 KB of Static Random Access Memory (SRAM), 1 KB of Electrically Erasable Programmable Read-Only Memory (EEPROM) and 16 MHz clock speed [6]. This development board features six analog-to-digital converter (ADC) pins, six pulse width modulation (PWM) pins, 14 digital input/output (IO) pins, 5VDC voltage output pin and 3VDC voltage pin. The small physical dimensions; 68.6 mm × 53.4 mm and 25 g in weight enhance the device portability.

2.3 SIM900 Quad-band GSM GPRS Module

The SIM900 Quad-band GSM GPRS module delivers GSM/GPRS 850/900/1800/1900 MHz performance and it provides service for voice, SMS and data [7]. AT commands are used to interface, configure and control the GSM GPRS shield. The GPRS Shield consists of 12 GPIOs, embedded TCP/UDP stack, speaker jack, headphone jack, SIM Card holder, GSM Antenna and ADC [7]. This module powered by 9VDC.

2.4 Passive Infrared (PIR) Motion Sensor

Passive infrared (PIR) sensors detect motion by comparing the amount of infrared radiation that reaches on a pair of detectors. When the two detectors read different values, the sensor indicates it as movement of an object, such as a person. The sensitivity of the PIR motion sensor can be up to 7m. The angle of detection is about 120°.

2.5 RF Link Transmitter and Receiver

RF link transmitter and receiver are used as a medium of data communication between the sensor and the microcontroller. It consists of two ranges of frequency mode, which are 315 MHz and 433 MHz. The radio frequency operate by applying Amplitude Shift Keying (ASK) modulation technique. These modules have the capability to communicate up to 100m range for open air and 40m for indoor uses. The transmitter operates from 3-12V.

3. System Design

The microcontroller at transmitter system (show at Figure 2 (a)) will connect to the network and then calibrate the PIR motion sensor to avoid false trigger. When PIR motion sensor detect any physical motion, the LED indicator will turn on and the signal will be transmitted from the RF transmitter to the RF receiver at the microcontroller. Next, camera will capture image of the intruder. The image will be saved on the SD card and will be uploaded to the database via Wi-Fi. The image quality is depends on the internet access speed and Wi-Fi signal quality. The user enable to access the intruder image on the website database. The microcontroller at receiver system (show at Figure 2 (b)) processes the received signal, turn on LED and triggers alarm for 15 seconds. The microcontroller will send a short messaging service (SMS) text to inform the user via mobile phone to notify invasion. The buzzer and LED will be turn off if there is any signal received by the microcontroller.

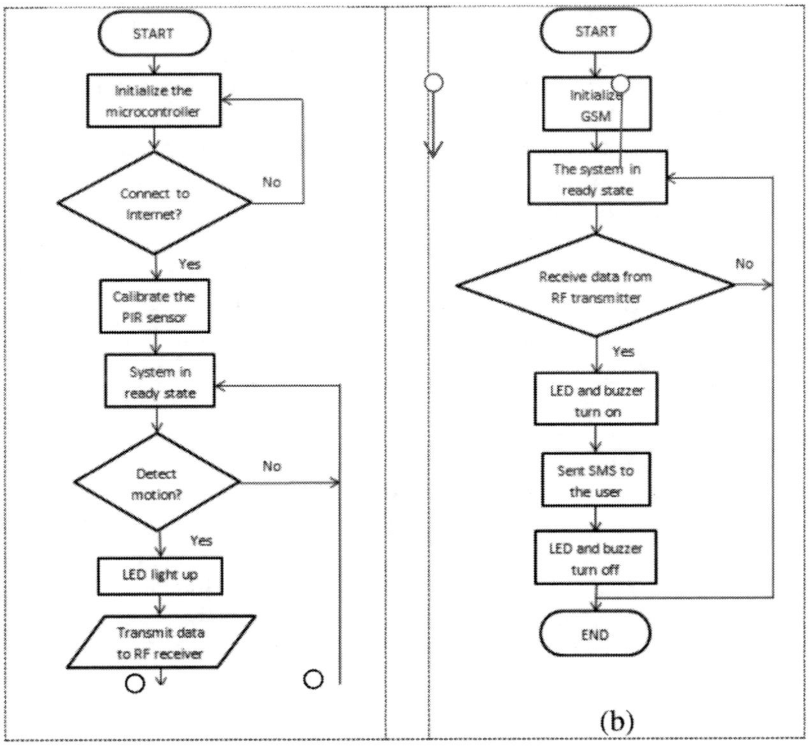

(b)

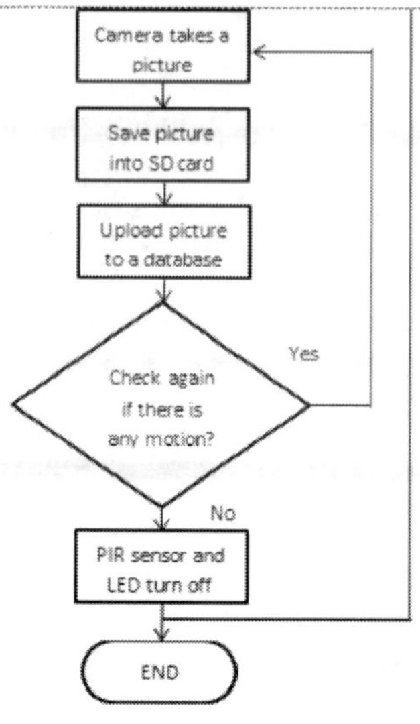

(a)

Figure 2(a). System operation for transmitter system.

Figure 2(b). System operation for receiver system.

4. Methodology

4.1 Project Development

At the beginning of the project, the background of the related field has been studied to collect data and generate the ideas. Based on the survey on the current home security in market today, there are still have some loophole and could be filled. The idea in here to design and implement a wireless sensor home security system using clean renewable energy. The next step is designing the overall system including the system functionality and specification. The components of hardware that will be used for this project is studied and chose according to the specification, flexibility, cost and user friendly, which comply with engineering economic management.

After one month performing system evaluation, the system finally works properly and meet the project objectives. Then, the system undergoes laboratory analysis and measurement to obtain the result for technical paper writing. After the laboratory evaluation done, the system was installed at Taman Cahaya Alam Section U12 residential area. After the site testing done, the feedback from the owner of the tested house was collected for system upgrading and improving. From here, we could know the ideal home security system.

4.2 Testing and Troubleshooting

First problem was the GSM did not send the SMS to the user when the sensor was triggered. The solution of the problem is by using the GSM_GPRS_GPS_Shield_GSMSHIELD Library since the GSM shield don't have its own firmware c-language library. The second problem is the camera did not capture the image when sensor was triggered. This issue was happened because there was a loose wire connection between camera and microcontroller. The third problem happened when the buzzer was not working properly. The issue was overcame by changing the buzzer.

4.3 Installation

Figure 3 shows that the sensor device had been installed at the residential area. The sensor was installed at the height of 4.2m to avoid intruder easily terminate the sensor. The sensor will be triggered if someone located at the range of the angle α which is 85°.

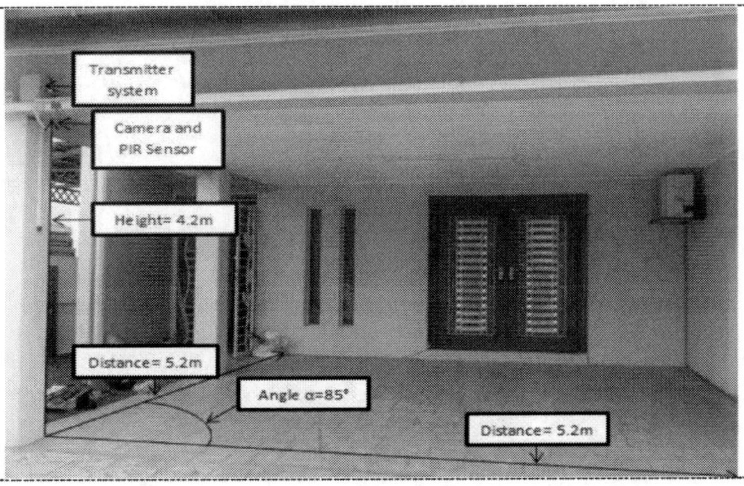

Figure 3. Horizontal detection angle and distance

In Figure 4, the blind spot is at the angle β and the minimum distance of detection is 1.5m.

Hence, the sensor disable to detect movement at 1.5 m distance and 35° angle. The sensor should be installed at the angle γ of 55°. The angle β can be reduced by facing the sensor toward the ground to increase the angle of detection. However, this procedure will decrease the distance of detection, the maximum distance of detection is 5.5m from the sensor. Any motion in this distance will be detected by the sensor.

Figure 4. Vertical detection angle and distance

The solar panel is installed on the house roof. The distance limit of transmitter and receiver must not be exceed 40 m.

5. Result and Discussion

5.1. Testing and Measurement

Figure 5 shows the graph of the voltage versus time of the PIR motion sensor and RF transmitter. The X-axis represent time and Y-axis represent voltage of the PIR sensor and RF transmitter within duration of 45s. The graph show the voltage value increase from 0V to 3.27V when the sensor is triggered. The data will be sent from RF receiver for 2 seconds duration. The voltage will be drop to 0V after 40s. This indicates that the sensor is now at the off state or no movement is detected.

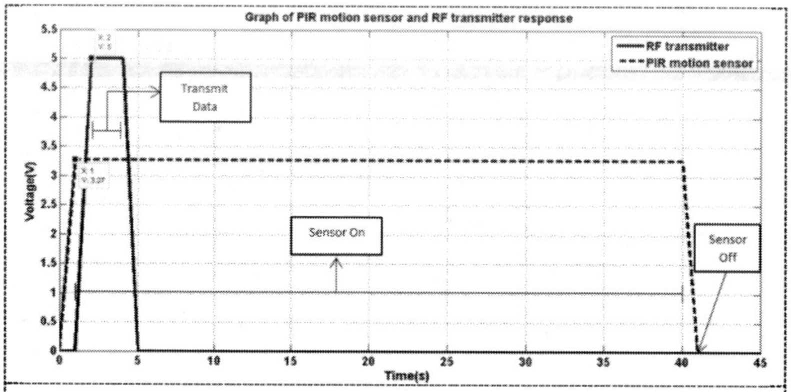

Figure 5. PIR motion sensor and RF transmitter response graph

Figure 6 show the graph of voltage versus time of the RF receiver and alarm. The X-axis represent time and Y-axis represent voltage of the RF receiver and alarm within duration of 45s.. During 5s to 6s, the graph indicates that there will be a sending data from the RF transmitter. When the motion sensor detects the motion, the signal will be transmit to the microcontroller. The graph shows 4.8V indicates that the alarm will be turned on by the microcontroller. After 15s, the voltage will be dropped to 0V. This indicates that the system is on the standby mode.

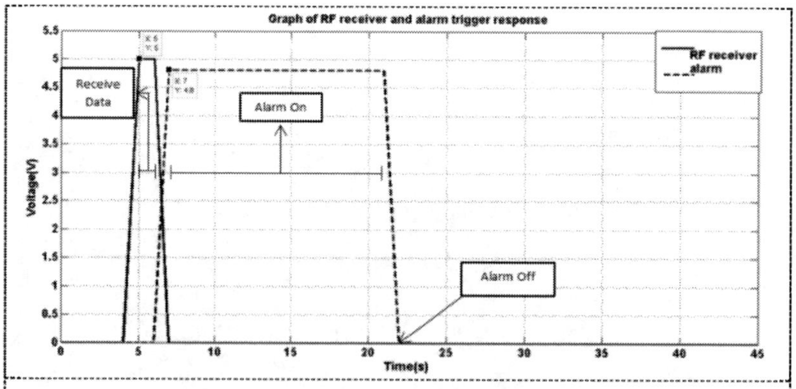

Figure 6. RF receiver and alarm response graph

The PIR motion sensor is chose as the main proximity sensor due to low cost, low power consumption, and small form [8]. The angle of blind spot of the sensor can be reduced by facing the sensor towards the ground to increase the angle of detection. However, this procedure will decrease the distance of detection. Wi-Fi is used to upload the captured image into database compared to GPRS due to size of the picture which is about 80KB to 100KB, depending on the quality, contrast and brightness of the image.

Increasing the sensitivity of the sensor will result the increase of the distance of detection but it will lead the sensor to consume more power. The level of security system will be increased by installing more sensors and camera will result in increasing in power consumption, cost, size and weight of the device. Besides, it wills faster the time on the SD card to reaches full capacity.

6. Conclusion

As the conclusion, the Wi-Fi and GSM Based Motion Sensor for Home Security System has been designed and tested with real time monitoring capability based on renewable energy. With limited budget, it is possible to develop and implement the system. This system will help to increase the safety of the private premises because the user can monitor premises from anywhere in real time. The user can get alert anywhere through GSM technology as long as the mobile communication network coverage is present. Besides, the user is also possible to identify the captured image through database system in real time. With implementation of this system, the security system can be improved to a higher level and educate the community to implement renewable energy concept in daily life. The security could be improved more by considering more detection features. However, this is only a crime prevention and detection mechanism. In order to fully eliminate crime issues, the development on positives human capital and attitudes is the best way of crime prevention solution.
this project.

7. References

[1] Assaf M.H., Mootoo R, Das S.R., Petriu E.M., Groza V., Biswas S., *Sensor based home automation and security system"* Instrumentation and Measurement Technology Conference (I2MTC), 2012 IEEE International, DOI: 10.1109/I2MTC.2012.6229153, Publication Year:2012, Page(s): 722-727

[2] M.Meyer, M. Hotter and T. Ohmacht, *A new system for Video-based Detection of moving objects and its integration into digital networks,* Security Technology 1996, 30th Annual 1996 International Carnahan Conference (1996), pp.105-110

[3] Y. Zhao and Z.Ye, *Low cost GSM/GPRS based wireless home security system*, IEE Trans. Consumer Electron, vol.56, no.4, (2007) January, pp. 546-567

[4] Ming Xu, Longhua Ma, Feng Xia, Tengkai Yuan, Jixin Qian, Meng Shao, *Design and Implementation of a Wireless Sensor Network for Smart Homes*, Beijing 2001 International Conference.

[5] Jayashri Bangali and Arvind Shaligram, *Design and Implementation of Security Systems for Smart Home based on GSM technology*, International Journal of Smart Home, Vol.7, No.6 (2013), pp. 201-208

[6] M. Banzi, M. Shiloh, *Getting Started with Arduino 3rd edition*, Maker Media Inc, Sebastopol, California, USA 2014

[7] *GSM/GPRS Shield Datasheet*, Tinysine (Tinyos) Electronics, Anhui, China, 2014

[8] P. Zappi, E. Farella, L. Benini, *Tracking Motion Direction and Distance With Pyroelectric InfraRed Sensors*, in IEEE Sensors Journal, Volume 10, no.9, pp.1486-1494, September. 2010.

8. Acknowledgement

The authors would like to thank Mr Shamry and his colleague from Lui(M) Sdn. Bhd. for participating in series of technology discussions and technical support. The authors also appreciate the cooperative from resident of Taman Cahaya Alam, Section U12 Shah Alam for allowing the premises to be used as test site. Lastly, the authors would like to acknowledge Research Grant 600-RMI/DANA 513/REI and Faculty of Electrical Engineering, UiTM Shah Alam for financial and facilities support in developing this project.

IOP Conf. Series: Materials Science and Engineering **99** (2015) 012011 doi:10.1088/1757-899X/99/1/012011

Real-Time Remote Monitoring with Data Acquisition System

Ahmad Faizal Zainal Abidin[1,2], Mohammad Huzaimy Jusoh[1,2], Elster James[2], Syed Abdul Mutalib Al Junid[2] and Ahmad Ihsan Mohd Yassin[2]

[1]Applied Electromagnetic Research Group, Advance Computing and Communication Communities of Research, Faculty of Electrical Engineering, Universiti Teknologi MARA, Shah Alam, Selangor, Malaysia
[2]Faculty of Electrical Engineering, Universiti Teknologi MARA, Shah Alam, Selangor, Malaysia

Email: huzaimy@salam.uitm.edu.my

Abstract. The purpose of this system is to provide monitoring system for an electrical device and enable remote monitoring via web based application. This monitoring system allow the user to monitor the device condition from anywhere as the information will be synchronised to the website. The current and voltage reading of the monitored equipment, ambient temperature and humidity level are monitored and recorded. These parameters will be updated on the web page. All these sensor are connected to the microcontroller and the data will saved in micro secure digital (SD) card and send all the gathered information to a web page using the GPRS service connection synchronously. The collected data will be displayed on the website and the user enable to download the data directly from the website. The system will help user to monitor the devices condition and ambient changes with ease. The system is successfully developed, tested and has been installed at residential area in Taman Cahaya Alam, Section U12, Shah Alam, Selangor, Malaysia.

1. Introduction

Real time data monitoring is an important support application in order to monitor supported electrical device conditions especially when the observed parameters influence the supported equipment electrical device operation such as temperature, humidity, voltage, current and wind condition [1]. Internet embedded technology makes data transferring and accessibility around the globe possible where machine could communicate with computer in performing its operation [2]. The idea of wireless data transmission is to provide device simplicity instead of wired system and lower cost for long range communication [3]. In some cases, the frequent human site visit is not permissible due to a number of factors such as safety, rough terrain, huge cost per visit, weather condition and danger wildlife. To overcome the issues, a long-term long-range wireless monitoring system, which required low maintenance essential to be established [4]. In nowadays Internet of Things (IoT) era, the sensor data sampling can be live fed into website and can be accessed anywhere as long as internet access is permissible [5]. In taking the advantage of nowadays technology achievement, an unmanned monitoring system can be established in order to overcome the stated difficulties. On top of that, by establishing the real time monitoring system, the human site visits for configuration and maintenance could be minimized. Hence, project and manpower costs could be also minimized. In this research project, a real time remote data monitoring sensors device is developed along with web based data acquisition (DAQ) system for user friendly data access.

2. System Design and Operation

The real-time data monitoring device is consists of six inputs and two outputs. All electronics operation is controlled by a single microcontroller unit, which is programmed on how the all peripherals should be behaved. In this device, it consists of five sensors; temperature and humidity (integrated in a single module), two current and two voltage sensors. The temperature and humidity sensors are responsible to detect ambient changes. One of the current sensors is used to detect current of equipment to be monitored while the other one is used to detect the real-time data monitoring device. The same operation also applied to both voltage sensors, where one of the voltage sensors is used for the real-time data monitoring device and the other one is used for main equipment monitoring. All of the sensors data reading will register into the microcontroller unit along with date and time stamp synchronously. The data will be uploaded over GSM/GPRS module to remote monitoring database to provide user the real time data, at the same time the data is logged into microSD module. The data is accessible anywhere through web base application where the data stored inside cloud storage application. The web base application serve as remote data acquisition application which shows real-time numerical data and graphical data plotting. Hence, the data is accessible anywhere and any form of internet capable electronic gadgets such as lap top, desktop and smartphone. The data collected by the sensors will be uploaded into website server and these data will be updated at the website specific channel and displayed for viewing [10]. The data will be updated every 30 minutes. The system design architecture can be referred in Figure 1 as follow.

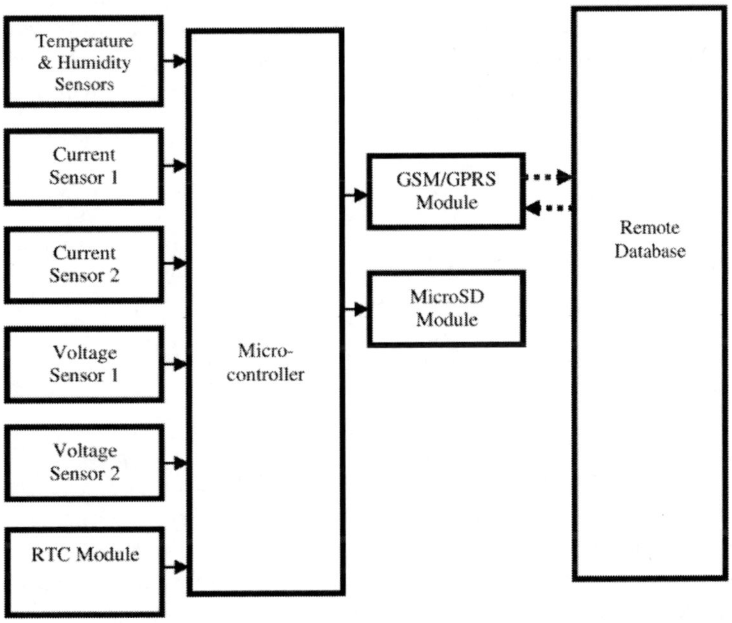

Figure 1. Real-time monitoring device system architecture

2.1 Microcontroller Unit

A microcontroller behaves as a small capacity computer on electronic hardware architecture. The microcontroller will determine on how the electronics peripherals that attached to it operate and act as integrated system which therefore programmed and burn into the microcontroller flash memory. The electronic peripherals that attached to the microcontroller communicate with each other via serial communication either via Inter Integrated Circuit (I2C), Serial Peripheral Interface (SPI) or Universal Asynchronous Receiver/Transmitter (UART). In this project, Atmel Atmega2560 microcontroller is chosen to perform the hardware operation control and serial data processing task. The microcontroller consists of 54 digital input/output pins where 15 of it can be used as pulse width modulation (PWM), 4

UARTs interface, 16 10-bit analog digital converter (ADC) input pins, 256 kB flash memory (which 8 kB contain bootloader) and 8 kB Static Random Access Memory (SRAM). 16 MHz crystal oscillator is attached along with the microcontroller as clock speed generator.

2.2 GSM/GPRS Module

For long range remote duplex data communication, General Packet Radio Service (GPRS) communication is suitable candidate to execute the task[Amin, A. & Khan, M. N. A. 2014. A Survey of GSM Technology to Control Remote Devices. *International Journal of u- and e- Service, Science and Technology, vol. 7, no. 6, pp. 153-162*]. The data packet will be sending over GPRS and uploaded into data storage cloud. A quad-band 850/900/1800/1900 MHz GSM module is chosen to perform the data transmitter. The GSM module embedded with TCP stack which allows data upload into web server. With existing telco network service on wide spreading the network areas and satellite communication, the data could be sent and received most location on the globe.

2.2 Temperature and humidity sensor

DHT11 is a low cost temperature and humidity sensor that comprises a calibrated digital signal output for temperature and humidity [4]. The sensor has ±5% accuracy for 20-80% humidity range and ±2°C accuracy for temperature from 0-50°C [4]. 5 VDC required to operate this sensor.

2.3 Analog Voltage Divider

The Analog Voltage Divider V2 by DFRobot capable to detect voltage from 0.0245V to 25V. The module communicates with the microcontroller via ADC channel, which the Atmel Atmega2560 serves 10-bit ADC. The sensor is based on the concept of voltage divider, which the measured voltage is scale reduced by five times, converted into digital reading, ratio over 1024 (10-bit) and times with maximum 25V as perform in equation (1) as follows.

$$V_m = {}^{D_v}/_{1024} \times 25\,V \tag{1}$$

Where, V_m = measured voltage
D_v = converted digital in decimal value

The module communicates with the microcontroller via ADC channel, which the Atmel Atmega2560 serve 10-bit ADC.

2.4 Current Sensor (AC/DC)

In this project, the ACS758 current sensor based on Hall Effect magnetic sensing capable to measure from -50A to 50 A AC and DC current with the peak voltage measuring up to 1000 V for AC and 500V for DC. This module will produce analog voltage output based on magnetic perturbation onto the Hall Effect sensor. The output analog voltage will supply into ADC channel of microcontroller unit and the current reading will be interpreted based on the equation (3).

$$I = \left({}^{(D_i - D_r)}/_{D_f} \times V_f \times sensitivity \right) - offset\ value \tag{2}$$

Where, D_i = converted detected decimal digital value
D_r = raw decimal digital value when input = 0 A, in this case D_r equal to 510
D_f = full digital value, in this case 10-bit ADC equal to 1024
V_m = maximum voltage at full ADC bit
Sensitivity = 0.04 V/A

Offset value = 0.04 A

$$I = \left({}^{(D_i - 510)}/_{1024} \times 5\,V \times 0.04\,V/A \right) - 0.04\,A \tag{3}$$

3. Methodology

First, the GSM/GPRS module will attempt to establish the connection with GPRS network. Once GPRS connection is established, the GSM/GPRS module will verify the established connection by receiving the network information of the signal strength and network registration through the user interface window. Then, real time clock (RTC) module is accessed through I2C interface to obtain the current time. The SD card is accessed via SPI interface and a new csv spreadsheet file is created. If the file is already existed, the data inside the file will be overwritten. All five sensors started taking measurement on the indicated parameter; temperature and humidity, monitored equipment voltage and current and the project device voltage and current. The recorded measurements are saved in the csv file along with date and time stamp. The recorded measurement will be sent over GPRS connection to the remote monitoring database. If the data sending fail, the microcontroller will retransmit the data to the remote monitoring server. The remote monitoring database server will received the obtained data and display these data accordingly based on the parameter assigned. The website is public accessed but the data from the remote monitoring database can be accessed by the authorized user only. The real-time device monitoring system flow can be referred in Figure 2.

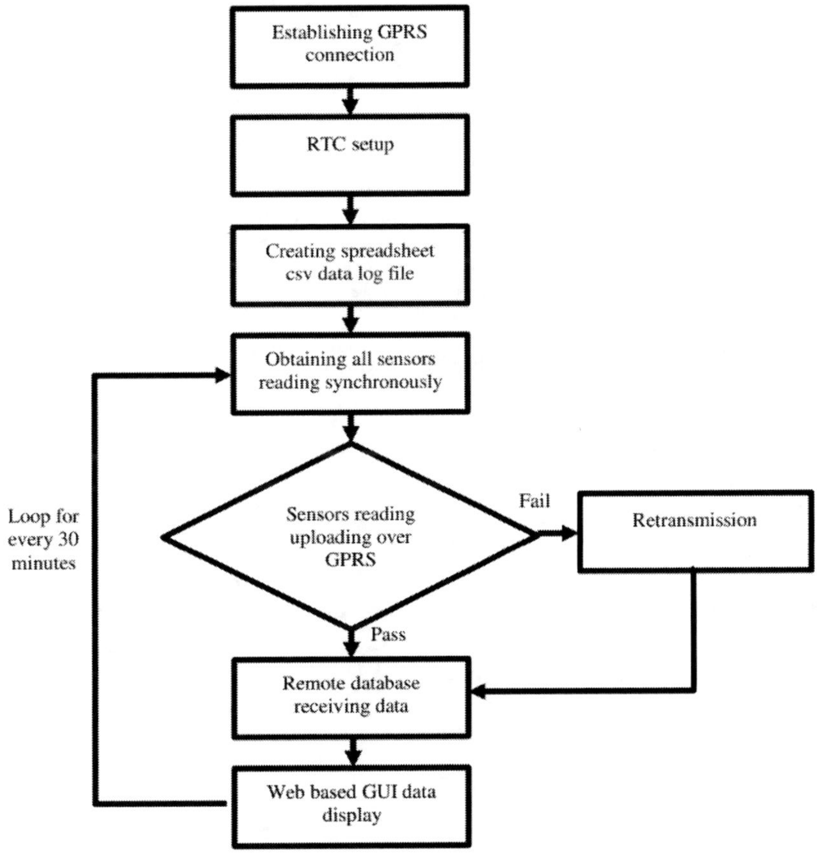

Figure 2. Real-time device monitoring system flow

IOP Conf. Series: Materials Science and Engineering **99** (2015) 012011 doi:10.1088/1757-899X/99/1/012011

4. Results

A device running and testing had been performed on 2nd May 2015 at 3.19 p.m. until 4th May 2015 at 6.56 a.m. at Cahaya Alam, Section U12, Shah Alam, Selangor, Malaysia residence area to evaluate the device workability in real situation. For the results, the measurements are categorised into three parameters: humidity and temperature, voltage and current analysis.

4.1 Temperature and Humidity

The Figure 3 shows the temperature (Figure 2 (top)) and humidity (Figure 2 (below)) reading samples taken once for every 30 minutes. The measurements had been ran for over 79 samplings, which the device had been ran for 39 hours and 30 minutes continuously. The temperature reading shows 31°C max and 30°C min, where the temperature during night is obviously drop by 1°C at first day, constantly at 31°C during days and roughly during night for second day. For humidity sensing, the humidity is higher during night and lower during day generally. During night, the humidity percentage is varies in the range of 63% to 74%. The humidity percentage drop reach at 59% and varies in the range of 59% to 68%.

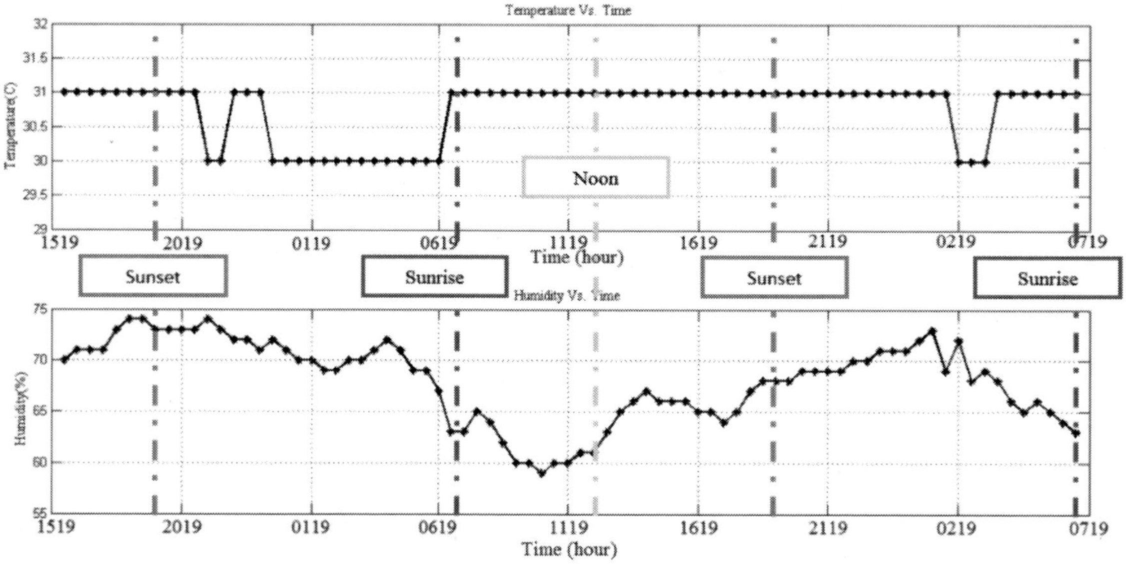

Figure 3. Humidity (top) and temperature (below) versus time. Samplings continuously taken for over 39 hours and 30 minutes

4.2 Voltage

Voltage reading monitoring is vital to identify the real time power supply voltage, where the power storage left can be identified. In voltage monitoring, the Figure 4 (top) shows the voltage monitoring on power supply of supported device and Figure 4 (below) shows the power supply of real-time monitoring voltage magnitude. For device testing, the voltage measurement was done on battery power supply of portable security system device that was installed at the site. The real-time monitoring device is powered up by 9V battery, the Figure 4 (below) shows the power supply voltage magnitude taken for every 30 minutes.

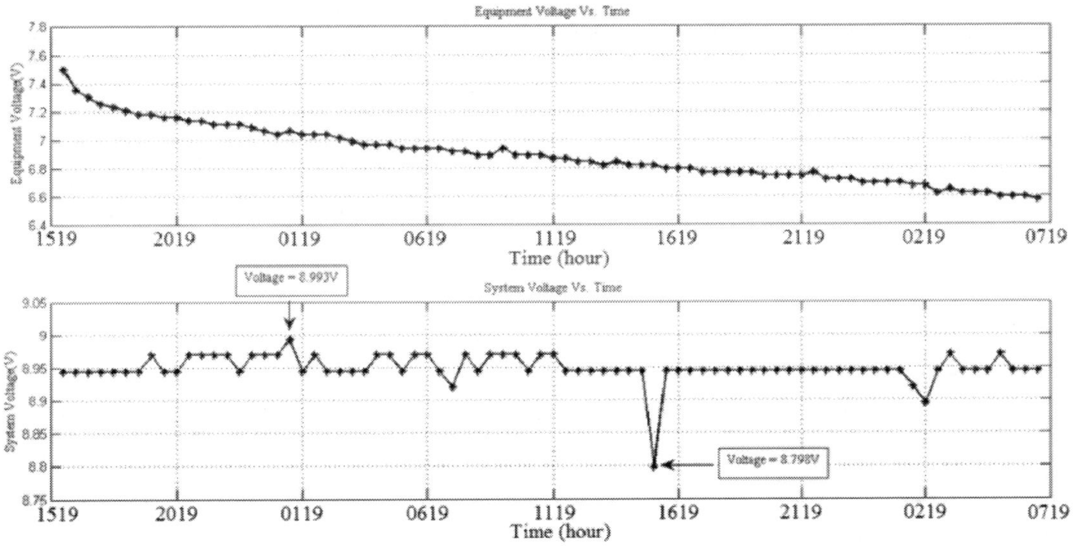

Figure 4. Supported equipment (top) and real-time device monitoring (below) power supply voltage reading versus time

4.3 Current

The Figure 5 shows the current consumption sampling taken on supported device (Figure 5 (top)) and real-time monitoring device (Figure 5 (below)) versus local time. The supported device has sleep mode features which will be partially hibernated during inactivity, where it will be fully activated when triggered by passive infrared sensor (PIR) pulse. The 0 A current consumption in Figure 4 (top) shows the inactivity of supported device, which from the figure it can give the details of device activity. For the real-time device monitoring current consumption, the device constantly consume 200 mA from 1519 hour until 1119 hour, fluctuate between 100 mA and 200 mA from 1149 hour until 0019 hour and constantly consume 99 mA from 0049 hour until 0649 hour.

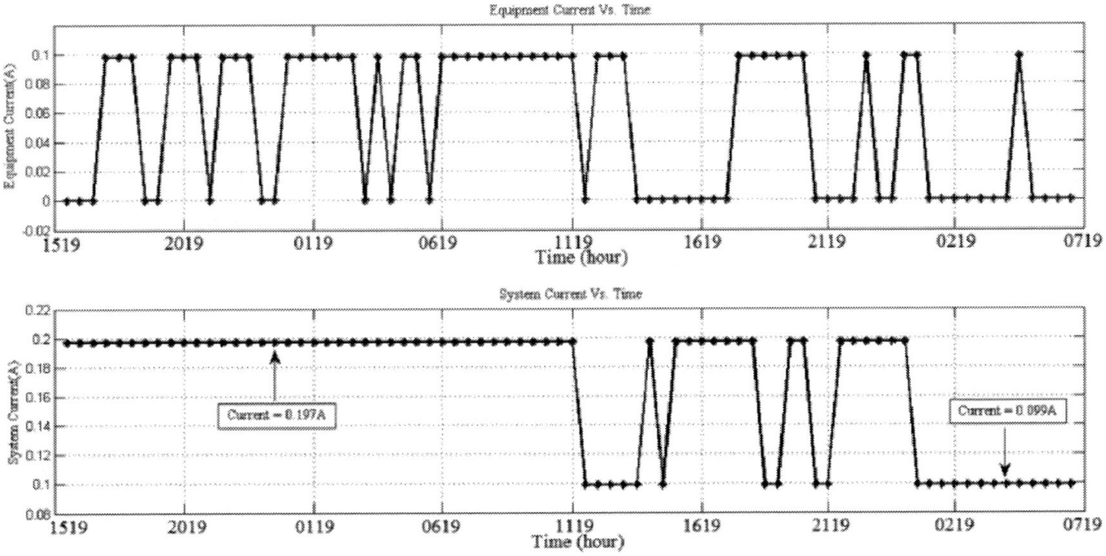

Figure 5. Supported electrical device (top) and real-time monitoring device current consumption (below) versus time

5. Conclusion

The idea of this system is to enable unmanned monitoring remotely. The system is designed to monitor any electrical device voltage and current magnitude, the system self-voltage, self-current, surrounding temperature and humidity magnitude. User enable to acknowledge the electrical device conditions anywhere and anytime without site visit. Thus the device could give time and cost beneficial. Furthermore, the device could assist user in planning for future upgrade or maintenance of the equipment. Overall, the system has been received positive remarks from industrial panel from electronic engineering solution company, which who also provide remote monitoring technology to North-South Expressway (PLUS) Malaysia.

References

[1] Nhivekar, G. S. & Mudholker, R. R. 2011. Data Logger and Remote Monitoring System for Multiple Parameter Measurement Applications. *e-Journal of Science & Technology, vol.3, no.6, pp. 55-62*

[2] Singh, J. S. & Padmalatha, L. 2013. Development of Http Server for Remote Data Monitoring and Recording System. *International Journal of Computer & Technology, vol. 11, no. 4, pp. 2440-2445*

[3] Ghayvat, H., Mukhopadhyay, S., Gui, X., & Suryadevara, N. 2015. WSN- and IOT-Based Smart Homes and Their Extension to Smart Buildings. *Sensors 2015, vol. 15, no., pp. 10350-10379*

[4] Lazarescu, M. T. 2015. Design and Field Test of a WSN Platform Prototype for Long-Term Environmental Monitoring. *Sensors 2015, vol.15, no., pp. 9481-9518*

[5] Ling, N. & Mei-Xia, D. 2015. Design of a Remote Data Monitoring System based on Sensor Network. *International Journal of Smart Home, vol. 9, no. 5, pp. 23-30*

Acknowledgements

The authors would like to acknowledge the chief executive officer of Luimewah (M) Sdn. Bhd., Mr. Shamry Mubdi Subra Mullisi and his team for providing us very brief technical consultancy and support. The authors would like to say thanks for the cooperation from resident of Taman Cahaya Alam, Section U12, Shah Alam, Selangor, Malaysia. Last but not least, the authors would like to acknowledge 600-RMI/DANA 5/3/REI (8/2014) and Faculty of Electrical Engineering, UiTM Shah Alam for financial and facilities support.

An Overview of Hand Gestures Recognition System Techniques

Farah Farhana Mod Ma'asum[1]**, Suhana Sulaiman, Azilah Saparon**
Center for Electronics Engineering Studies (CEES)
Faculty of Electrical Engineering, Universiti Teknologi Mara Shah Alam,
40450 Shah Alam, Selangor, Malaysia

E-mail: farah.farhana88@yahoo.com

Abstract. Hand gesture recognition system has evolved tremendously in the recent few years because of its ability to interact with machine efficiently. Mankind tries to incorporate human gestures into modern technology by searching and finding a replacement of multi touch technology which does not require any touching movement on screen. This paper presents an overview on several methods to realize hand gesture recognition by using three main modules: camera and segmentation module, detection module and feature extraction module. There are many methods which can be used to get the respective results depending on its advantages. Summary of previous research and results of hand gesture methods as well as comparison between gesture recognition are also given in this paper.

1. Introduction

Massive improvement in computer development can be clearly seen involving both interaction between human and computer itself. Human interaction with computer has becoming a core component in our daily life as we interact with it throughout the day. Technology nowadays incorporate flat panel displays as an output device of images being transmitted electronically. For an instance, most cars are equipped with computerized systems for navigation and entertainment which needs to be operated and displayed by using a screen panel. However, controlling a computer by gesturing in the air will be everyone wishes as this will ease down our tasks. Hand gesture has been visualized as the new evolvement of human computer interaction which will replace the functions and usage of touch screen displays.

Human computer interaction (HCI) is defined as the relation between the human and computer (machine), represented with the emerging of computer itself [1]. Vital endeavour of hand gesture recognition system is to create a natural interaction between human and computer when controlling and conveying meaningful information. There are two main characteristics should be considered when designing a HCI system as mention in [1]: functionality and usability.

System functionality referred to a set of actions or services that are provided to its users [1], while usability of a system is defined by the level and scope of the system in which the system can be used efficiently in order to obtain certain specific user purposes [2]. By having an equal balance between both functionality and usability of a system, an effectiveness and powerful performance system can be achieved. However in order to create a useful hand gesture recognition system, differences between hand postures and hand gestures need to be distinguished first. Hand posture is a static hand configuration which is represented by a single image without any involvement of movements [3][4], while hand gesture is defined as a dynamic movement consisting of sequences of hand postures over a short span of time [4]. For example, making thumbs up and holding it in a certain position is a hand posture while, waving good-bye is an example of hand gesture.

Gestures by using sign language are usually used among deaf community for communication between human and computer [5]. Hand gestures may represent compressed information [6] which can be decoded between both sender and receiver in order to obtain and understand the gesture meaning. This paper intends to provide an overview of hand gesture recognition system which addresses different approaches that can be used to acquire information of hand gestures recognition, which are segmentation, features extraction, and recognition.

Most of the studies have defined specific gestures for a specific task based on its own purpose of studies. In research done by S. Billinghurst *et. al* [7], natural gestures for common web browsing tasks are tested by examining differences between postures, screen sizes and user's educational background. S. Billinghurst claims that there are quite distinctive differences between the psychology and engineering participants. The psychology and engineering groups has different ways of swiping touch screen panel. Psychology group preferred to swipe from right to left while the engineering group preferred the opposite directions.

2. System components and framework

In order to analyse and process hand gesture recognition, it needs to undergo three main steps which are: camera module, detection module and feature extraction module, as shown in Figure 1. Camera module acquires images from input source while detection module is where the quality of the threshold images is determined. Feature extraction module on the other hand acts as extraction of user's gesture according to its purposes. Comparison between each researcher methods is further explained in Section 2.1.

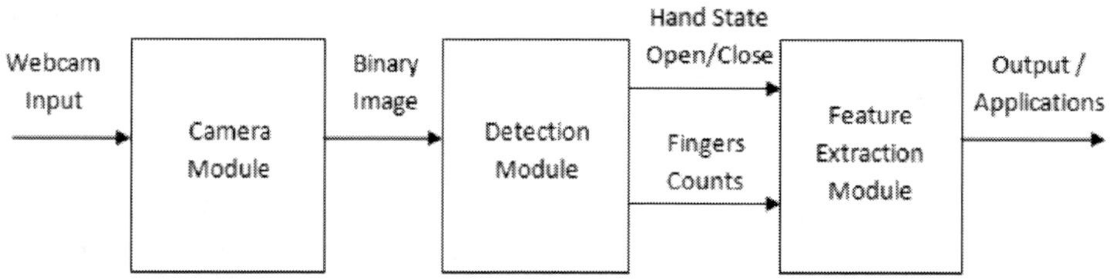

Figure 1. Framework of hand gesture recognition system.

2.1. Camera & Segmentation Module

In this camera module, image of hand can be retrieved from various sources such as single camera (web camera), depth camera and infrared camera with sensors. The image of hand is then need to be segmented so that it could divide input image into required regions. It dealt with separating the user's hand with background in order to capture the gestures without noises.

However, both hand postures (static) and hand gestures (dynamic) do required different segmentation process depending on the type of gestures. If it is dynamic gesture, it needs to undergo two steps before it is being segmented. Firstly, the hand gesture needs to be located and tracked using a bounding box, then the video of gestures will have to be divided into frames and each frame have to be processed alone [8].

A perfect segmentation is difficult to obtain due to overlapping of skin colours with complex background. There are various researches which has been done using different type of colour space technique in segmentation process such as HSV colour model, YCbCr colour space [9] and RGB colour model [10]. However, due to sensitivity of lighting changes in colour spaces, many researches tend to eliminate the luminance component [11].

The most important step for hand segmentation is thresholding which it is used to separate hand from the background. Thresholding can be used as an extraction method from the background by assigning certain values for each pixel as a threshold value[10]. Thresholding will produce a binary image with all pixels of 1 representing the hand. This occurs when pixel with intensity less than the threshold value is set to 0, while any pixel which has the intensity more than the threshold value will be set to 1. Figure 2 shows the results of hand segmentation by using RGB colour space with different threshold values, 70 and 20, by A. Dhawan *et. al* [10] .

Figure 2. Segmentation using RGB color space with different threshold value (left: original image, center: threshold value of 70, right: threshold value of 20) by A. Dhawan [9]

Figure 3 shows the results when H.S. Yeo *et. al* [9] applied YCbCr colour model with different wide and narrow YCbCr threshold value. It can be seen clearly that background with similar colours as skin is not extracted when using a narrow threshold.

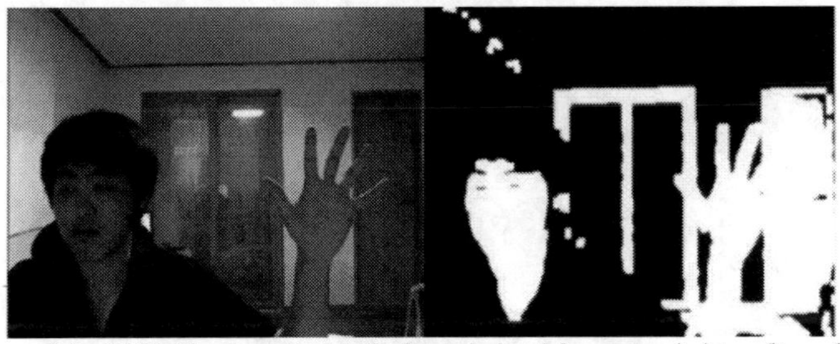

Figure 3. Using wide YCbCr threshold (left: original image, right: after skin extraction), by H.S. Yeo [8]

2.2. Detection Module

Detection module is a module where is responsible for detecting and tracking hand and fingers. Fingers and hand locations, hand close/open state and fingers direction need to be extracted in order to gain more information about the gestures.

2.3. Feature Extraction Module

Features vector of the segmented image can be extracted in many ways according to particular application. Some methods used the shape of the hand such as hand contour which is then being calculated for convex hull and convexity defects [10], while M.M. Hassan *et. al* [5] prefer to divide the segmented image into different blocks representing its brightness measurement of the image. Center of gravity (COG) [12] of segmented hand and the distance from COG to the farthest point of the fingers, is another method which can be used as an feature extraction.

3. Literature Review on Gesture Recognition Systems

M.M. Hassan *et. al* [5] applied scaled normalization for gesture recognition based on brightness factor matching. Initially, the input gesture image is divided into 25x25 blocks (each of 5x5 block size), which is then colour segmented by HSV colour model and thresholding technique. Each segmented images are normalised by using centre of mass technique as shown in Figure 5. In Figure 5, the coordinates of red line are shifted to match the centroid of the hand object at the origin X and Y axis. After being normalized according to its respective normalization technique, these images are then being analyzed by using local brightness value mathematical equation[5]. Comparison between database features and tested gesture features are then being calculated using equation (1) for the percentage between feature image and database, as in [5],[13]. The database consists of 6 different gestures, while 10 samples per gesture are used (5 samples for training and 5 samples for testing). And from his outcome, it has shown that the recognition rate for center mass normalization, as in [5] achieved better results compared to classical normalization technique, as in [13]. This is because of its ability to extract more data from hand information itself. However, time limitation is the disadvantage of this research since it consume more time in interpreting and comparing each gestures and its database.

$$Recognition\ (\%) = \frac{number\ of\ matched\ features}{total\ number\ of\ blocks} \times 100\% \qquad (1)$$

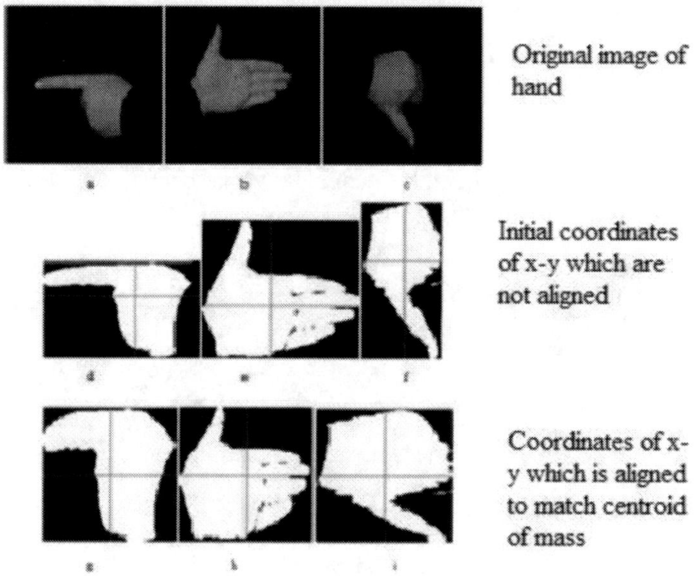

Figure 4. Applying scaling normalization on the input image (a,b,c: Original Image. d,e,f: Image after being segmented. g,e,f: image after scaling normalization process) [5]

G. Thomas *et. al* [14] on the other hand had compared three different types of methods which can be used to recognize hand gestures. These three methods are pixel-by-pixel comparison, edges method and orientation histogram. Pixel-by–pixel comparison is a method where each and every frame is being compared with the image in the database. Edges method [14] on the other hand is used to find out which portion of the image produce the highest gradient value. In order to determine edges of the hand image, two filters named X-Y filter is used – see Figure 6. These two filters represent different directions respectively, for the x direction, X= $[\,1 \quad -1 \quad 0\,]$ while for the y direction, Y= $\begin{bmatrix} -1 \\ 1 \\ 0 \end{bmatrix}$.

Orientation histogram method is actually a method where it combines both edges method and pixel-to-pixel comparison method. Each pixels of the image (x,y) has its own gradient. Gradient direction and gradient magnitude are given by arctan (dx,dy) and $\sqrt{dx^2 + dy^2}$ respectively, as in [14]. If the gradient magnitude is greater than threshold, the gradient direction is detected and the frequency is incremented. This histogram is then saved as a training pattern database. As the last steps, all three methods (pixel-by-pixel comparison, edges method and orientation histogram) used Euclidean Distance to determine the differences between image and its database. And from this finding, it shows that orientation histogram produced the highest percentage of recognition since it is able to recognize gesture fast and robust. However, it does have disadvantage of being rotation dependent.

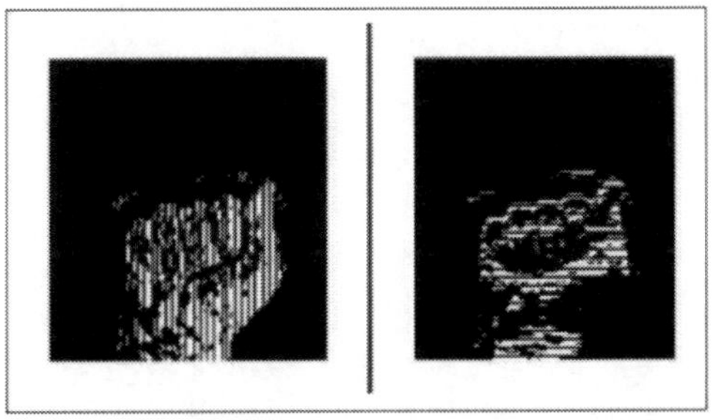

Figure 5. Edge detection using X-Y filter [13]

H.S. Yeo *et. al* [9] and A. Dhawan *et. al* [10] on the other hand implies quite similar technique of feature extraction in their research. Both researchers employed convex hull and convexity defects method to determine both hand and fingertips contours. However, they used difference type of input sources in their researches. A. Dhawan *et. al* [10] only focuses on single camera as an input source, while H.S. Yeo *et. al* [9] compares the results of image recognition using single camera and depth camera. In H.S. Yeo *et.al* [9] studies, he has proved that user can interact with computer without using any physical controllers. Another findings by H.S. Yeo *et. al* [9], single camera does not works well under low light condition. In order to overcome this issue, a large range of YCbCr threshold value has to be used. Nevertheless, this limitation does not occur when using Kinect depth camera due to its stability to keep the hand contours, by applying a fixed depth threshold.

Table 1 shows summary of some hand gesture recognition system: comparison between methods in hand gesture recognition methods are used, while Table 2 tabulates the summary of recognition system which used other tracking devices such as gloves etc.

Table 1. Summary of Hand Gesture Recognition Systems Using Different Types of Camera

Research	Method	Detection	Feature Extraction	Recognition (%)
M.M. Hassan, P.K. Mishra [13]	Vision-based using single camera	HSV colour space/ Threshold method	Classical normalization: Divide the hand image into blocks of intensity features, which is then being extract by using edge information technique	Classical normalization 91%
M.M. Hassan & P. K. Mishra [5]	Vision-based using single camera	HSV colour space/ Threshold method	Comparison between classical normalization technique and block scaling using center of mass normalization technique	Classical normalization 83.3% Block scaling normalization using center of mass 96.6%
Ginu Thomas [14]	Vision-based using single camera	RGB colour space/ Otsu threshold method	Image captured are being tested in few methods: a) Pixel by pixel comparison b) Edges method	Pixel by pixel method 86%, Edges method 92% Orientation

			c) Orientation histogram	histogram 93.6%
H.S. Yeo, B.G. Lee, H. Lim [9]	Vision-based comparison using single camera and depth camera	YCrCb colour space/ Threshold method	Fingertips recognition by using convexity defects method and convex hull method around hand contours	Depth camera produced better result compare to single camera
A.Dhawan, V. Hanrao [10]	Vision-based using single camera	RGB colour space/ Threshold method		

Table 2. Summary of Hand Gesture Recognition Systems Using Tracking Devices

Research	**Method**	**Detection**	**Feature Extraction**	**Recognition (%)**
Y. Yao, Y. Fu [15]	14- patches colour glove	RGB-D using Kinect sensor	Database indexing technique is used by collecting training samples. Similarity between training samples and gesture are tested	51.87%
J. Nagi, A. Giusti, L. Gambardella, G. A. Caro [16]	Coloured Glove	HSV colour space	Support Vector Machine (SVM) is trained in a cascaded multi-binary-class configuration, where the spatial gestures are effectively learned and recognized by a swarm of UAVs.	Spatial gesture recognition is robust and scales well with swarm sizes of up to 20 robots
Kılıboz, Nurettin Çağrı Güdükbay, Uğur [17]	Six-degree-of-freedom (DOF) magnetic motion tracking device	Collected motion data is converted into relative position data in gradient form (x-y axis)	Finite State Machine (FSM) is used as a gesture recognizer method. Needleman-Wunsch sequence matching algorithm is applied in order to produce similarity scores between two sequences.	Recognition rate: 73%. The outcome can be improved by using a more powerful tracking approach which has better sensor capability

4. Conclusion

In this paper, three main steps of hand gesture which are camera module, detection module and feature extraction module are discussed. A robust hand and finger gesture tracking recognition system is viable to be used in human computer interaction system. The purpose of this paper is to review other research studies on different types of detection module and feature extraction methods, as tabulated in Table 1 and Table 2. Each method has its own advantages and disadvantages to be considered. The selection of specific recognition extraction depends on the application needed. It can be concluded that depth camera do produced the best result in hand gesture recognition, and this is because it can be easily calibrated to suit any skin colour and it can work against almost any background.

5. Proposed Future Work

For future work, we would like to focus on implementing a robust marker-less hand gesture recognition system using a low-cost hardware. A sliding tracker will be used for the user to determine the threshold value so that it can easily suits any user's skin color by controlling RGB color segmentation. Convex hull and convexity defects method on the other hand will be used as extraction methods to get the hand contours. Various gestures types will be tested in the future work such as pointing, clicking, swiping and scrolling.

5. Acknowledgement

Thank you to Kementerian Pelajaran Malaysia and UiTM for Tenaga Pengajar Muda (TPM) scholarship for funding this research.

References

[1] F. Karray, M. Alemzadeh, J. a Saleh, and M. N. Arab, "Human-Computer Interaction: Overview on State of the Art," *Int. J. Smart Sens. Intell. Syst.*, vol. 1, no. 1, pp. 137–159, 2008.

[2] J. Carey and M. Tremaine, "Integrating Human-Computer Interaction Development into the Systems Development Life Cycle: A Methodology," *Commun. Assoc. Inf. Syst.*, vol. 15, pp. 512–543, 2005.

[3] G. Murthy and R. Jadon, "A review of vision based hand gestures recognition," *Int. J. Inf. Technol. Knowl. Manag.*, vol. 2, no. 2, pp. 405–410, 2009.

[4] P. Garg, N. Aggarwal, and S. Sofat, "Vision based hand gesture recognition," *World Acad. Sci. Eng. Technol.*, pp. 972–977, 2009.

[5] M. M. Hasan and P. K. Misra, "Brightness Factor Matching For Gesture Recognition System Using Scaled Normalization," *Int. J. Comput. Sci. Inf. Technol.*, vol. 3, no. 2, pp. 35–46, 2011.

[6] S. Mitra and T. Acharya, "Gesture recognition: A survey," *IEEE Trans. Syst. Man Cybern. Part C Appl. Rev.*, vol. 37, no. 3, pp. 311–324, 2007.

[7] S. S. Billinghurst and K.-P. L. Vu, "Touch screen gestures for web browsing tasks," *Comput. Human Behav.*, vol. 53, pp. 71–81, 2015.

[8] N. Ibraheem, R. Khan, and M. Hasan, "Comparative Study of Skin Color based Segmentation Techniques," *Int. J. Appl. Inf. Syst.*, vol. 5, no. 10, pp. 24–38, 2013.

[9] H.-S. Yeo, B.-G. Lee, and H. Lim, "Hand tracking and gesture recognition system for human-computer interaction using low-cost hardware," *Multimed. Tools Appl.*, May 2013.

[10] A. Dhawan and V. Honrao, "Implementation of Hand Detection based Techniques for Human Computer Interaction," *Int. J. Comput. Appl.*, vol. 72, no. 17, pp. 6–13, 2013.

[11] R. Zaman Khan and N. A. Ibraheem, "Hand Gesture Recognition : A Literature," *Int. J. Artif. Intell. Appl.*, vol. 3, no. 4, pp. 161–174, 2012.

[12] A. A. M. Faudzi, M. H. K. Ali, M. A. Azman, and Z. H. Ismail, "Real-time hand gestures system for mobile robots control," *Int. Symp. Robot. Intell. Sensors*, vol. 41, no. Iris, pp. 798–804, 2012.

[13] M. M. Hasan and Pramod K. Misra, "HSV Brightness Factor Matching for Gesture Recognition System," *Int. J. Image Process.*, vol. 4, no. 5, pp. 456–467, 2011.

[14] G. Thomas, "A Review of Various Hand Gesture Recognition Techniques," *VSRD Int. J. Electr. Electron. Commun. Eng.*, vol. 1, no. 7, pp. 374–383, 2011.

[15] Y. Yao and Y. Fu, "Contour Model-Based Hand-Gesture Recognition Using the Kinect Sensor," *IEEE Trans. Circuits Syst. Video Technol.*, vol. 24, no. 11, pp. 1935–1944, 2014.

[16] J. Nagi, A. Giusti, L. M. Gambardella, and G. A. Di Caro, "Human-Swarm Interaction Using Spatial Gestures," *IEEE Int. Conf. Intell. Robot. Syst. (IROS 2014)*, no. Iros, pp. 3834–3841, 2014.

[17] N. Ç. Kılıboz and U. Güdükbay, "A hand gesture recognition technique for human–computer interaction," *J. Vis. Commun. Image Represent.*, vol. 28, no. January, pp. 97–104, 2015.

IOP Conf. Series: Materials Science and Engineering **99** (2015) 012013 doi:10.1088/1757-899X/99/1/012013

Capacitance Regression Modelling Analysis on Latex from Selected Rubber Tree Clones

A D Rosli[1], H Hashim[2], N A Khairuzzaman[2], A F Mohd Sampian[2], R Baharudin[1], N E Abdullah[2], M S Sulaiman[3] and M Kamaru'zzaman[2]

[1]Faculty of Electrical Engineering
Universiti Teknologi MARA Malaysia-P.Pinang

[2]Faculty of Electrical Engineering
Universiti Teknologi MARA Malaysia-Selangor

[3]Faculty of Electrical Engineering
Universiti Teknologi MARA Malaysia-Terengganu

E-mail: anis.diyana@ppinang.uitm.edu.my

Abstract. This paper investigates the capacitance regression modelling performance of latex for various rubber tree clones, namely clone 2002, 2008, 2014 and 3001. Conventionally, the rubber tree clones identification are based on observation towards tree features such as shape of leaf, trunk, branching habit and pattern of seeds texture. The former method requires expert persons and very time-consuming. Currently, there is no sensing device based on electrical properties that can be employed to measure different clones from latex samples. Hence, with a hypothesis that the dielectric constant of each clone varies, this paper discusses the development of a capacitance sensor via Capacitance Comparison Bridge (known as capacitance sensor) to measure an output voltage of different latex samples. The proposed sensor is initially tested with 30ml of latex sample prior to gradually addition of dilution water. The output voltage and capacitance obtained from the test are recorded and analyzed using Simple Linear Regression (SLR) model. This work outcome infers that latex clone of 2002 has produced the highest and reliable linear regression line with determination coefficient of 91.24%. In addition, the study also found that the capacitive elements in latex samples deteriorate if it is diluted with higher volume of water.

1. Introduction

Latex is a cytoplasm that has suspension rubber and non-rubber aqueous serum. The basic components of freshly tapped natural rubber latex are 50-80% water, 15-45% rubber hydrocarbon and approximately 2-4% non-rubber constituents. This composition of latex are influenced by weather, soil condition, clone, tapping system, season and so on [1]. The natural latex is obtained from major rubber trees, known as Hevea Brasiliensis, which grown in ASEAN region namely Malaysia, Indonesia, Thailand, Vietnam and also Cambodia [2].

In the past, the Rubber Research Institute of Malaysia (RRIM) focused on planting rubber with improved latex yield. With the new demand in rubber wood, RRIM initiates breeding activities that emphasize a selection of quick growing clones with a good latex yield as well as timber production.

The selection of these new latex-timber clones (LTC) are based on the important genotypes of the rubber tree such as latex yield, resistance to wind damage, resistance to diseases, girth increment during tapping, respond to wounding, resistance to dryness and so on [3].

Accurate identification of rubber clones is crucial since each clone has different interests in rubber industry. Currently, the rubber clones type can only be discriminated by expert persons as they observe tree features such as shape of leaf, trunk, branching habit and pattern of seeds texture. This approach is time consuming [4]. Therefore, in order to overcome this constraint, a series of agronomic research on rubber clone identification methods have been proposed.

Dry rubber content (DRC) is one of the possible latex's features that can be further investigated for clone classification. Various methods that have been reported in literatures for the determination of DRC are Metrolac, Spot, Viscosity, Latex film dialysis, Titration, Microwave attenuation, Low resolution pulsed NMR technique, Spin eco-technique and etc. [5].

Another possible method to determine the DRC of latex is by measuring the capacitance of the latex [6]. This method infers a hypothesis that DRC and dielectric constant of the latex have a linear correlation, and hence, the capacitance of the latex also will vary accordingly. Other investigations that based on their dielectric properties are on the quality of an apple, watermelon, honeydew melons and also a ripeness of banana [7-11].

This paper discusses the initial development of designing a complete sensor for clone detection. It focuses on constructing a capacitance sensor via Capacitance Comparison Bridge (known as capacitance sensor) to measure an output voltage of different latex samples representing selected clones. The capacitance regression modelling on latex is then statistically analyzed with respect to the selected clones.

2. Methodology

The main part of this paper is the development of the capacitance sensor using the Capacitance Comparison Bridge. The workflow of this paper is illustrated in Figure 1.

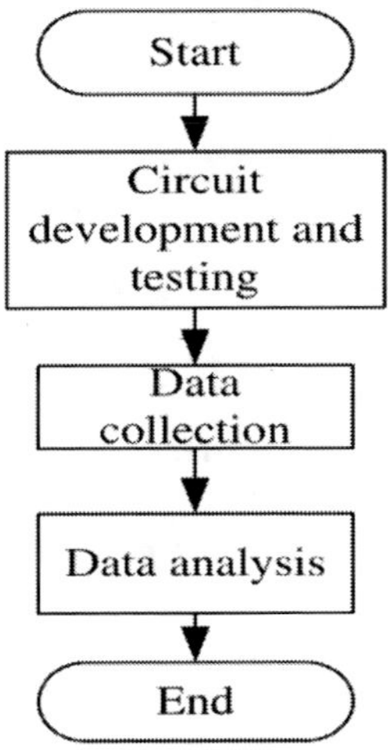

Figure 1. Flowchart for capacitance regression modelling analysis

IOP Conf. Series: Materials Science and Engineering **99** (2015) 012013 doi:10.1088/1757-899X/99/1/012013

2.1 Circuit development and testing

2.1.1 Circuit development

The AC bridge circuit is used to measure the capacitance value of the different latex samples based on the output voltage. Figure 2 shows the Capacitance Comparison bridge circuit [12].

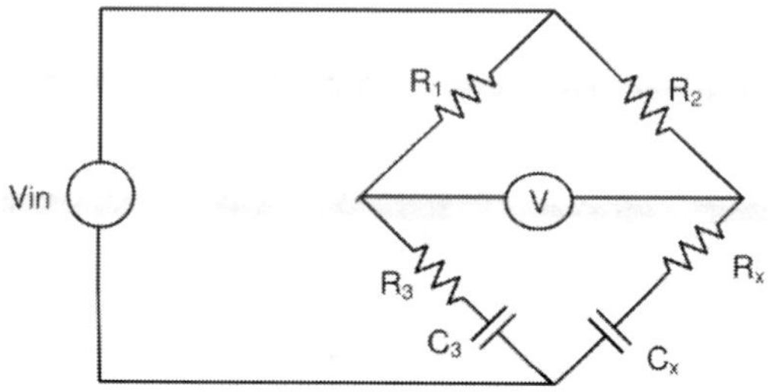

Figure 2. Capacitance Comparison Bridge Circuit

Theoretically, the equation for Capacitance Comparison Bridge can be derived as follows:

$$Z_1 = R_1 \tag{1}$$
$$Z_2 = R_2 \tag{2}$$
$$Z_3 = R_3 - jXC_3 \tag{3}$$
$$Z_X = R_X - jXC_X \tag{4}$$

During balance condition, there is no output voltage produced, resulting in equation (5).

$$Z_1 Z_X = Z_2 Z_3 \tag{5}$$

By substituting equation (1), (2), (3), and (4) into equation (5), equation (6) is produced.

$$\begin{aligned} R_1(R_X - jXC_X) &= R_2(R_3 - jXC_3) \\ R_1 R_X - jR_1 XC_X &= R_2 R_3 - jR_2 XC_3 \end{aligned} \tag{6}$$

Equation (6) is separated into two parts which are real and imaginary part, indicated by equation (7) and (8) respectively.

$$R_X = \left(\frac{R_2}{R_1}\right) \cdot R_3 \tag{7}$$

$$C_X = \left(\frac{R_2}{R_1}\right) \cdot C_3 \tag{8}$$

From the derivation, the prototype circuit is constructed as depicted in Figure 3.

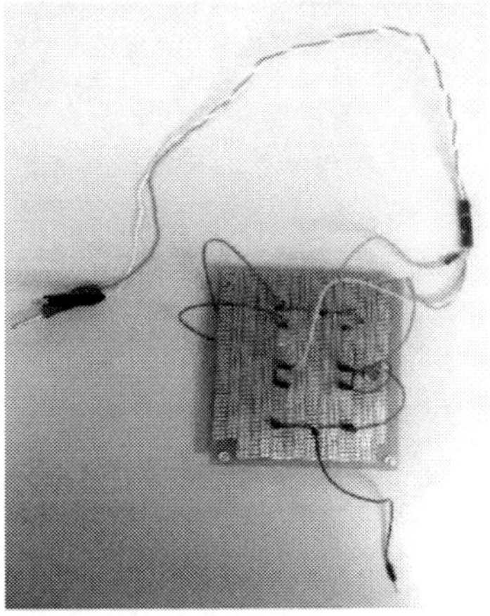

Figure 3. Prototype of Capacitance Comparison Bridge circuit

2.1.2 Sensor development

Capacitor is an electrical component which consist of two conducting plate of metal separated by the distance or gap called dielectric. The value of the capacitance depends on the area of the plate, the distance between two plates and also the relative permittivity of the dielectric material. Their relationship is explained by equation (9).

$$C = \varepsilon \frac{A}{D} \qquad (9)$$

Where

C = capacitor in farad
A = area of the plate
D = distance between two plates
ε = relative permittivity of the dielectric material between the plates

Figure 4 depicted a capacitance sensor designed by using the copper plate with the same size, length of 3cm and width of 3.3cm.The distance between two plates is fixed to 0.7cm throughout the experiment.

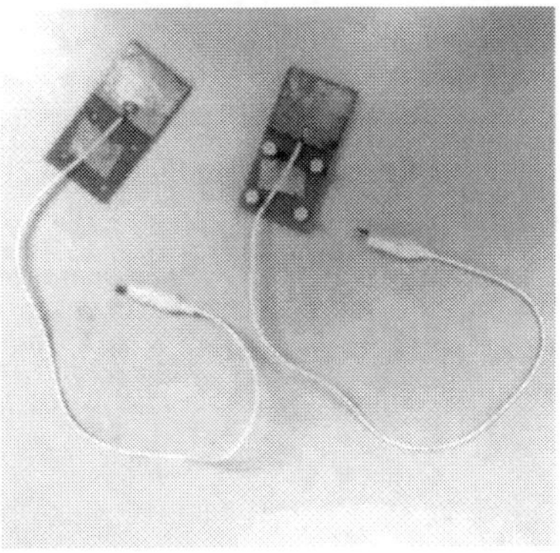

Figure 4. Capacitance plate design

The sensor development begins with selecting the value of capacitor, C_3, and resistors R_1, R_2, R_3 and R_x that satisfy equation (7) and (8). These resistors are selected based on the resistance in the latex, ranging from 50Ω to 100Ω. However, from the testing conducted, 500Ω resistor gives the best variation of the output voltage. Meanwhile, a 10pF capacitor is selected as the value of C_3 due to the more reliable output voltage produced compared to other capacitor's value.

2.2 Data collection

Four rubber tree clones of 2002, 2008, 2014 and 3001 are used in this work. The samples of these clones are obtained from the Rubber Research Institute of Malaysia (RRIM) located at Permatang Station, Kota Tinggi, Johor, Malaysia. All the latex sample has been added with preservative recommended by the Rubber Research Institute of Malaysia (RRIM), Kota Tinggi, Johor and each clone have two samples of latex.

The frequency, distance of capacitance plate and volume of latex are fixed at 100Hz, 0.7cm and 30ml respectively. The frequency is set to 100Hz based on previous research work. The work suggested that 100Hz frequency produces the best effect of capacitance value on the dry rubber content (DRC) [6].

Initially, the capacitance plate sensor is tested in the distilled water to obtain its initial value. Then, the sensor is used to measure both the output voltage and capacitance of 30ml latex sample. The measurements obtained via LCR meter are recorded. This process is repeated for every 10ml of diluted water added to the latex sample until the mixture reaches 60ml.The mixture of latex and diluted water is stirred prior to the measurement for more accurate results. The experiment set-up for data collection is illustrated in Figure 5.

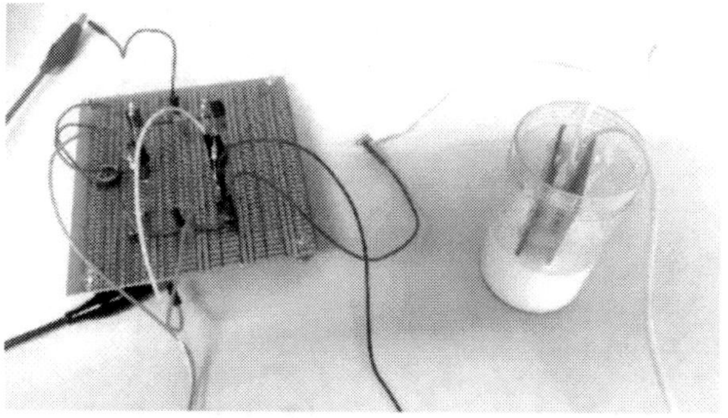

Figure 5. Data collection process using bridge circuit

2.3 Data analysis

Analyzing data collected for four different rubber tree clones are through observing the effect of diluted water on the capacitance of the latex. The collected data is interpreted into a line graph to obtain the voltage and capacitance relationship. Equations (10)-(11) are used to compute the linear regression equation and the coefficient of correlation, R.This parameter explains the linear relationship between voltage and capacitance. Linear relationship between this two independent variable is considered strong if the value of being greater than 0.8. Whilst, if R less than 0.5, it can be described as a weak linear relationship. The perfect correlation is equal to 1 where all the data lies exactly on a straight line.

$$y = a + bx \tag{10}$$

$$R = \frac{n\Sigma xy - \Sigma x \Sigma y}{(n\Sigma x^2 - \Sigma x^2)^{1/2} \cdot (n\Sigma y^2 - \Sigma y^2)^{1/2}} \tag{11}$$

Where,

$a =$ The intercept point of the regression line and the y-axis.

$b =$ The slope of the regression line.

$x, y =$ Independent variable.

$n =$ Number of samples.

3. Result and Discussion

The voltage and capacitance of four different clones, namely 2002, 2008, 2014 and 3001 are analyzed and discussed in this section.

3.1 Clone 2002

Table 1 shows the voltage and capacitance relationship when latex sample of clone 2002 was diluted with different volumes of water.

Table 1. Voltage and capacitance of 2002 clone in different water content.

Latex (ml)	Water (ml)	Voltage (V)	Capacitance (µF)
30	0	0.931	16.776
30	0	0.935	18.221
30	10	0.926	14.796
30	10	0.926	13.733
30	20	0.919	11.526
30	20	0.919	11.402
30	30	0.911	9.885
30	30	0.910	9.675

While Figure 6 shows the regression between the capacitance and the output voltage for the respective clone 2002. The value of the correlation coefficient, R is equal to 0.9552 which indicates strong positive correlation between the voltage and the capacitance values.

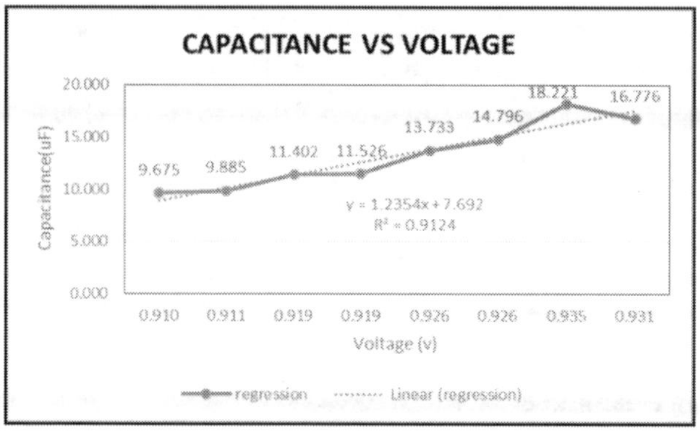

Figure 6. Regression between capacitance and voltage for 2002 clone

3.2 Clone 2008

Measurements for latex sample of clone 2008 when diluted with different volumes of water are described in Table 2.

Table 2. Voltage and capacitance of 2008 clone in different water content.

Latex (ml)	Water (ml)	Voltage (V)	Capacitance (μF)
30	0	0.925	16.865
30	0	0.935	18.195
30	10	0.920	13.890
30	10	0.925	15.300
30	20	0.916	12.546
30	20	0.920	13.413
30	30	0.909	6.517
30	30	0.913	11.753

Hence, Figure 7 shows the regression between the capacitance and output voltage for clone 2008. From the graph plotted, value of the correlation coefficient, R is 0.8023. It represents a moderate positive correlation between the voltage and capacitance values.

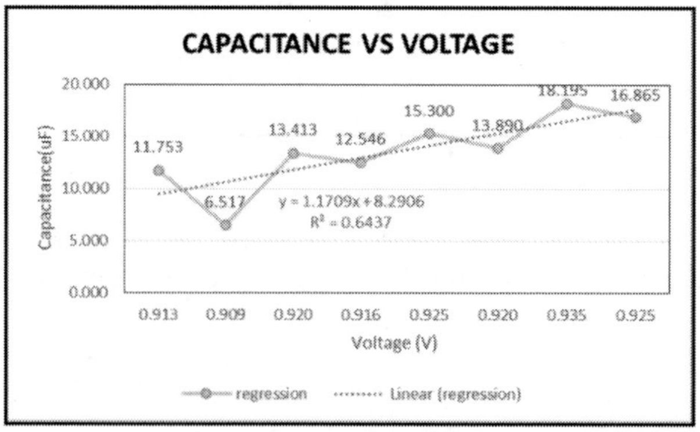

Figure 7. Regression between capacitance and voltage for 2008 clone.

3.3 Clone 2014

Table 3 shows the voltage and capacitance variation when latex sample of clone 2014 was diluted with different volume of water.

Table 3.Voltage and capacitance of 2014 clone in different water content.

Latex (ml)	Water (ml)	Voltage (V)	Capacitance (μF)
30	0	0.938	19.254
30	0	0.932	12.302
30	10	0.927	15.854
30	10	0.918	9.973
30	20	0.920	13.257
30	20	0.910	7.948
30	30	0.910	9.673
30	30	0.901	7.133

From the graph in Figure 8, the value of R obtained is 0.8385. This infers that the capacitance and voltage of clone 2014 has moderate positive correlation.

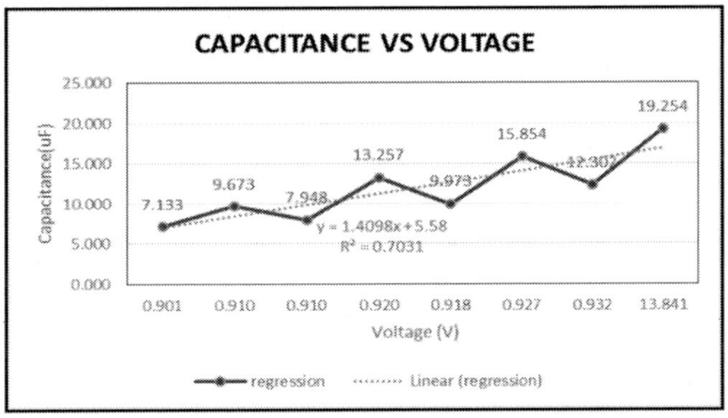

Figure 8. Regression between capacitance and voltage for 2014 clone.

3.3 Clone 3001

Table 4 depicted the voltage and capacitance variation when latex sample of clone 3001 was diluted with different volumes of water

Table 4. Voltage and capacitance of 3001 clone in different water content.

Latex (ml)	Water (ml)	Voltage (V)	Capacitance (μF)
30	0	0.927	14.869
30	0	0.928	16.627
30	10	0.923	12.585
30	10	0.920	13.959
30	20	0.915	9.770
30	20	0.918	13.557
30	30	0.911	9.219
30	30	0.913	11.246

Figure 9 shows that clone 3001 has weak correlation between its voltage and capacitance value with R equal to 0.7301. When comparing with other clones, the variation of the output voltage of clone 3001 can be claimed as less dependent on the value of capacitance.

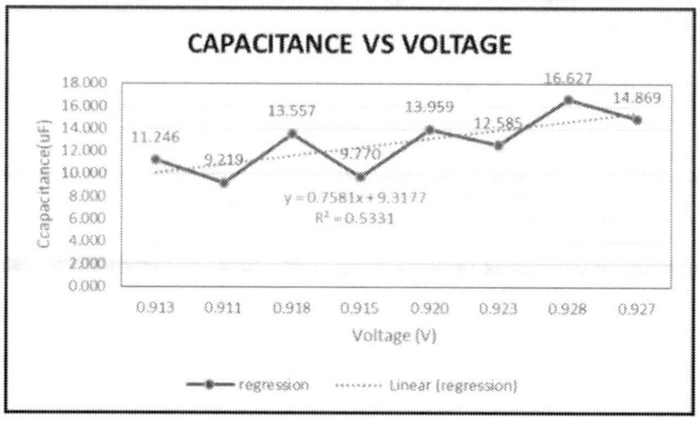

Figure 9. Relationship between capacitance and voltage for 3001 clone.

Hence, the general model of capacitance value of each clone (in microfarad) can be determined by rewriting equation (11) to the linear regression equation as shown in equation (12).

$$Capacitance = b \cdot Voltage + a \qquad (12)$$

Table 5 shows the summarization of regression modelling with respect to the clone types. It can be observed clone 2002 produces the highest coefficient of determination, R^2 values compared to the clone 2008, 2014 and 3001. This implies that with every change in 1 volt, leads to an increment in capacitance value a factor of 1.2354. This work also proved that the output voltage of each clone on its latex can be directly proportional to its capacitance values.

Table 5. The coefficient determination and linear regression model for each latex clone.

Clone	Simple Linear Regression Model (SLM)	Coefficient of determination (R^2)
2002	Capacitance = 1.2354 · Voltage + 7.692	0.9124
2008	Capacitance = 1.1709 · Voltage + 8.2906	0.6437
2014	Capacitance = 1.4098 · Voltage + 5.5800	0.7031
3001	Capacitance = 0.7581 · Voltage + 9.3177	0.5331

4. Conclusion

The motives of this work is to formulate a regression model that can provide capacitance values in terms of micro farad by measuring voltage of latex using Capacitance Comparison bridge circuit. Samples of latex from four selected clones (2002, 2008, 2014 and 3001) commonly used in the rubber industry, were measured using the designed and developed circuit system. These samples were diluted with different volumes of water in order to get variations in the voltage measurement since latex naturally contains water in its composition. Thus, its dielectric would change if water is being added to it. The outcome of the experiment has produced capacitance regression modelling for each of the clone with their respective coefficients of determination. It can be concluded that the capacitance of latex from clone 2002 is strongly correlated with voltage measured from the Capacitance Comparison bridge circuit. This work also proved that by using this capacitance sensor circuit, the output voltage of the latex from these selected clones can be directly proportional to its capacitance values.

Therefore, a lookup table given in Table 5 can be used as a reference for any further investigation work using designed and developed capacitance sensor.

Acknowledgement

The authors would like to thank all RRIM staff especially the Director of Rubber Research Institute of Malaysia (RRIM) at Sg. Buloh, Dr.Mohd Nasaruddin Mohd Aris and also the Crop Improvement and Protection Unit staff such as Mr. Amran Saari and Mr. Razak for their valuable information, suggestions, guidance and also for providing the data prior to the experiment of this research. Not to forget to Faculty of Electrical Engineering, Universiti Teknologi MARA Shah Alam, and also to Research Management Institute (RMI), UiTM and Ministry of Education for funding our research work under Research Acculturation Grant Scheme (RAGS) under the code 600-RMI/RAGS 5/3 (27/2014).

REFERENCES

[1] Ansarudin F , Abbas Z, Ali A H and Ismail M A 2012 *Reflection technique for the determination of moisture content in hevea rubber latex* Progress in Electromagnetics Research Symposium Proc. (Kuala Lumpur, Malaysia, 27-30 March 2012) pp 210–3

[2] Shuhaimi N H H, Othman N, Ismail H and Sasidharan S 2013 Effects of natural rubber origin on mechanical properties of rubber compound in the presence of natural and commercial antioxidants *Adv. Mater. Res.* **747** 703–6

[3] Kiam T S 2000 *Forest plantation development in malaysia and the potential of rubber wood as an important source of timber in the future* Proc. of the International Conference on Timber Plantation Development (Manila, Philippines, 7 - 9 November 2000)

[4] i-klon, RITes bantu pekebun kecil getah 2013 *Utusan Online* .Retrieved from http://ww1.utusan.com.my/utusan/Sains & Teknologi/20131103/st 01/i-klon-RITeS-bantu-pekebun-kecil-getah

[5] Kumara P H S 2006 *Methods of estimation of dry rubber content in natural rubber latex* Bulletin of the Rubber Research Institute of Sri Lanka 47 pp 65–9

[6] Kumar R R 2009 *Design and development of instrumentation systems to determine the dry rubber content in natural rubber latex (Doctoral dissertation)* Cochin University Of Science And Technology

[7] Guo W C, Nelson S O, Trabelsi S and Kays S J 2007 10-1800-MHz Dielectric properties of fresh apples during storage *J. Food Eng.* **83** 562–9

[8] Nelson S O, Guo W C, Trabelsi S and Kays S J 2007 Dielectric spectroscopy of watermelons for quality sensing *Meas. Sci. Technol.* **18** 1887-92

[9] Nelson S O, Trabelsi S and Kays S J 2006 *Dielectric spectroscopy of honeydew melons for quality sensing* Instrumentation and Measurement Technology Conf. (Sorrento, Italia, 24 - 27 April 2006) pp 180-2

[10] Nelson S O, Trabelsi S and Kays S J 2008 Dielectric spectroscopy of melons for potential quality sensing *American Society of Agricultural and Biological Engineers* **51** pp 2209-14

[11] Soltani M, Alimardani R and Omid M 2011 Evaluating banana ripening status from measuring dielectric properties *J. Food Eng.* **105** 625-31

[12] Johnson C D 2006, *Process Control Instrumentation Technology*, 8th ed. (Ohio: Prentice Hall) pp. 70.

A Rudimentary Optical System in Detecting Ripeness of Red Watermelon

Noor Ezan Abdullah[1], Hadzli Hashim[1], Muhammad Fathullah Sulaiman1,Nina Korlina Madzhi[1], Ahmad Faiz Mohd Sampian[1], Faridatul Aima Ismail[1]

Fakulti Kejuruteraan Elektrik, Universiti Teknologi MARA, 40450 Shah Alam, Selangor Darul Ehsan Malaysia

E-mail: noor_ezan@salam.uitm.edu.my, hadzli66@gmail.com

Abstract. The purpose of this project is to detect the ripeness and quality of the watermelon particularly for red watermelon. The ripeness of the watermelon will be evaluated by using near-infrared spectroscopy sensor (NRIS). The color wavelength will classify the ripeness of the watermelon. An infrared light will be used to get the appropriate wavelength from the watermelon either from the rind or inner of it and the signal received will be analyzed. An appropriate algorithm is used to extract the information of the inner of the watermelon. A microcontroller namely Programmable Interface Controller (PIC) will be used to execute the algorithm and the result will be displayed on Liquid Crystal Display (LCD). Based on the result obtain from the device, the data is computed by using Statistical Package for the Social Sciences (SPSS). This approach is vital to verify the relationship between unripe and ripeness of red watermelon. The objective of this project is to produce an efficient system to detect the ripeness of the watermelon.

1. Introduction

Normally, determination process of fruit and vegetables quality is believed to be very crucial for both producers and processors[1]. In this work, a research has been conducted to detect the watermelon ripeness via optical system. Apparently, previous research has stated that it is very difficult to judge the ripeness of the watermelon just by examining the outward characteristics of the fruit such as the size or its external color in which different limitations are considered except with aid of experience person [2]. Hence, it is a necessity for the farmers to constantly maintain the good quality of watermelon and deliver a good supply of watermelon to the market due to the great demand on this fruit[2]. In modern agriculture, the need to maintain the quality of fruit is very important in order to improve the industry. Current and brand new technologies are being developed and projected to fully utilized the Near Infrared spectroscopy (NIR) which can offer better and non-destructive methods for measuring the quality of fresh fruit and vegetable[3].Thus, the techniques of measuring fruit and vegetables by using nondestructive methods are highly desirable[4]. As a result, researchers have actively been focused on developing non-destructive techniques in order to measure fruit quality attributes[5]. According to Nilsson(2005), NIR is becoming a routine used in agriculture , pharmaceuticals, combustion products, astronomy, food stuffs and in research[6]. In particular, fruits are categorized and sorted according to their attributes such as size, colour, and surface defects but other attributes such as firmness and total soluble solids content are also important internal quality of fruit products[5]. Various methods for measuring fruit quality non-destructively are being evaluated by using different technologies which include surface reflectance, diffuse reflectance, transmittance, interactance (body transmittance), fluorescence, and delayed emission from acoustical, ultrasonic, vibrational, and deformation properties[3]. Amongst these techniques, practically, the principle operation of NIR uses surface reflectance and diffuse reflectance in order to measure the ripeness of a watermelon. According to a journal published by Dull, infrared spectrometry is currently a technique that is widely used by researchers to analyze food and agricultural products, with most of the applications are using the near infrared (NIR) region which is between 700 to 2500nm [7]. The NIR region was exploited due in part to the availability of strong radiation sources and detectors with high sensitivity. Thus, NIR has been used in the non-destructive quality evaluation of a different types of fresh fruit and vegetables [7]. The main objective of this study is to discover the most appropriate algorithms to detect information of the inner content of the red watermelon. In this work, NIR technique based on surface reflectance is used to determine the watermelon ripeness based on the calculation from the surface of the watermelon. All the acquired data was analyzed by using statistical method in order to grade and evaluate the ripeness of watermelon.

2. Principle of NRIS

The electromagnetic (EM) fields that exist in a transmission medium are being gathered to create an electromagnetic spectrum[9]. The electromagnetic spectrum includes all type of electromagnetic radiation in which can be divided into the following band such as cosmic rays, gamma rays, x-rays, ultraviolet, visible light, infrared, and radio wave as shown in Figure 1.

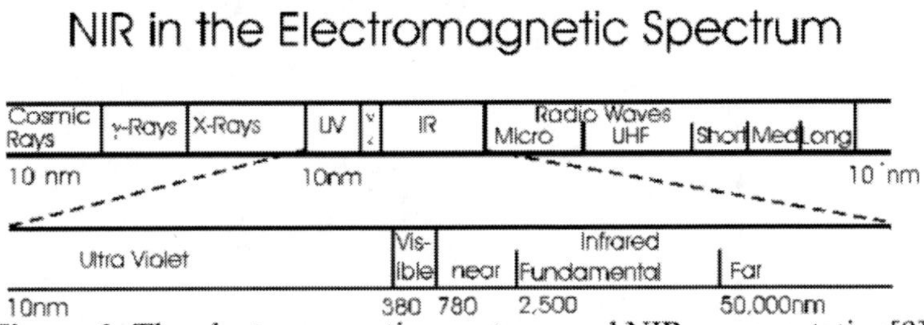

Figure 1. The electromagnetic spectrum and NIR representation[9].

In Figure 1, NIR range is between 750-2500nm. When the light from the source such as light emitting diode (LED) is emitted, a small region or part of the light will be reflected by the surface and also after it absorbs the colour of the object. This phenomenon is called a regular reflectance [10]. The remaining radiation is transmitted through the surface of the other part of the watermelon where the transmitted radiation can be transformed into other forms of energy such as heat as illustrated in Figure 2.

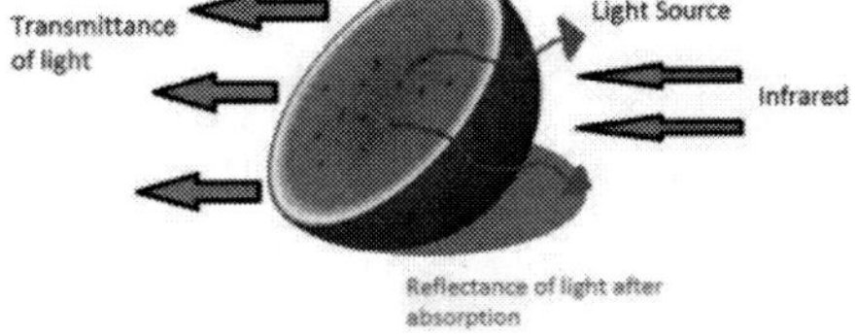

Figure 2. Radiation of light source

3. Methodology

3.1. Introduction

This section will briefly explain the samples collection process and the hardware development for designing the watermelon ripeness detector. The hardware development phase will discuss in brief the measurement setup for NIR sensor and will be followed by some explanations on the components used.

3.1.1. Samples Collection

There are about 30 red watermelon samples were evaluated from Hypermarket and Night Market in Shah Alam Selangor.A total of 10 samples from each shop were chosen in which 10 samples from Giant supermarket, and 20 other were from 2 shops at Pasar Malam Seksyen 7 Shah Alam. A total of 30 unripe watermelon were evaluated in which from Giant supermarket and Pasar Malam Seksyen 7 Shah Alam.

3.1.2. Hardware Development

The circuit development is started by creating a schematic design by using Proteus software. All of the components were undergoing troubleshooting process to ensure the circuit is working. The programming of the PIC will be the starting of the simulation of the project. Figure 3 shows the flow chart of the hardware development.

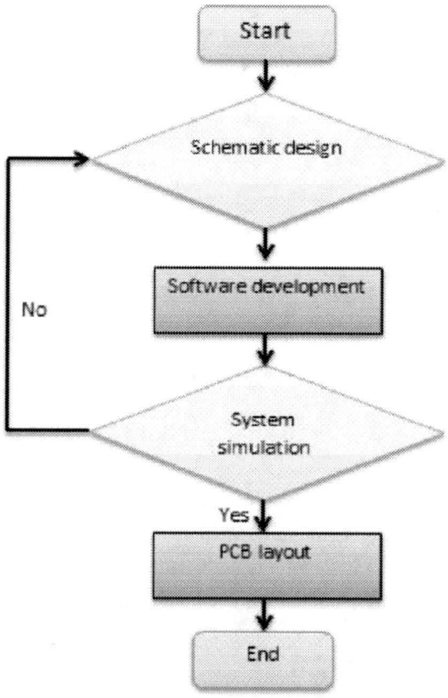

Figure 3. Hardware Development Flowchart

3.1.3. Measurement Setup

There are several methods to obtain the measurement of near infrared spectra as shown in Figure 4. In reflectance mode (a), the light source or infrared light emitting diode (IRLED) and detector or phototransistor are fixed under a specific angle for example 45° in order to avoid specular reflection. In transmittance mode (b), the IRLED is pointed at the opposite of the phototransistor so that the light is able to pass through the fruit and then is detected by the phototransistor. While in interactance mode (c), the IRLED and phototransistor are placed parallel to each other in such a way that light cannot directly enter the phototransistor due to a barrier is placed between them which physically prevented the light from reaching the detector. Both reflectance and transmittance technique may also be applied in order to collect light and increase the signal to noise ratio[11].

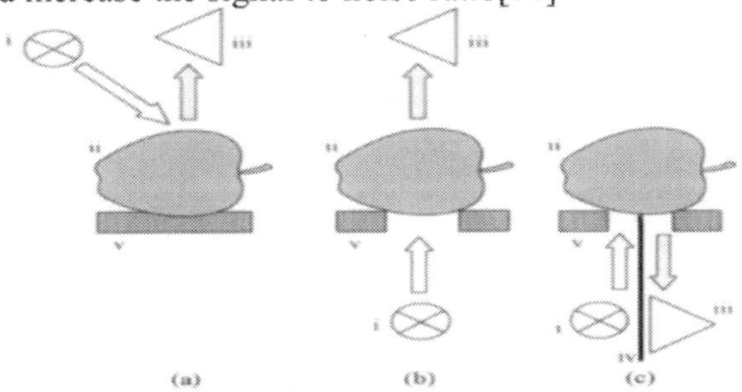

Figure 4. Setup for the acquisition of (a) reflectance, (b) transmittance, and (c) interactance spectra, with (i) the light source, (ii) fruit, (iii) detector, (iv) light barrier, and (v) support. In interactance mode, light due to specular reflection is physically blocked from entering the detector.[10]

The principle of the reflectance technique used in this project is indicated as in Figure 5 where the incident radiation of the IRLED is projected to the sample and the radiation will pass through the rind of the watermelon and detect the red part of the inside of the watermelon. The radiation is then reflected and then detected by a phototransistor to measure its voltage[11].

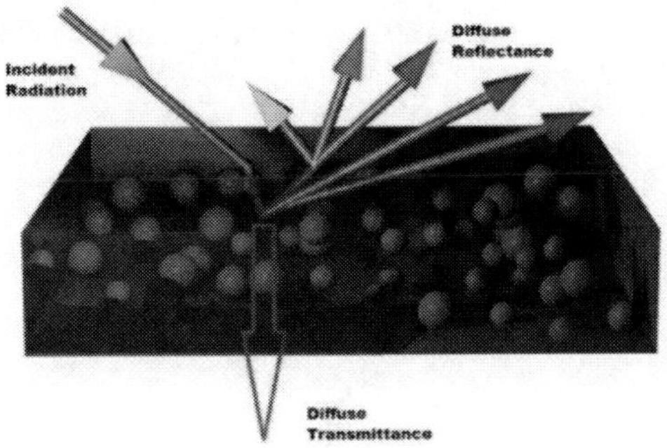

Figure 5. The principle of reflectance beam diffuses into a sample

For the measurement setup, it is known that the penetration of NIR radiation into fruit tissues decreases exponentially with the depth found a penetration depth of up to 4mm in the 700-900 nm range between 2 and 3mm in the 900-1900 nm range[11]. Based on the Figure 6, the angle of reflectance used was 45° so that the IR phototransistor is able to receive the infrared radiation that was reflected from the red sample of the watermelon. The range between the sensor and the surface of the watermelon was kept at 2.5cm in order for it to fully utilized the reflectance technique so that the phototransistor able to receive large amount of infrared radiation[11].

IOP Conf. Series: Materials Science and Engineering **99** (2015) 012014 doi:10.1088/1757-899X/99/1/012014

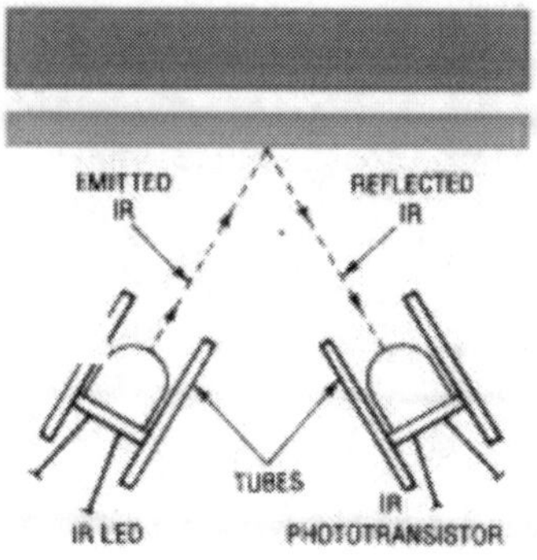

Figure 6. The angle of reflectance

When a sample is illuminated by the beam light, the reflected radiation is measured by the phototransistor. The intensity of the radiation will be measured by the receiver in the units of voltage. However, the reflectance value may be varied depends on the condition of the watermelon whether it is ripe, unripe or overripe

3.1.4. Measurement Point

The detector used in this project is a phototransistor. The collector emitter voltage, V_{CEO}, is measured either on a transistor curve tracer or statically using the circuit shown in Figure 7 where voltmeter is placed between the collector and emitter[12].

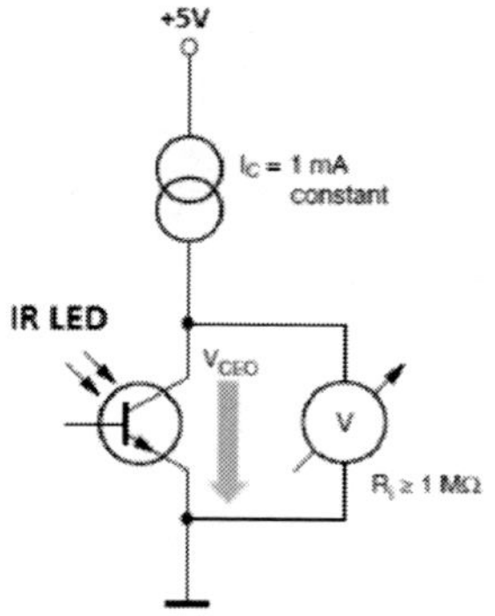

Figure 7. Measurement point of V_{CEO}[12]

3.1.5. Near Infrared Sensor

The emitter used in this project is Honeywell SE5455-003 SE series gallium Arsenide (GaAs) Infrared emitting diode. The SE3455/5455 series consists of a GaAs infrared emitting diode mounted in a To-46 metal can package. The SE5455 series has glass lensed cans providing narrow beam angle between 20°-90°. It has the wavelength of 935nm which in the range of near infrared radiation. It is also ideal for high pulsed current applications. Figure 8 shows the structure of SE5455 GaAs Infrared emitting diode[13].

IOP Conf. Series: Materials Science and Engineering **99** (2015) 012014 doi:10.1088/1757-899X/99/1/012014

Figure 8. SE5455 GaAs Infrared Emitting Diode[13]

The phototransistor used in this project is from SD3443/5443 series which consists of NPN silicon mounted in a TO-46 metal can package. The SD3443 has flat window cans providing a wide acceptance angle, while the SD5443 has glass lensed cans providing a narrow acceptance angle. It is also compatible and spectrally matched to SE5455 GaAs Infrared Emitting Diode. Figure 9shows the structure of phototransistor used[14].

Figure 9. SD3443/5443 silicon phototransistor[14]

3.1.6. Microcontroller PIC16F877A

This project is using microcontroller of PIC16f87XA family which has 40 pins and can be executed by using C programming language. The sensor which is the SE5443 phototransistor is connected to RA0 which is the Analog-Digital converter inside of the PIC[15] via its analog output from the emitter. The value obtained is in bit value and an algorithm is used inside the programming to display the actual voltage. From the output, the voltage of ripeness of red watermelon can be seen. Figure 10 shows the features of PIC16F877A. Figure 11 shows the PIC microcontroller kit used to program the PIC chip.

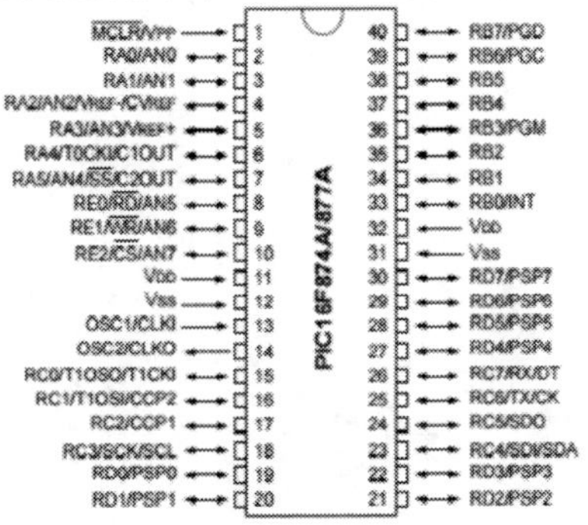

Figure 10. PIC16F877A chip[15]

Figure 11. PIC Microcontroller Kit

3.1.7. Liquid Crystal Display (LCD) 16x2

Liquid Crystal Display (LCD) screen is a flat panel display that uses the light modulating properties of liquid crystals. There are several types of LCD display such as 16x2 and 16x4. This project is using LCD 16x2 which is a very basic module and very commonly used in various projects and devices. Since the output of this project is only voltage, this LCD 16x2 is very suitable to be used to display watermelon ripeness. Figure 12 shows the features of LCD 16x2[16].

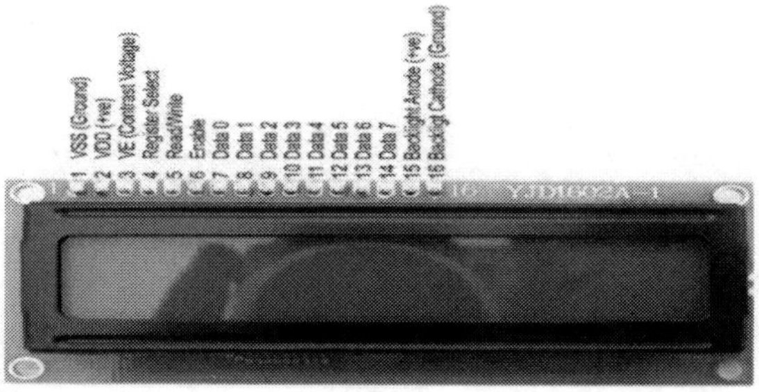

Figure 12. 16x2 Liquid Crystal Display (LCD)[16]

3.1.8. Project Algorithm

Based on Figure 13, Light Emitting Diode (LED) is being projected to the rind from outer and also from the inner. From the outer and inner layer, the value of light intensity is converted into voltage in order to get the value of α and β.

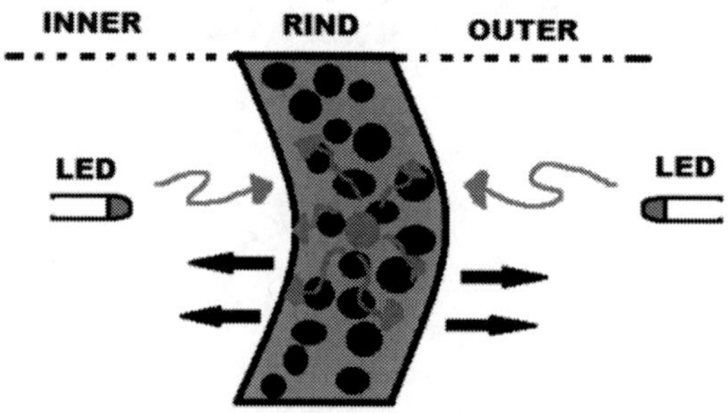

Figure 13. Presentation of how incident light is being reflected by the inner, rind & outer.

Based on the fundamental of light properties on surface material[17], the following equations are extracted.

$$i_o = \alpha + \delta_{ro} \tag{1}$$

$$i_i = \beta + \delta_{ri} \tag{2}$$

Where
$i_o = i_i =$ incident light due to LED
$\alpha =$ diffuse reflectance light due to i_o
$\delta_{ro} =$ diffuse transmitted light due to i_o
$\beta =$ diffuse reflectance light due to i_i
$\delta_{ri} =$ diffuse transmitted light due to i_i

Since $i_o = i_i$, equation (1) and (2) can be generalized as

$$\beta + \delta_{ri} = \alpha + \delta_{ro} \tag{3}$$
$$\therefore \beta = \alpha + \delta_{ro} - \delta_{ri} \tag{4}$$

Then, the measurement of any diffuse reflectance light at the inner section of watermelon, for any given sample of $n=1, 2, 3, \ldots\ldots N$ can be summarized as

$$\therefore \beta_{act}(n) = \alpha_{act}(n) + \delta_{ro}(n) - \delta_{ri}(n) \tag{5}$$

Equation (5) is validated by making comparison with the measurements of diffused reflectance gathered from raw data of 200 ripe and 200 unripe watermelons. Figure 14 shows the correlation between these parameters. It is shown that the correlation is 100% (Pearson's factor, r =1)[18].

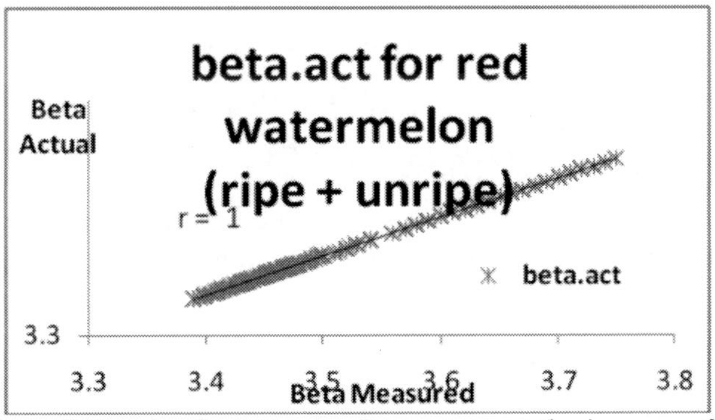

Figure 14. Correlation bet. diffused reflectance measurements at the inner section of watermelon

Since the motive of this project is to determine ripe and unripe of red watermelon, it is viable to produce an estimation of reflectance measurement (β_{est}) of the inner section based on the available measured raw data parameters such as α, δ_{ro} and δ_{ro}. The equation can be formulated by firstly, finding the average diffuse transmitted light from the measured raw data as shown below.

$$\delta_{ri}^{avg} = \sum_{n=1}^{N} \frac{\delta_{ri}(n)}{N} \quad (6)$$

$$\delta_{ro}^{avg} = \sum_{n=1}^{N} \frac{\delta_{ro}(n)}{N} \quad (7)$$

Finally, the estimation of reflectance measurement of the inner section of watermelon can be derived as in equation (8). It is also known as *HF* formula representing the contributed mathematical ideas from Hadzli (Reseacher) – Faiz/ Fathullah (Investigator).

$$\therefore \beta_{est}^{HF}(n) = \alpha_{act}(n) + \delta_{ro}^{avg} - \delta_{ri}^{avg} \quad (8)$$

Equation (8) is also validated by making comparison between the measurements of diffused reflectance gathered from raw data of 200 ripe and 200 unripe watermelons. Figure 15 shows the correlation between these parameters. This time, the correlation is about 79% (Pearson's factor, r =0.789)[18].The result is considered to be reliable and can be used for further statistical inference analysis in comparing ripe and unripe of red watermelon.

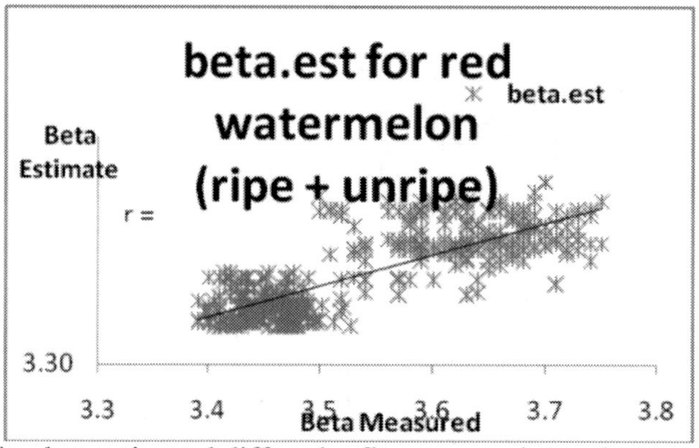

Figure 15. Correlation bet. estimated diffused reflectance and actual measurements at the inner section of watermelon

3.1.9. Validation Formula For Watermelon Ripeness Detector

In order to validate the samples of data collectedfrom the red watermelon, equation (9), (10), and (11) are extracted from Table 1 in order to determine the sensitivity, specificity and accuracy of the device respectively[19].

Table 1. Validation of Watermelon Ripeness Detector

	Ripe	Unripe
Display "Ripe"	TP	FP
Display "Unripe"	FN	TN

$$Sensitivity = \frac{TP}{TP + FN} x\,100\% \quad (9)$$

$$Specificity = \frac{TN}{TN + FP} x\,100\% \quad (10)$$

$$Accuracy = \frac{TP + TN}{TP + TN + FN + FP} x\,100\% \quad (11)$$

4. Result and Discussions

4.1. Test of Normality

The data collected were transferred to SPSS statistical software in order to identify whether the voltage readings for the raw data could be categorized as parametric samples or not. Based on Table 2, it shows the result of Kolmogorov-Smirnov (one of Normality test result) to test its normality. It can be observed that all of the samples have p-value less than 0.05 which indicate that the distributions of data are not normal. Hence, the suitable test for this population data is by using non-parametric test.

Table 2. Kolmogorov-Smirnov Test of Normality

Tests of Normality

		Kolmogorov-Smirnov[a]		
	group	Statistic	df	Sig.
beta actual	ripe	.067	200	.028
	unripe	.080	200	.003
beta estimate	ripe	.101	200	.000
	unripe	.142	200	.000

a. Lilliefors Significance Correction

4.2. Non-Parametric Test

A non-parametric test was done to ensure that the data can be validated. Table 3 shows the hypothesis between 2 groups of samples which include actual ripe and unripe data and estimate ripe and unripe data. Independent sample test of median, Mann-Whitney U and Kruskal-Wallis Test were conducted to validate the distribution of data. Based on Table III, there is overwhelming evidence that the two group of ripe and unripe are significantly different.

Table 3. Comparison Between Ripe and Unripe of Beta Actual and Beta Estimate

Hypothesis Test Summary

	Null Hypothesis	Test	Sig.	Decision
1	The medians of ripe are the same across categories of group.	Independent-Samples Median Test	.000	Reject the null hypothesis.
2	The distribution of ripe is the same across categories of group.	Independent-Samples Mann-Whitney U Test	.000	Reject the null hypothesis.
3	The distribution of ripe is the same across categories of group.	Independent-Samples Kruskal-Wallis Test	.000	Reject the null hypothesis.
4	The medians of unripe are the same across categories of group.	Independent-Samples Median Test	.000	Reject the null hypothesis.
5	The distribution of unripe is the same across categories of group.	Independent-Samples Mann-Whitney U Test	.000	Reject the null hypothesis.
6	The distribution of unripe is the same across categories of group.	Independent-Samples Kruskal-Wallis Test	.000	Reject the null hypothesis.

Asymptotic significances are displayed. The significance level is .05.

4.3. Error Bar and Box Plot

Comparison between ripe and unripe value of actual and estimated value has been done as shown in error bar in Figure 16 whereas Figure 17 shows the box plot of the sample taken between actual and predicted ripe watermelon. It shows that there is difference as can be seen visually between the value of ripe and unripe of red watermelon for both figures.

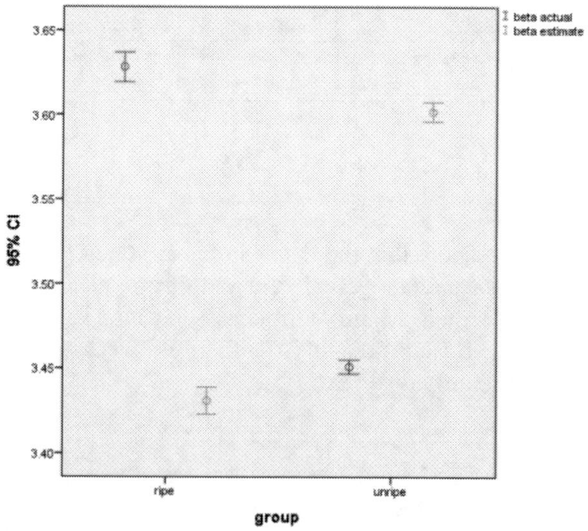

Figure 16. Error bar of ripe and unripe watermelon (actual versus estimate)

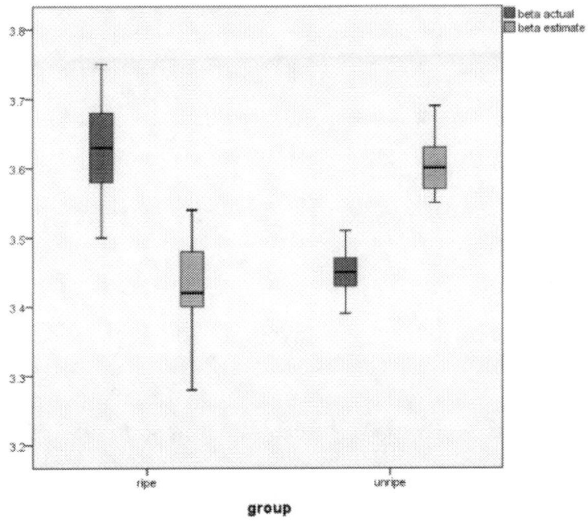

Figure 17. Boxplot of ripe and unripe watermelon (actual versus estimate)

Based on the boxplot, the range of voltage for ripe and unripe can be estimated as shown in Table 4.

Table 4. Estimate range of voltage of ripe and unripe for watermelon

	Min(V)	Max(V)
Ripe	3.30	3.56
Unripe	3.57	3.71

4.4. Validation of Watermelon Ripeness Detector

Table 5 shows the resultant of using the watermelon ripeness detector to test its accuracy. In this work, there were a total 60 red watermelons which include 30 ripe and 30 unripe watermelonstested by using the device. From Table 5, its sensitivity and specificity of Watermelon Ripeness Detector are 80% and 73.3% respectively. The accuracy of the device, thus can be specified where only 76.7% of the tested watermelon gave accurate result.

Table 5. Validation of Watermelon Ripeness Detector

Result from LCD	Ripe	Unripe	Total
Display “Ripe”	24	8	30
Display “Unripe”	6	22	30
Total	30	30	60

Sensitivity of Watermelon Ripeness Detector:

$$\frac{24}{30} \times 100 = 80\%$$

Specificity of Watermelon Ripeness Detector:

$$\frac{22}{30} \times 100 = 73.3\%$$

Accuracy of Watermelon Ripeness Detector :

$$\frac{24 + 22}{60} \times 100 = 76.7\%$$

5. Conclusion and Discussion

The outcome of the project has shown that the developed device which is Watermelon Ripeness Detector (WRD) that uses near infrared sensor is successfully able to measure the ripeness of the watermelon with the accuracy of 76.7% which is enough to justify the device effectiveness. As a conclusion, this project still need to be observed and adjustment to the device is very important in order to achieve a better validation about the ripeness of watermelon. This project could act as a stepping stone in order to develop a better watermelon ripeness detector in order to maintain the quality of fruits in market.

6. Acknowledgement

The authors would like to acknowledge Industrial Based Instrumentation & Technology (I-BIT) Group, Faculty of Electrical Engineering, UiTM Shah Alam for sharing their knowledge and providing equipment in hardware developing. Not to forget, a special thanks to Faculty of Electrical Engineering Universiti Teknologi MARA Shah Alam, Research Management Institute (RMI), UiTM and Ministry of Higher Education (MOHE) for funding our research work under RAGS grant code 600-RMI/RAGS 5/3 (22/2012).

References

[1] H. Q. Tian, Y. B. Ying, H. S. Lu, X. P. Fu, and H. Y. Yu, 2007 Measurement of soluble solids content in watermelon by Vis/NIR diffuse transmittance technique, *J Zhejiang Univ Sci B*, **8** 2 p 105–110

[2] N. E. Abdullah, H. Hashim, Y. Wati, M. Yusof, F. N. Osman, A. S. Kusim, and M. S. Adam, 2012 ,*Characterization of Watermelon Leaf Diseases using Fuzzy Logic*. IEEE Symposium on Business, Engineering and Industrial Applications (ISBEA 2012), Bandung, Indonesia.

[3] A. E. Watada, 1989 Non-Destructive Methods of Evaluating Quality of Fresh Fruits and Vegetables, *Acta Horticulturae*, **258**. p. 321–329.

[4] K. Choi, K. Lee, and G. Kim, 2006 Non-destructive quality evaluation technology for fruits and vegetables using near-infrared spectroscopy, *Int. Semin. Enhancing Export Compet. Asian Fruits.*

[5] Y. He, Y. Zhang, A. G. Pereira, A. H. Gómez, and J. Wang, 2005 Nondestructive Determination of Tomato Fruit Quality Characteristics Using Vis / NIR Spectroscopy Technique, *Int. J. Inf. Technol.*, **11** 11 p. 97–108.

[6] R. Nilsson, Applications of Near Infrared Spectroscopic Analysis in the Food Industry and Research, *Food Saf. Cent.*, p 1-12.

[7] K. H. S. Peiris Dull, G. G., Leffler, R. G., Kays,S.J., 1998 Near-infrared spectrometric method for nondestructive determination of soluble solids content of peaches, *Journal of the American Society for Horticultural Science*, **123** 5 p 898–905.

[8] A. H. Feller, 1979 *Role of an Electromagnetic Environment Model in Spectrum Management.*

[9] J. Workman, 2005 *An Introduction to Near Infrared Spectroscopy.*

IOP Conf. Series: Materials Science and Engineering **99** (2015) 012014 doi:10.1088/1757-899X/99/1/012014

[10] B. M. Nicolaï, K. Beullens, E. Bobelyn, A. Peirs, W. Saeys, K. I. Theron, and J. Lammertyn, 2007 Nondestructive measurement of fruit and vegetable quality by means of NIR spectroscopy: A review, *Postharvest Biol. Technol.*, **46** 2, p. 99–118.
[11] J. Lammertyn, A. Peirs, J. De Baerdemaeker, and B. Nicolaï,2000 Light penetration properties of NIR radiation in fruit with respect to non-destructive quality assessment, *Postharvest Biol. Technol.*, **18** 2 p. 121–132.
[12] V. Semiconductors, 2001 Vishay Semiconductors, *Health Phys.*, p. 1–3.
[13] S. E. Series, G. Infrared, E. Diode, M. C. Package, T. Se, T. Se, T. To-, S. E. Series, G. Infrared, E. Diode, and M. C. Package, 2001 "SE5455-003," p. 1–5.
[14] S. Phototransistor, 2005 Sd3443/5443, p. 108–111
[15] P. Enhanced and F. Microcontrollers, 2013 Pic16F87Xa Pic16F87Xa.
[16] H. Max, 2012 Vishay 16 x 2 Character LCD ITEM SYMBOL STANDARD VALUE UNIT ITEM SYMBOL CONDITION FUNCTION, p. 31–33.
[17] M. S. Brown and C. B. Arnold,2010 Laser Precision Microfabrication, *Springer Ser. Mater. Sci.*, vol. 135, no. 0933–033X, p. 91–120
[18] J. Hauke and T. Kossowski, 2011 Comparison of Values of Pearson's and Spearman's Correlation Coefficients on the Same Sets of Data, *Quaest. Geogr.*, **30** 2 p. 87–93
[19] W. Zhu, N. Zeng, and N. Wang,2010 Sensitivity, Specificity, Accuracy, Associated Confidence Interval and ROC Analysis with Practical SAS Implementations, *Nesug*, p. 1–9

IOP Conf. Series: Materials Science and Engineering **99** (2015) 012015 doi:10.1088/1757-899X/99/1/012015

Impact of different NBTI defect components on sub-threshold operation of high-k p-MOSFET

H. Hussin[1, 2], N. Soin[1], S. Wan Muhamad Hatta[1] and M. F. Bukhori[3]
[1] Department of Electrical Engineering, University of Malaya, Kuala Lumpur 50603, Malaysia
[2] Faculty of Electrical Engineering, Universiti Teknologi MARA, Shah Alam 40450, Malaysia
[3]Dept. of Electrical, Electronics & Systems Engineering, Faculty of Engineering and Built Environment, Universiti Kebangsaan Malaysia, Bangi 43000, Malaysia

hanimh@salam.uitm.edu.my

Abstract. A study of the NBTI reliability of high-k p-MOSFET device for application in sub-threshold operation based on different defect mechanism is presented. The impact of the different defect mechanism is studied based on modelling the sub-threshold operation using Two-Stage NBTI model and NBTI-induced positive charges based on energy profiling approach. The time exponent of 0.1 is observed in sub-threshold operation modelled based on the Two-Stage NBTI model while time exponent of 0.3 is observed in sub-threshold operation modelled based energy profiling approach. Considerable threshold voltage shifts are observed during sub-threshold operation based on both defect mechanisms. Extraction of E'centres and E'/Pb H Complex as well as positive charges was found to be temperature dependence hence the degradation is also thermally activated during subthreshold operation for both defect mechanisms.

1. Introduction

Operations in a transistor's sub-threshold regime was critically essential for meeting a system's stringent power desires in ultra low-power uses. However, seeing the aggressive downscaling of complementary metal-oxide-semiconductor (CMOS) transistors, stress-induced Negative Bias Temperature Instability (NBTI) has appeared as a reliability risk even in low-power systems. As a result of scaled oxide thickness in sub-nanometre regime devices, the increased electric field further enhanced the NBTI degradation effects. Many studies that used circuit-level simulation used the Reaction-Diffusion (R-D) model to investigate the effects of NBTI degradation on a circuit's performance [1], [2]. However, as the oxide trapping/de-trapping effect mechanism affect device performance considerably [3], [4], [5] a device-level analysis of the effect of NBTI, which involved both oxide and interface traps under sub-threshold operation regime, is necessary. From the device-level perspective, operation in sub-threshold regime for Ultra – Effective Oxide Thickness, (UT-EOT) devices were contributed by the minority carrier and holes under the gate and diffuses due to the application of V_{DS}. However, different mechanisms occurred given the very thin EOT. Direct tunnelling of holes from substrate to bulk oxide will be trapped in the E'center defect precursor, which further degraded the device performance [6]. Previous study on sub-threshold effect due to NBTI was discussed based on hydrogen transport as the underlying mechanism [7]. Here, we present a study of the NBTI reliability of SiO_2/Hf (2/0.6 nm) with gate length of 32 nm under sub-threshold operation.

The modelling of NBTI-induced positive charges for degradation in sub-threshold operation is based on energy profiling approach.The subthreshold effects based on defect mechanism of interface trap and hole trapping effect with positive charges induced are presented.

2. Device and Simulation Conditions

2.1. Device process

The testbed PMOSFET devices with high-*k*/SiO_2 gate stacks simulated in this study were based on the foundry-standard 32-nm CMOS process. The fabrication process incorporated shallow trench isolation, deposition of high-*k* dielectrics with metal gate, stress engineering using epi-SiGe pockets, silicidation, and dual-stress liner [8], [9]. The device fabrication process flow in this simulation is presented in figure 1. This simulation process, which is based on the gate-first scheme, was adopted to overcome the process-related problems (such as ultra-shallow junction formation), suppress the leakage current, and improve the on-and-off state drain current. Incorporating the laser annealing process can help to suppress transient-enhanced diffusion. To experimentally validate the testbed device, the electrical characteristics of IdVg obtained using this process flow was compared with experimental data in [9] and shown in figure 2. The main performance characteristics based on this study are compared in table 1. Given the information shown this table, it can be concluded that the process flow for the testbed devices used in this work was approximately within the range of the characteristics of the device fabricated by [9]. The validation of testbed device method is similar to other simulation work which confirms that the Id-Vg simulation results matched the experimental data [10]. This is to ensure that the impact of NBTI degradation in this study is realistically assessed.

- Shallow trench isolation formation
- Well implantation
- High-*k* gate dielectric and metal gate deposition
- HALO implantation
- S/D extension
- SDE extension
- Sidewall spacer
- Two-step recess SiGe pocket
- Epitaxial SiGe growth
- HDD implantation
- Spike and laser annealing
- Nickel silicide
- Dual-stress liner deposition

Figure 1. The process flow of the simulated advanced-process 32 nm high-*k* PMOS transistor [11]

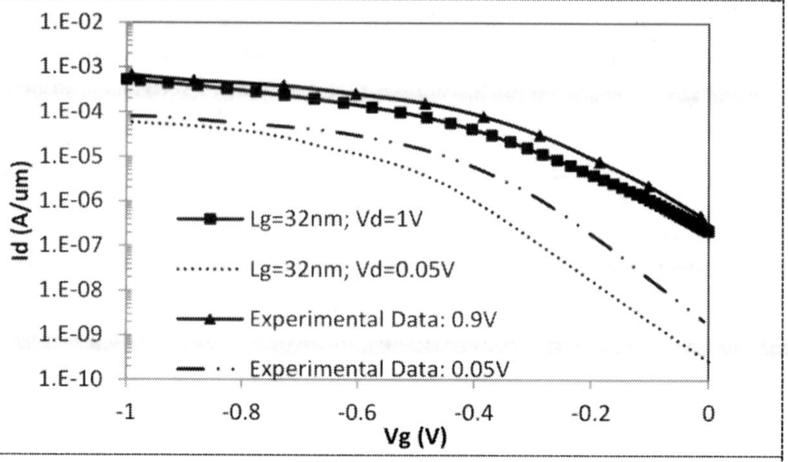

Figure 2. Comparison Id-Vg characteristics between simulation and experimental data in [9]

Table 1. Performance summary of simulated devices

	This work (Lg = 32 nm)	(Lg = 24 nm) [9]
Vdd (V)	1	0.9
Ion (A/um)	5.16 × 10−4	5.25 × 10−4
Ioff (A/um)	2.22 × 10−7	3.70 × 10−7

2.2. Simulation conditions

For the 32 nm technology high-k metal gate devices, both drift diffusion and hydrodynamic models were used in the simulation which solved the Poisson equation and carrier continuity equations for 2D simulation involved in high-k devices. The drift diffusion model was activated for the entire region, while the hydrodynamic model was only activated in the silicon region. The models were used selectively according to the region to ensure the convergence in the simulation is achieved. In addition to that, the hydrodynamic transport was restricted to only one type of carrier for a particular device. This helps to increase the simulation speed that guaranteed the accuracy.

In this work, the hydrodynamic transport model is used due to the effects of carrier heating in the largely-varying electric fields can be accurately captured inside a sub-micron device subjected to the extreme conditions of NBTI stresses. The relatively simpler drift-diffusion simulation model neglects the carrier heating effect, hence more physically-accurate simulation results is yielded using the hydrodynamic transport model as shown in figure 3. The drift-diffusion model underestimated the NBTI-induced ΔVth by as much as 12% where this underestimation appears to be rising with prolonged stress time [5].

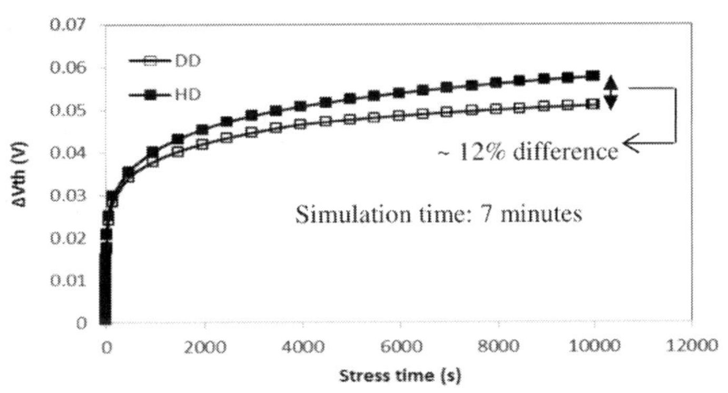

Figure 3. Underestimation by drift diffusion model after a 10000 s stress by 12% and expected to increase at extended stress time[5].

2.3. NBTI Simulation

The threshold voltage degradation in this simulation was determined using the widely adopted on-the-fly (OTF) method [12]–[15]. Application of pre-stress voltage was needed in this model to equilibrate the occupancy of different states and thus ensuring that all states were not empty at the early phase of stress period [16].

The investigation, based on different defect mechanism is conducted using default Two-stage NBTI models and redefine the energy level using an energy profiling approach model. Table 2 shows the different energy level for each defects used in this work.

Table 2. Energy levels for each defect mechanism

	Energy level (eV)	
Defect	Two-stage NBTI model	Redefine energy model using energy profiling approach
Pre-existing defect	Below valence band	Below valence band
E'center	$0 < E < 0.3$	Within energy band gap
E/Pb H complex	$0 < E < 0.5$	Above energy band gap

The energy profiling approach is based on the probing of energy distribution of positive charges in the gate dielectric [4]. The defects energy level as defined in a default model is redefined in order to have defect with characteristics based on positive charges. Hence, the pre-existing defect, E'center and E/Pb H Complex are the positive charges below, within and above energy band gap which known as grown hole traps (AHTs), cyclic positive charges (CPCs) and antineutralization positive charges (ANPCs) respectively based on [4]. The features of pre-existing defect, E'center and E/Pb H complex are similar to AHT, CPC and ANPC [4], [17], [18] hence the redefining of energy levels is important to accurately access the device degradation due to NBTI.

3. Results and Discussion

3.1. Simulation validation

To validate the simulation approach, figure 4 shows comparison between simulation result and experimental data in [4], [19]. The features of each charges contributed in the degradation was proven to be in agreement with the previous experimental works [4], [19]–[22]. The graph shows a simulation trend which agrees with the experiment for the region below Ev, within and above energy band gap.

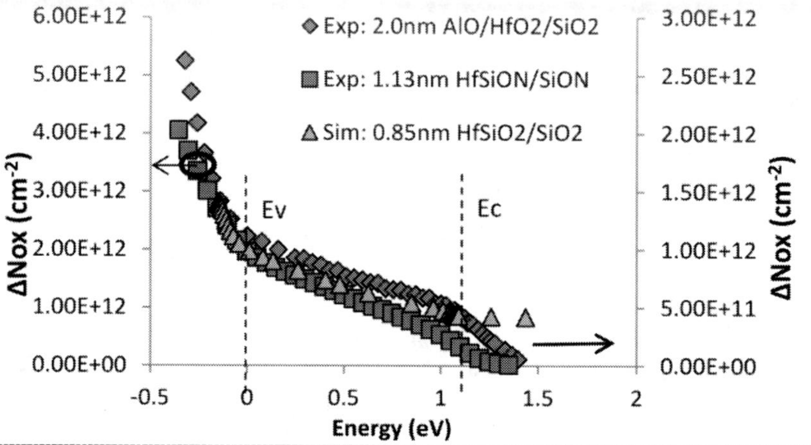

Figure 4. Simulation of $HfSiO_2/SiO_2$ compared with experimental data of Hf-based devices

The time exponent, n for redefine (new) and default (old) energy level varied significantly during the sub and super-threshold operations as observed in figure 5. The old energy level showed that the time exponent, n was around 0.1. However, the super-threshold voltage gave a different n. The n for new define energy level can be varied up to 0.3 as observed in most experimental works involving defects associated with DLHT or ANPC [23] which shown to be distributed closer or above E_c.

The total positive charges contributed in the degradation for new and default energy level is shown in figure 6. Significant differences can be seen in the new define energy level based on the applied stress temperature. Higher temperature gave more total positive charges contributed towards the degradation during sub and super-threshold operation regime for the new defines energy level. However, the default energy level showed no temperature effects upon the generated total positive charges.

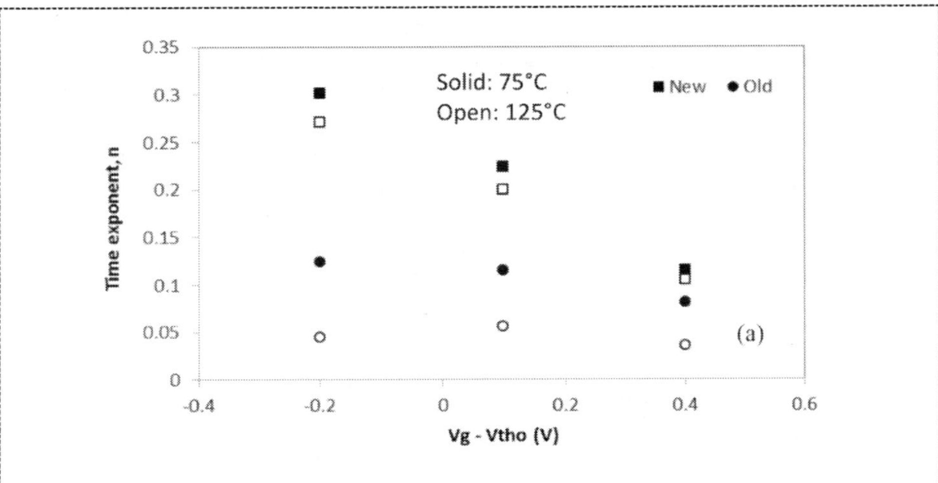

Figure 5. Plot for new define and default energy level under different stress temperature for time exponent versus $V_g - V_{tho}$

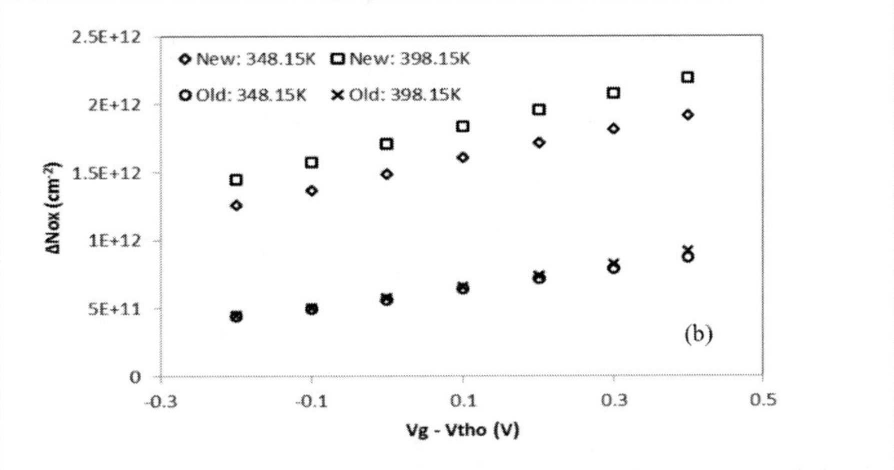

Figure 6. Plot for new define and default energy level under different stress temperature for total positive charges concentration versus Vg – Vtho

3.2. Comparison of defect component characteristics

The behavior of defect components which contributed in the degradation for different stress voltages as shown in figure 7 (a) and (b) for E'centres or CPCs and E'/Pb H complex or ANPCs concentration respectively is compared. As observed in figure 7 (a), the sub-threshold operation voltage shows that the generation of CPCs was almost saturated, but slightly increased, approaching 10000s stress time while the super-threshold voltage shows the increasing of the CPCs generation when the stress time increased. In contrast, the E'centers concentration saturated and slightly reduced approaching 10000s stress time. The generation of E'centres concentration was smaller compared to CPCs for both sub and super-threshold voltage.

In figure 7 (b), first impression on the concentration of E'/Pb H complex and ANPCs increased as the stress time increased for both sub and super-threshold voltage. Closer observation indicated that the generated of E'/Pb H complex and ANPCs during super-threshold operation regime increased throughout the stress duration. In contrast, the generated E'/Pb H complex and ANPCs during the sub-threshold operation regime only increased when the stress time was approaching above 1000s. Another closer observation indicated that the generated ANPCs was shown to be as compared to E'/Pb H complex.

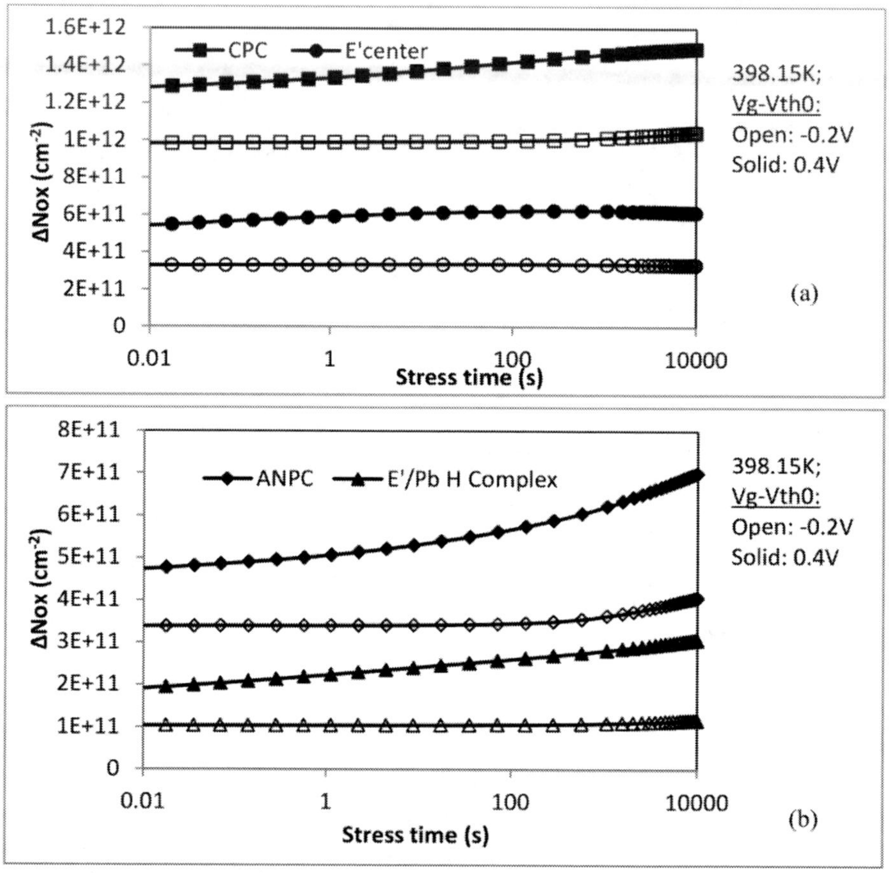

Figure 7. Defect components generated during NBTI stress for (a) E'center and CPC (b) E'/Pb H Complex and ANPC

Figure 8 (a) and (b) shows the CPC and ANPC complexes under different stress voltages and stress temperatures using new and default energy level respectively. As shown in figure 8 (a), the concentrations of generated CPC were higher than generated ANPC during the sub and super-threshold operation regime and under higher stress temperature. This is in agreement with previous experimental works which highlighted that the generated of positive charges were thermally activated [24]. Both components of positive charges were found to be temperature dependence. Therefore, the degradation during the sub-threshold operation regime was also thermally activated. This is not in agreement with [7] because the threshold voltage shift observed was not dependent on temperature during sub-threshold operation which is suspected due to the contributed defect mechanism studied based on interface trap.

For comparison, the plot of default energy level presented in figure 8 (b) gave different kinetics based on different stress temperatures for the sub and super-threshold operation regime. Above 1000s stress time duration, the E'center concentration was higher for the lower stress temperature during the super-threshold voltage. In contrast, for sub-threshold operation, similar generated E'center concentration was observed for both stress temperatures. Upon approaching the 1000s, the generated E'center concentration started to reduce more for higher stress temperature, hence during the 10000s stress duration, more generated E' centers concentration was observed during smaller stress temperature. However, the generated E'/Pb H Complex concentration for both sub and super-threshold is stress time and temperature dependence. As the stress time and stress temperature increased, the generated E'/Pb H Complex concentration too increased.

3.3. Lifetime prediction

The lifetime of sub-threshold operation based on a hundred-year lifetime was estimated. The ΔV_{th} as a function of device lifetime which can be used in simulation study for circuit level degradation projection based on application-specific criterion [7] where it was readily understandable. Figure 9 (a) and (b) shows the level of device degradation in terms of ΔV_{th} as a function of device lifetime for new and default energy levels respectively for sub-threshold and super-threshold operation regime. The ΔV_{th} lines for each voltage were extrapolated to 10 and 100 years. The level of degradation in NBTI effect was assessed based on failure criterion is ΔV_t = 30mV [7]. In the first impression for sub-threshold operation regime, based on the failure criterion target, the device can satisfy a hundred-year lifetime at sub-threshold operation voltage which was suitable for biomedical application which require extended device lifetime [7] for both new and old energy level. A closer observation as presented in figure 9 (c) shows that the threshold voltage shifted during the 10 and 100 years operation. For biomedical application devices, the sub-threshold operation regime which is for $V_g - V_{tho} = -0.2$ V and up to $V_g - V_{tho} = 0.2$ V meet the failure criterion target. The percentage difference for threshold voltage shift extracted based on new define and default energy level is shown in the figure. There was a significant difference in the prediction using default and new define energy level.

Conclusion

For sub-threshold operation, the time exponent observed for default and new define energy level was 0.1 and 0.3 respectively. The time exponent of 0.1 was referred to the pre-existing defect available in the device while the 0.3 was referred to the total generated defect contributed to the degradation consist of the CPC and ANPC which located near and above conduction energy level respectively. The degradation which characterized based on the redefines energy level found that the degradation was significant when stressed at higher temperature as compared to lower temperatures during the sub-threshold regime. However, the characterization using default energy level found that the degradation was not influenced by the stress temperature. More degradation was observed based on redefine energy level where the concentrations of generated CPC were higher than generated ANPC during sub and super-threshold operation regime and under higher stress temperature.

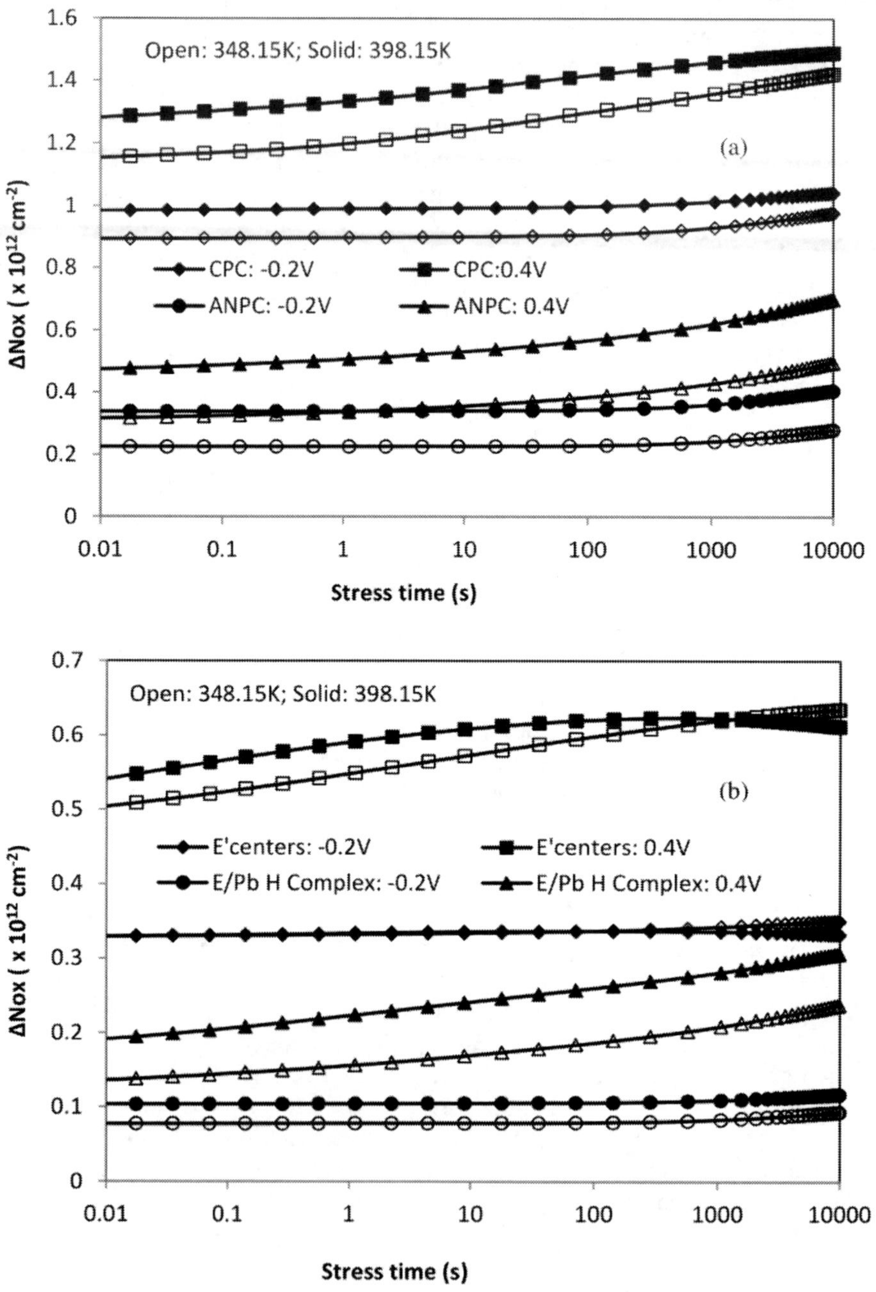

Figure 8. Plot of defect components during sub and super-threshold voltage under different stress temperatures for (a) Redefine energy levels (b) default energy level

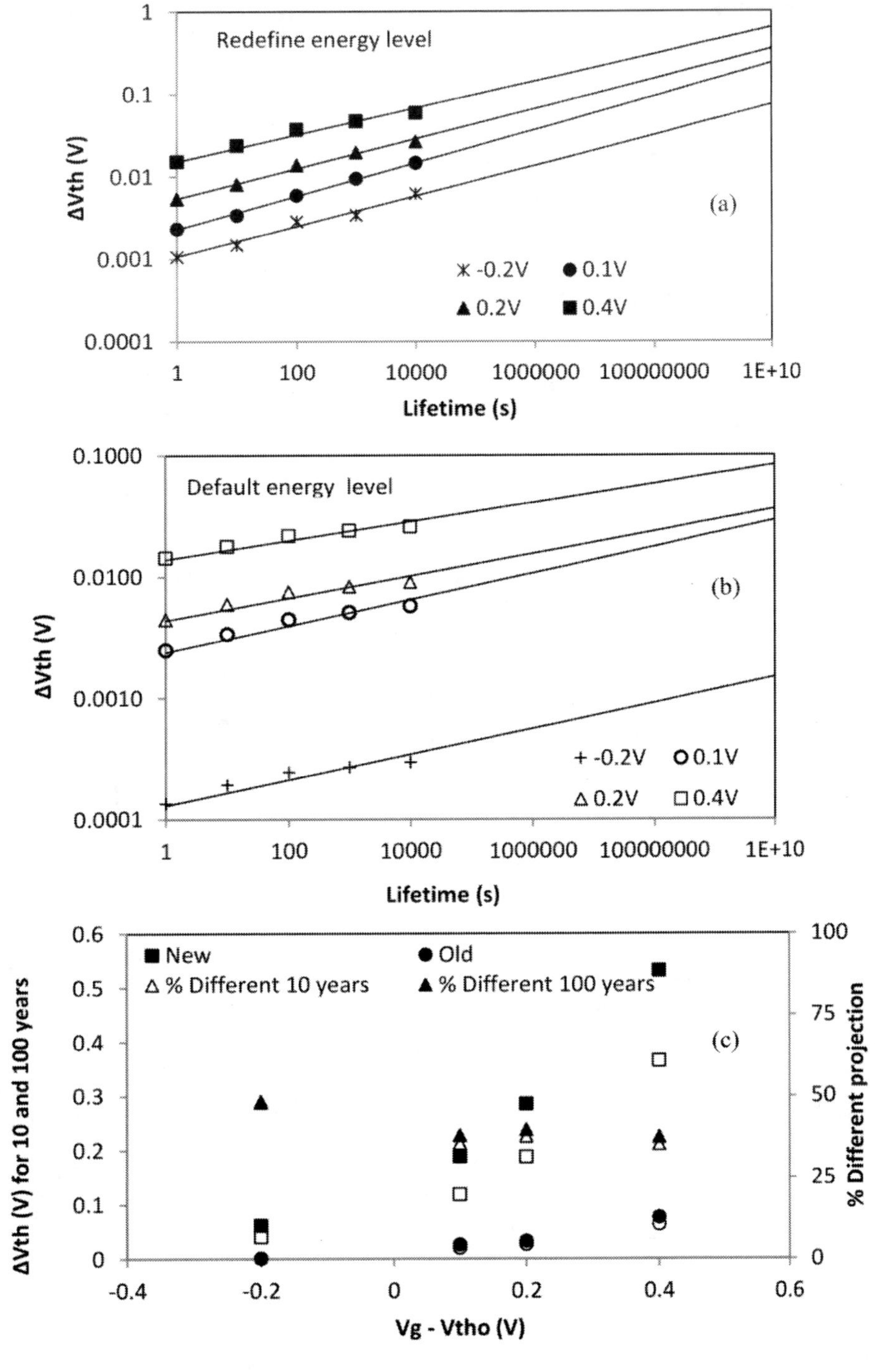

Figure 9. Power-law time dependence of extrapolated ΔV_{th} with respect to lifetime during sub and super-threshold operating regime for (a) New energy level (b) default energy level and (c) Estimation of ΔV_{th} after 10 and 100 years operation with the percentage different of estimation using new define and default energy level

IOP Conf. Series: Materials Science and Engineering **99** (2015) 012015 doi:10.1088/1757-899X/99/1/012015

References

[1] W. Wang, S. Yang, S. Bhardwaj, S. Vrudhula, F. Liu, and Y. Cao, "The Impact of NBTI Effect on Combinational Circuit :," *IEEE Trans. Very Large Scale Integr. Sytems*, vol. 18, no. 2, pp. 173–183, 2010.

[2] J.-S. Yuan, W.-K. Yeh, S. Chen, and C.-W. Hsu, "NBTI reliability on high-k metal-gate SiGe transistor and circuit performances," *Microelectron. Reliab.*, vol. 51, no. 5, pp. 914–918, May 2011.

[3] D. S. Ang, Z. Q. Teo, T. J. J. Ho, and C. M. Ng, "Reassessing the Mechanisms of Negative-Bias Temperature Instability by Repetitive Stress / Relaxation Experiments," *IEEE Trans. Device Mater. Reliab.*, vol. 11, no. 1, pp. 19–34, 2011.

[4] S. W. M. Hatta, Z. Ji, J. F. Zhang, M. Duan, W. D. Zhang, N. Soin, B. Kaczer, S. De Gendt, and G. Groeseneken, "Energy Distribution of Positive Charges in Gate Dielectric: Probing Technique and Impacts of Different Defects," *IEEE Trans. Electron Devices*, vol. 60, no. 5, pp. 1745–1753, May 2013.

[5] S. F. W. M. H. and Y. A. W. H. Hussin, N. Soin, M. F. Bukhori, "Effects of Gate Stack Structural and Process Defectivity on High- Dielectric Dependence of NBTI Reliability in 32 nm Technology Node PMOSFETs," *Sci. World J.*, vol. 2014, pp. 1–13, 2014.

[6] M. Cho, J.-D. Lee, M. Aoulaiche, B. Kaczer, P. Roussel, T. Kauerauf, R. Degraeve, J. Franco, L.-Å. Ragnarsson, and G. Groeseneken, "Insight Into N/PBTI Mechanisms in Sub-1-nm-EOT Devices," *IEEE Trans. Electron Devices*, vol. 59, no. 8, pp. 2042–2048, Aug. 2012.

[7] J. Franco, S. Graziano, B. Kaczer, F. Crupi, L.-Å. Ragnarsson, T. Grasser, and G. Groeseneken, "BTI reliability of ultra-thin EOT MOSFETs for sub-threshold logic," *Microelectron. Reliab.*, vol. 52, no. 9–10, pp. 1932–1935, Sep. 2012.

[8] S. F. W. M. Hatta, N. Soin, D. A. Hadi, and J. F. Zhang, "NBTI degradation effect on advanced-process 45nm high-k PMOSFETs with geometric and process variations," *Microelectron. Reliab.*, vol. 50, no. 9–11, pp. 1283–1289, Sep. 2010.

[9] N. Yasutake, T. Ishida, K. Ohuchi, N. Aoki, N. Kusunoki, S. Mori, I. Mizushima, T. Morooka, K. Yahashi, S. Kawanaka, K. Ishimaru, and H. Ishiuchi, "A High Performance pMOSFET with Two-step Recessed SiGe-S/D Structure for 32nm node and Beyond," in *Proceeding of the 36th European Solid-State Device Research Conference*, 2006, pp. 77–80.

[10] M. F. Bukhori, "Simulation of Charge-Trapping in Nano-Scale MOSFETs in the Presence of Random-Dopants-Induced Variability," 2011.

[11] "Sentaurus Technology Template : 32-nm Gate-First Flow and CMOS Processing." TCAD Sentaurus Application Note, Version F-2011.09, 2011.

[12] S. Mahapatra and M. A. Alam, "Defect Generation in p-MOSFETs Under Negative-Bias Stress : An Experimental Perspective," *IEEE Trans. Device Mater. Reliab.*, vol. 8, no. 1, pp. 35–46, 2008.

[13] T. Yang and W. J. Liu, "Understand NBTI Mechanism by Developing Novel Measurement Techniques," *IEEE Trans. Device Mater. Reliab.*, vol. 8, no. 1, pp. 62–71, Mar. 2008.

[14] V. D. Maheta, E. N. Kumar, S. Purawat, C. Olsen, K. Ahmed, and S. Mahapatra, "Development of an Ultrafast On-the-Fly IDLIN Technique to Study NBTI in Plasma and Thermal Oxynitride p-MOSFETs," *IEEE Trans. Electron Devices*, vol. 55, no. 10, pp. 2614–2622, 2008.

[15] V. Huard, M. Denais, and C. Parthasarathy, "NBTI degradation: From physical mechanisms to modelling," *Microelectron. Reliab.*, vol. 46, no. 1, pp. 1–23, Jan. 2006.

[16] S. Gupta, B. Jose, K. Joshi, A. Jain, M. A. Alam, and S. Mahapatra, "A Comprehensive and Critical Re-assement of 2-Stage Energy Level NBTI Model," in *IEEE Int. Reliab. Phys. Symp*, 2012, pp. 1–6.

[17] T. Grasser, B. Kaczer, W. Goes, H. Reisinger, T. Aichinger, P. Hehenberger, P. Wagner, F. Schanovsky, J. Franco, M. T. Luque, and M. Nelhiebel, "The Paradigm Shift in Understanding the Bias Temperature Instability : From Reaction – Diffusion to Switching Oxide Traps," *IEEE Trans. Electron Devices*, pp. 1–15, 2011.

[18] J. F. Zhang, "Defects and instabilities in Hf-dielectric/SiON stacks (Invited Paper)," *Microelectron. Eng.*, vol. 86, no. 7–9, pp. 1883–1887, Jul. 2009.

[19] S. W. M. Hatta, Z. Ji, J. F. Zhang, W. D. Zhang, N. Soin, B. Kaczer, S. D. Gendt, and G. Groeseneken, "Energy distribution of positive charges in high-k dielectric," *Microelectron. Reliab.*, vol. 54, no. 9–10, pp. 2329–2333, Aug. 2014.

[20] J. F. Zhang, "Defects and instabilities in Hf-dielectric/SiON stacks (Invited Paper)," *Microelectron. Eng.*, vol. 86, no. 7–9, pp. 1883–1887, Jul. 2009.

[21] J. F. Zhang, C. Z. Zhao, A. H. Chen, G. Groeseneken, and R. Degraeve, "Hole Traps in Silicon Dioxides — Part I : Properties," *IEEE Trans. Electron Devices*, vol. 51, no. 8, pp. 1267–1273, 2004.

[22] C. Z. Zhao, J. F. Zhang, M. H. Chang, a. R. Peaker, S. Hall, G. Groeseneken, L. Pantisano, S. De Gendt, and M. Heyns, "Stress-Induced Positive Charge in Hf-Based Gate Dielectrics: Impact on Device Performance and a Framework for the Defect," *IEEE Trans. Electron Devices*, vol. 55, no. 7, pp. 1647–1656, Jul. 2008.

[23] S. W. M. Hatta, "Probing Technique For Energy Distribution Positive Charges In Gate Dielectrics and Its Application to Lifetime Prediction. PhD. Thesis.," LJMU/UM, 2013.

[24] A. A. Boo, D. S. Ang, Z. Q. Teo, and K. C. Leong, "Correlation Between Oxide Trap Generation and Negative-Bias Temperature Instability," *IEEE Electron Device Lett.*, vol. 33, no. 4, pp. 486–488, Apr. 2012.

Influence of Growth Time and Temperature on the Morphology of ZnO Nanorods via Hydrothermal

Syahirah Shaziman, Ahmad Syakirin Ismail@rosdi, Mohamad Hafiz Mamat, Ahmad Sabirin Zoolfakar
Faculty of Electrical Engineering, MARA University of Technology, 40450 Shah Alam, Selangor, Malaysia.

Emails: ahmad074@salam.uitm.edu.my

Abstract. This research investigates the effect of temperature and growth duration on the morphology of ZnO. ZnO nanostructure can be synthesized *via* hydrothermal method using Zinc Acetate and hexamethylenetetramine (HMT) as precursor and glass slide as a substrate. The temperature and growth time vary from 50-95℃ and 50-100 minutes, respectively. The substrates with ZnO nanostructures are annealed at different temperatures which are 200°C and 500°C. The result indicates that different temperature and growth time produces different morphologies in size. Thus, the morphology of ZnO nanostructure affects by the parameters such as temperature and growth time *via* hydrothermal grown.

1. Introduction

Zinc oxide (ZnO) is frequently used in several areas of technology due to its unique properties such as large excitation binding energy of 60meV direct wide band gap (Eg~3.3 eV at 300 K) which is widely used for production of green, blue-ultraviolet, and white light-emitting devices [1] and can be effectively grown at low temperature (less than 700 C) [1]. Related to the remarkable features and has large surface-to-volume ratio of ZnO nanostructures [2], it has been widely used in various applications, such as solar cells[3], photocatalysts [4], memristor[5], gas sensors[6] and piezoelectric devices [7].

The properties of such materials of ZnO nanostructures are strongly dependent on their microstructures [8], including crystal size, specific surface [9] and especially on morphologies [10]. Different application may need different types of ZnO nanostructure, size, shape, morphology and growth orientation [2]. There are several types of dimensions in ZnO such as dimensional (1D), two-dimensional (2D) and three-dimensional (3D) ZnO nanostructures. However, most of the researches are mainly focused on one-dimensional ZnO nanostructures, such as nanowires, nanorods and nanotubes[11] due to its unique features such as less grain boundary, surface defect, and dislocation [12]. These nanostructures can be synthesized through various techniques. The synthesis techniques include thermal evaporation[13], chemical vapour deposition (CVD) [14] , physical vapour deposition[15], hydrothermal[16] , sol–gel [17], and electrochemical deposition [18]. However hydrothermal synthesis is chosen as the best method among these because of its low-temperature (60-100 C), low cost, high yield [19,20] and the growth parameters are much easily controlled the shape and dimension of ZnO nanostructures, compared to other techniques[21,22].

There are several parameters that affect the morphology of ZnO nanostructures in hydrothermal synthesis. Mudan Wang et al.[23] studied that parameters such as seed layer, ZnO colloid concentration, substrate, and precursor concentration are hardly effect on the alignment of ZnO nanorods array. In this study, it clearly shown that different sizes of ZnO nanorods can be obtained by manipulating all these parameters. Thus, this research investigates the effect of time and temperature on the morphology of ZnO nanorods *via* hydrothermal method.

IOP Conf. Series: Materials Science and Engineering **99** (2015) 012016 doi:10.1088/1757-899X/99/1/012016

2. Methodology

2.1. A. Preparation of ZnO seed layers

Glass slides were cleaned by using acetone, ethanol and distilled water. The seed layer was prepared by mixing solution of Zn Acetate ($(CH_3COO)_2.2H_2O$, 0.4 M), Aluminium Nitrate Nanohydrate ($Al(NO_3)_2$) and Monoethanol Amine(C_2H_7NO) in 2- methyoxyethanol ($C_3H_8O_2$). The prepared solution was stirred and heated at 250 rpm and 80 C, respectively for 3 hours. It was continuously stirred on the hot plate without heating for 24 hours. The solution was then spin coated on glass substrate at 2500 rpm for 1 minute. The coated substrate was annealed at 200 C for 10 minutes to obtain ZnO seed layer. The spin-coating method was repeated for 5 times to obtain the surface thickness on the substrate. The coated glasses were post annealed at 500 °C for 1 hour to introduce nucleation sites that give contribution for absorbing zinc nutrients during the growth process and for forming nanorods [2].

2.2. Growth of ZnO nanorods *via* hydrothermal method

7.01 g of hexamethylenetetramine (HMT)($C_6H_{12}N_4$,0.1M) solution will be mixed with Zn Nitrate ($Zn(NO_3)_2$,0.1M). Both of the solutions were immersed in ultrasonic bath for 30 minutes. The precursor was stirred well on a hot plate for 3 hours. The glass slides coated with the annealed seed layer which act as a substrate was immersed in the precursor and heated at C for 50 and 80 minutes respectively to study the effect of ZnO at different growth time. After 50 and 80 minutes, the substrates were rinsed with deionized water. However, the same procedure was carried out for at 95 C to investigate the effect of temperature on the morphology of ZnO nanorods. The substrates with ZnO nanorods were dried at 1 C for 10 minutes and post-annealed with different temperature at 200 and °C for 1 hour. The morphology of ZnO nanorods were observed by FESEM images.

3. RESULT AND DISCUSSION

3.1 Effect of growth time on ZnO nanorods

In the formation of ZnO nanorods, thin film ZnO seed layer is required to promote the epitaxial nucleation on the glass substrate [24]. Fig. 1 shows the effect of growth time on the size of nanorods at different annealing temperature. Based on the Fig. 1(a), at immersion and post-annealing temperature at 80 °C and 95 °C, respectively, the size of nanorods increases from 46.25 nm to 162.50nm as the growth time increases from 50 to 80 minutes. Overall, it can be observed that the size of nanorods increase as the growth rate increases at fixed immersion temperature.

Theoretically, the chemical reaction of the formation of ZnO nanorods can be summarized as below [10,25,26]

$$(CH)_6N_4 + 6H_2O \rightarrow 6HCHO + NH_3 \quad (1)$$

$$NH_3 + H_2O \rightarrow NH_4^{+} + OH^{-} \quad (2)$$

$$Zn^{2+} + NH_3 \rightarrow Zn\ NH_3)_4^{2+} \quad (3)$$

$$Zn^{2+} + 4OH^{-} \rightarrow Zn(OH)_4^{2-} \quad (4)$$

$$Zn(NH_3)_4^{2+} + 2OH^{-} \rightarrow ZnO + 4NH_3 + H_2O \quad (5)$$

$$Zn(OH)_4^{2-} \rightarrow ZnO + H_2O + 2OH^{-} \quad (6)$$

IOP Conf. Series: Materials Science and Engineering **99** (2015) 012016 doi:10.1088/1757-899X/99/1/012016

In the initial growth of ZnO nanorods, the HMT begins to decompose into ammonia (Eqs.(1)) and provide the hydroxide ions, OH^- (Eqs.(2)) [10]. The Zn^{2+} cation further reacts with NH_3 and OH^- anion to form ZnO nuclei (Eqs.(3) and Eqs.(4)) [25]. The formed crystal nuclei will grow, and then degraded into ZnO nanorods nuclei with the influence of more OH ions under a certain temperature according to Eqs. (5) and (6) [25]. Hence, as the time increases, the ZnO nuclei will continuously growing. Zn on the surface of the Zn substrate is further oxidized from the oxygen to form ZnO nanorods on the substrate [27]. While Zn atoms attached to the ZnO nuclei edges, the oxidization process causes lateral growth of ZnO nuclei [28,29]. Based on the Fig. 1(b), at the immersion and post-annealing temperature of 95 °C and 500 °C, repectively, the size of nanorods decreases from 62.50 to 56.25. Hence, the ZnO nanorods may grow on the top of other layer of nanorods. This may due to the continuous consumption of Zn nanoparticles on the surface of the Zn substrate and ZnO are not enough for the growth of the ZnO nanorods [30].

Shuzi Li *et al* demonstrates that there are many small ZnO rods at the bottom of each ZnO hexagonal structure, this phenomenon means that the ZnO crystallite rods amalgamate each other at the later growth stage because the growth velocity of ZnO is faster along the c axis than that of other directions[31]. Based on the Fig. 2 (e) and (g), it can be observed that the densitiy of nanorods decreases as the temperature increases at the immersion temperature at 95 ˚C. This is a spontaneous process that occurs because larger crystals are more energetically favoured than smaller crystals[33]. In this case, kinetically favoured tiny crystallites nucleate first in supersaturated medium and are followed by the growth of larger particles (thermodynamically favoured) due to the energy difference between large and smaller particles of higher solubility based on the Gibbs-Thomson Law [34]

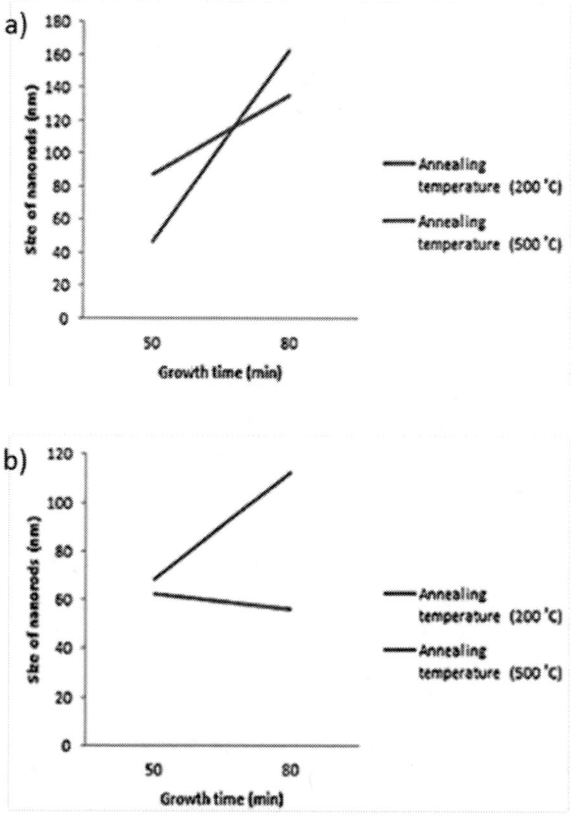

Fig. 1 Effect of growth time on ZnO nanorods at immersion temperature (a) 80 ˚C and (b)95 ˚C

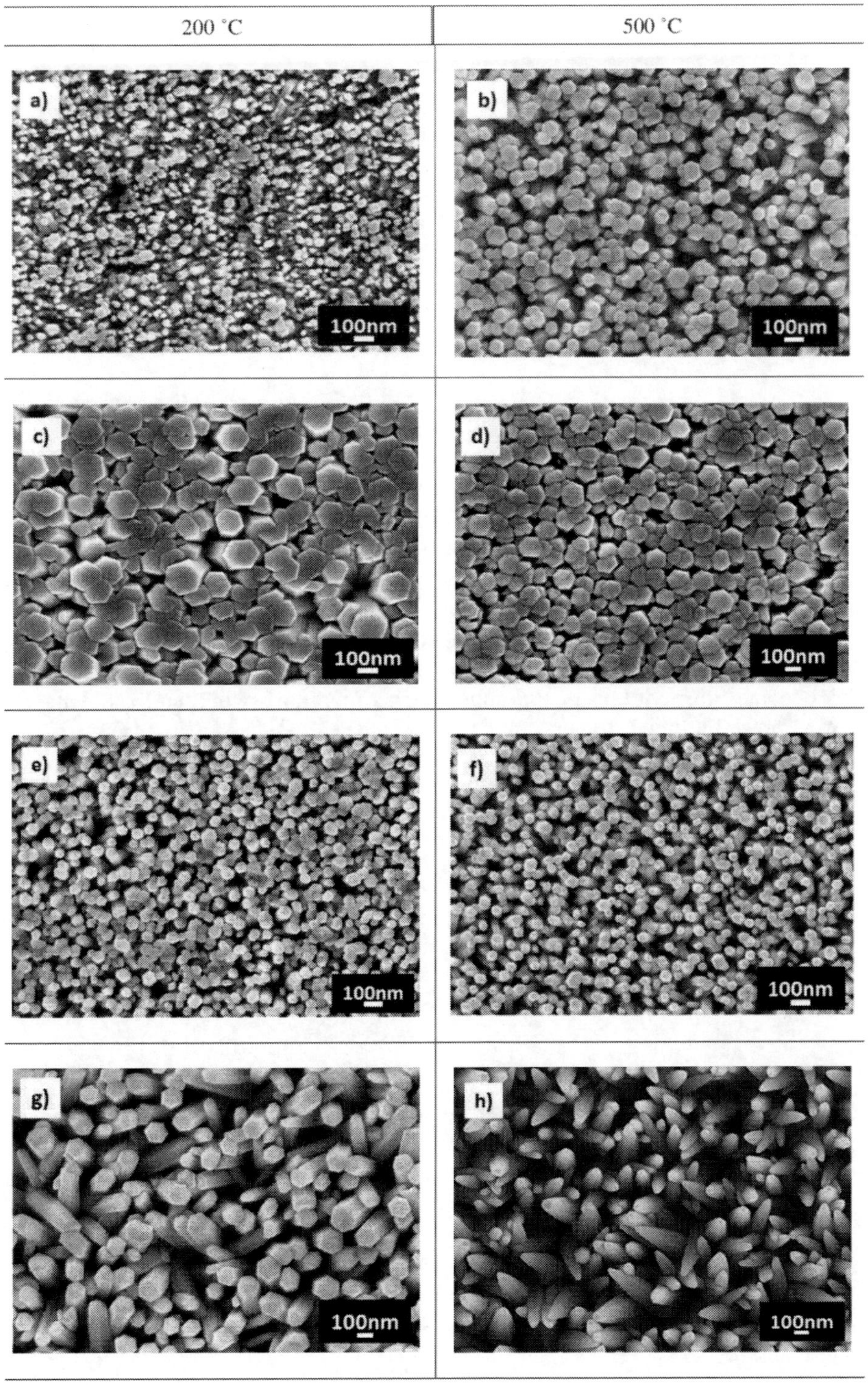

Fig. 2 FESEM images of ZnO nanorods were annealed at different immersion time and growth at different time at different annealing temperature (a-d) 80℃, (a-b) and (c-d) growth for 50 and 80 minutes;(e-h) 95℃,(e-f) and (g-h) growth for 50 and 80 minutes

3.2 Effect of immersion temperature on ZnO nanorods

Based on Fig. 2, the size of nanorods decreases when the immersion temperature increases. The length of nanorods is expected to be increased as the temperature increases. Based on the comparison between Fig. 2 (c)-(d) and (g)-(h), it shows that ZnO nanorods grow less densely packed as the temperature increases. This can be explained by that the increasing of reaction temperature that will enhanced the aggregation of the ZnO nanorods as well as accelerates the collision between nucleation atoms during the ZnO hydrothermal process [35]. This kind of collision may cause the dispersion and desorption of nucleation atoms, which reduces the probability for tight aggregation of the ZnO nanorods [35,36]. However, when the growth time is at 50 minutes with annealing temperature at 200 °C, the size increases from 46.25 to 68.75 nm. Increasing in temperature may due to nucleation atoms acquire enough thermal energy at a higher solution temperature which quickly elevates their surface mobility and makes them become more activated [37].

3.3. Effect of annealing temperature on ZnO nanorods

Based on the Fig. 1 and 3, it shows that the microstructures of ZnO nanorods change at a higher annealing temperature as the movement of atom increases [38]. However, it is observed that the size of nanorods is bigger at the annealing temperature of 200 °C than 500 °C . The ZnO nanorods at a higher annealing temperature has lesser crystallinity than the lower annealing temperature. It may occur due to the weak stability of glass that affects the stability and recrystallization of ZnO nanorods at a higher temperature[38]. The further increase of the annealing temperature may induce the desorption of ZnO from the substrate[38]. It is also accepted that as the temperature increases to higher, the movements of molecules become more violent as the annealing temperature is higher [39]. This will deteriorate the perfection of crystal lattice [39].

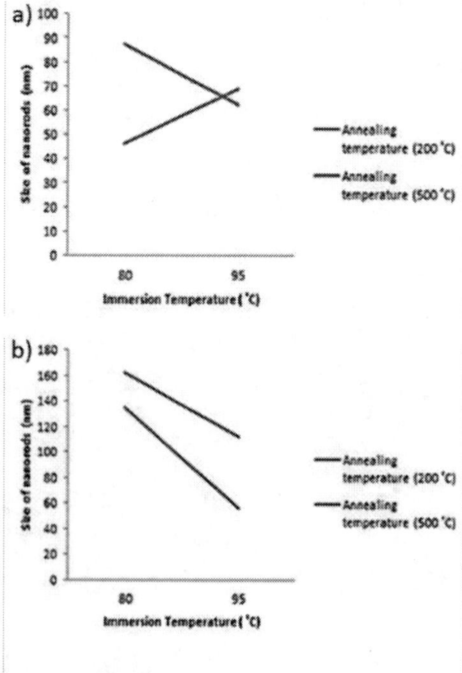

Fig 3 Effect of immersion temperature on ZnO nanorods at growth time (a) 50 minutes (b) 80 minutes

4. Conclusions

The effects of growth time and temperature on the morphology of ZnO nanorods have been successfully investigated. The ZnO nanorods are formed on the ZnO seed layers. As the growth time increases, the size of nanorods. However, the size of nanorods decreases at annealing and immersion temperature at 500 °C and 95 °C, respectively. The size of nanorods decreases as the immersion temperature increases. At the fixed growth time at 50 minutes, the size of nanorods increases at annealing temperature of 200 °C. It shows that different growth time and temperature gives a big impact on the size of nanorods. In addition, the size of nanorods is also affected by the post-annealed temperature at 200 C and 500 °C. Overall, the higher temperature results in smaller size of nanorods. The obtained results were characterized by FESEM.

References

[1] U. Ozgur, *et al.*, J. Appl. Phys., vol. 98, pp. 041301, 2005.

[2] K. A. Wahid, *et al.*, Appl. Surf. Sci., vol. 283, pp. 629-635, 2013.

[3] Zoolfakar A.S *et al.*, J. Mater. Chem., 22, pp. 21276, 2012.

[4] F.-H. Chu *et al*, Nanoscale, vol. 4, pp. 1471– 1475,2012.

[5] Zoolfakar A.S *et al.*, J. Phys. Chem. Chem. Phys., 15, pp. 10376, 2013.

[6] Brijesh Kumar *et al.*, Nano Energ., pp. 342–355,2012.

[7] M.-P. Lu *et al.*, Nano Lett., vol. 9 ,pp. 1223–1227, 2009.

[8] J. Zhang *et al.*, Int. J. Mater. Mech. Engin., vol. 1, pp. 38–43, 2012.

[9] M. Raula *et al.*, Langmuir, vol. 26, pp. 8769–8782, 2010.

[10] R. Shi, *et al.*, Appl. Surf. Sci., vol. 264, pp. 162-170, 2013.

[11] Y. Xia *et al.*, Adv. Mater, vol. 15, pp. 353–389, 2013.

[12] G.R. Li *et al.* , J. Phys. Chem. C 112 (2008) 11859–11864

[13] P.Hu *et al.*, Sens. Actuat. B:Chem, vol. 169, pp. 74–80, 2012.

[14] K.B. Sundaram *et al.*, Thin Solid Films, vol. 295, pp.87–91, 1997.

[15] H. Wang *et al.*, J. Mater. Sci. Techno, vol.127, pp. 153–158, 2011.

[16] K.F. Lin *et al.*, Chem. Phys. Lett, vol. .409 , pp. 208–211, 2005.

[17] J. Elias *et al.*, J. Electroanalytical Chem., vol. 621 pp. 171–177, 2008.

[18] L. Vayssieres *et al.*, J. Phys. Chem. B 105 (2001)3350.

[19] B. Liu, C.H. Zeng, J. Am. Chem. Soc. 125 (2003) 4430.

[20] Z. Qiu, K.S. Wong, M. Wu, W. Lin, H. Xu, Appl. Phys. Lett. 84 (2004) 2739.

[21] J.B. Shim *et al.*, J. Nanomater., 2011.

[22] S.H. Jung *et al.*, Cryst. Growth Des, vol. 8, pp. 265–269, 2008.

[23] M.W *et al.* Phys. Chem. Solids, 2014, pp.808-817.

[24] P. Li *et al.*, Mater. Chem. Phys., 106 ,2007,pp.63– 69.

[25] Y. Zhou *et al.* .Ceramics Int. , vol. 40,2014, pp.10415–10421.

[26] J. Xie *et al.*, Appl. Surf. Sci. vol. 257 ,2011, pp. 6358–636.

[27] L.Z. Pei *et al,* Physica E , vol. 42, 2010, pp. 1333–1337.

[28] K. Hou, C. Li, W. Lei, X.B. Zhang, X.X. Yang, K. Qu, B.P. Wang, Z.W. Zhao, X.W. Sun, Physica E 41 (2009) 470.

[29] S.R. Hejazi, H.R.M. Hosseini, M.S. Ghamsari, J. Alloys Compd. 455 (2008) 353

[30] J.P. Liu, X.T. Huang, Y.Y. Li, K.M. Sulieman, X. He, F.L. Sun, Cryst. Growth Des. 6 (2006) 1690.

[31] Shuzi Li et al, Mater. Lett.,Vol. 61,pp. 30–33, 2007.

[32] O. Krichershy, J. Stavan, Phys. Rev. Lett. 70 (1993) 1473.

[33] D. Polsongkrama *et al.*, Physica B, Vol. 403 pp.3713– 3717, 2008.

[34] J.W. Mullin, Crystallization, third ed, Butterworth/Heinemann, Oxford, 1997,pp. 1436.

[35] Y.J. Gao, W.C. Zhang, X.L. Wu, Y. Xia, G.S. Huang, L.L. Xu, J.C. Shen, G.G. Siu, P.K. Chu, Appl. Surf. Sci. 255 (2008) 1982– 1987.

[36] R. Sivakumar, T. Tsunoda, Y. Kuroki, T. Okamoto, M. Takata, Mater. Chem. Phys. 134 (2012) 345–349.

[37] Meng *et al.*, Appl. Surf. Sci., vol. 268 ,2013, pp. 561–565.

[38] Hainan Wu *et al.*, J. Alloy Comp., vol. 565 ,2013, pp. 85–89.

[39] Zh o *et al.,* Appl. Surf. Sci., Vol. 255, 15 March 2009, pp. 5861–5865.

The Strategy to Control the Morphology of ZnO Nanostructure UV Sensor

Q.Humayun[1*], U.Hashim[2], C.M. RUZAIDI[3], Kai Loong Foo[4]
[1]Department of Electronics Faculty of Engineering Technology, Universiti Malaysia Perlis (UniMAP),
[3]Faculty of Engineering Technology, Universiti Malaysia Perlis (UniMAP)
[2,4]Nano Biochip Research Group, Institute of Nano Electronic Engineering (INEE), Universiti Malaysia Perlis (UniMAP), Kangar, Perlis.

E-mail: qhumayun2@gmail.com

Abstract. The control morphology of ZnO nanostructures at specific area of electrodes by implemented a cost effective fabrication process, is extremely a challenging task. Rapid sensing, fast response and fast detection capability of the electronics devices is nowadays hot subject of keen interest. Our research is one of the successful attempts to achieve the desired goal at certain levels. Therefore in the current research article the ZnO thin film and ZnO nanorods were selectively deposited by low cost sol-gel and hydrothermal growth process at the selective area of microgap electrodes spacing and further the comparative study of ZnO thin film and ZnO nanorods were conducted electrically, for ultraviolet (UV) sensing application. On exposure to ultraviolet (UV) light the current gains, response/recovery times, repeatability, of the ZnO nanorods compared with ZnO thin film was improved probably due to the role of large surface area covered by the deposited nanostructures, and the most important is the bridging nanorods at the microgap electrodes tips. All the characterization including, surface, electrical and structural of the deposited nanostructures were completed by using SEM, sourcemeter and XRD respectively.

Keywords:Micro-gap, Sol gel, Hydrothermal, ZnO thin film, ZnO Nanorods, Ultraviolet light (UV).

1. Introduction

During the last few decades, nanomaterials have been the subject of extensive interest because of their potential use in a wide range of fields like, optoelectronics, catalysis and sensing applications [1-3]. Due to smaller size and larger surface area to volume ratio, nanomaterials exhibits unique properties (physical and chemical) compared to conventional and bulk counterparts [4]. Generally there are three types of nanomaterials (one-dimensional (1D), two-dimensional (2D) and three-dimensional (3D)) have been fabricated [4-6]. Presently, nanomaterials based research areas are rapidly expanding and growing [6]. Moreover metal oxide based nanomaterials have drawn a special attention because of attractive properties coupled with efficient flexibility in structures [7-8].
Physically ZnO ''Zinc Oxide'' is existed in white powder form. Naturally ZnO exhibit n-type conductivity, while its p-type conductivity can be generated by special synthesis process [8-9].

Generally the semiconductors defects and the numbers of oxygen vacancies plays an important role for the production n-type conduction [9].
There are various methods which has been implemented for the synthesis of nanostructures ZnO, in which the most common are sputtering [9-13], pulsed laser deposition (PLD) [11], chemical vapor deposition sol-gel [12] and metal organic chemical vapor deposition (MOCVD) [13-14] etc. The solution based methods is low cost, simple, need low temperature and also scalable for large areas. ZnO is considering good candidate for electronic device application. Properties of direct and wide band gap allow ZnO to function at high break down voltages, low noise and at high temperature and power [14]. Such characteristics could enhance the application of ZnO for light emitting application mainly blue and ultraviolet light emission diodes [15]. High exciton binding energy (60 meV) of ZnO shows the light emission efficiency at room temperatures [16-17]. Moreover for solar cell technology transparency properties of ZnO allow to use as transparent electrodes [17-18]. To control the nanostructures ZnO at selective area of electrodes is one of the challenging task and less reported.
Ultraviolet (UV) photo detector has an important role in military defence system, commercial communication, space communication, environmental pollution, water purification, fire alarm, environmental monitoring, and high temperature flame detection [18]. Therefore the serious conduction of research for the fabrication of low-cost UV sensors is quite necessary so that to control the effect of UV radiation [16]. The measurement of UV radiation is different from visible radiation, because direct optical instrument could not detect UV radiation properly [19].

2. Experimental

To align the nanostructures at desired area of electrodes spacing and, to achieve good resolution during pattern transformation process to wafer samples a chrome glass mask was chosen. The mask containing 22 dies of zerogap structures. The schematic of such designed can be seen in Figure 1 (a, b, c). The distance between two microgaps was maintained to 15000 µm. P-type 4 inch silicon wafer was selected for device fabrication process. And finally by low cost common mask imaging transfer technique namely photolithography, and by chemicals etching the structure of microgap were fabricated on wafer samples as have been reported previously [21-22, 27-28].

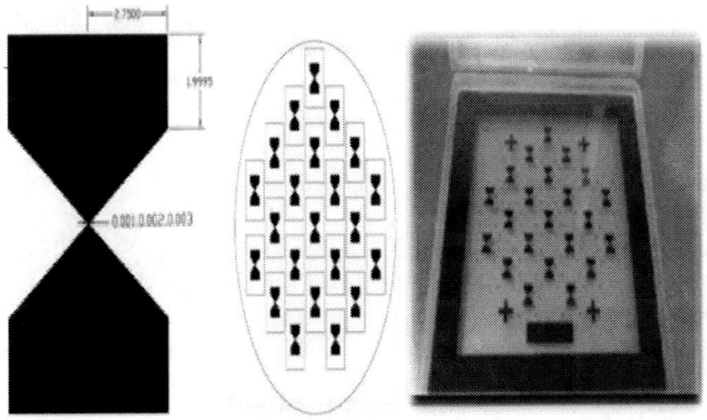

(a) (b) (c)

Figure 1. (a) The AutoCAD design of individual microgap scaled structure (b) Whole mask with different design microgap electrode (c) Original chrome mask with printed microgap structures.

The wet solution, namely sol–gel spin coating technique was implemented to synthesize on oxides substrate, the thin film ZnO. For this purpose the recipe for solution preparation was followed as; Dehydrate zinc acetate plus methoxyethanol-2 was dissolved, while monoethanolamine used as neutralizer, the molar ratio was fixed to 0.5 M. Using RCA1 and RCA2 the wafer samples were washed to remove the dust particles. To create the insulation layer an ~1 µm oxide layer was oxidized on samples by thermal process of oxidation. The metals titanium (Ti) followed by gold (Au) was deposited using evaporation process. The special designed zerogap mask was used for fabrication of microgap electrodes by low cost conventional photolithography process After the gold mirogap fabrication the solution of ZnO thin film was dropped on the surface of microcap electrodes by spin coated method with the spinning speed of 4000 rpm for 50 s. and then the coated samples were dried at 250 °C for 10 min on hot plates and were post heated up by using furnace until 350 °C to 1.5 h in the normal air atmosphere. Similarly for ZnO nanorods deposition the recipe for solution was followed as; Zinc nitrate hexahydrate (0.030 M) and hexamethylenetetramine (0.030 M) was mixed in 200 mL of deionised water, and the solution was poured into the 300 mL beaker with a continuous stirring of 30 minutes. Finally the samples were dipped into the mixer with the help of sample holder and were placed in oven for 7 h at 95 °C to achieve the well defined ZnO nanorods. At the end, finally at room temperature the samples were bring out from the oven and then washed out the residual salt from samples by fast nitrogen gas blower. In Figure 2 (a) the schematic of ZnO thin film and in (b) the ZnO nanorods deposition process at the selective area of microgap electrodes can be seen.

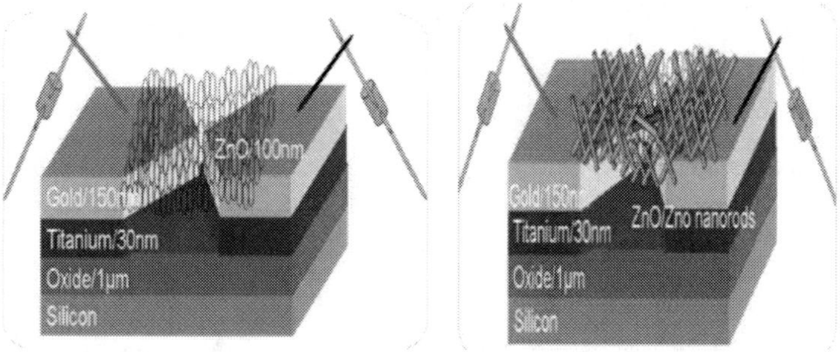

Figure 2. (a) Schematic of ZnO thin film and (b) ZnO nanorods deposition process at the selective area of microgap electrodes.

By using the SEM (JEOL 5320 LA) the nanostructures deposited samples surfaces were characterised and the structural analysis was conducted by using X-ray diffractometer (XRD Bruker Cu Kα, 1.5406 Å). Finally for electrical measurement the sensors and the meter (Keihtly 3400) were placed in series to conduct the current to voltage measurement, followed by switching the UV LED On and Off from 5 V to 5 V bias with respect to steps of 0.5 V. In this experiment the UV LED of wavelength 360 nm and 3 mW power was used for sensing.

3. Result and Discussion

The Figure 3 (a, b) demonstrate the XRD spectra thin film and nanorods ZnO, the spectrums have polycrystalline c-axis oriented peaks. The peaks (100), (002), (101), (102), and (103) shows the uniform perpendicular nature to the surface of substrate. Grain size was calculated from XRD data by using the formula reported in [10]:

$$D = (0.9\,\lambda)/Bcos\theta \tag{1}$$

FWHM value is "B" and the bragg angle is denoted by "θ" while "λ" is the wavelength. The grain size was calculated as 32.6 nm while for the ZnO nanostructure lattice constant "c" can be mathematically achieved by implemented the reported formula [11].

$$n\lambda = 2dsin\theta \tag{2}$$

The "n" and "λ" shows the diffraction and wavelength of X-rays, moreover for the miller indices (h, k, l) the spacing is denoted by "d" [12]. This spacing has good relation with the lattice constant "a" and "c" and the Miller indices, which has been reported in following relation [12]:

$$d_{hkl} = \sqrt{\frac{4}{3}\frac{(h^2 + hk + k^2)}{a^2} + \frac{l^2}{c^2}} \quad (3)$$

The value of lattice constant was calculated as 5.15 Å.

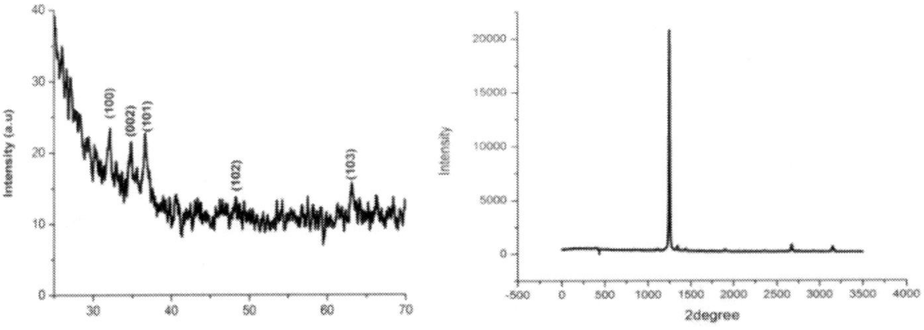

(a) (b)

Figure 3: (a) XRD spectra of deposited ZnO thin film and (b) for ZnO nanorods

The Figure 4 (a) shows the original samples having fabricated microgap structures on silicon substrate. Moreover from the SEM image as shown in Figure 4 (b) we can observe that the surface of the substrate having dense ZnO nano-particles and having roughness, porosity which creates the interconnected particles networks, while in SEM image in Figure 4 (c) the ZnO nanorods provides a bridging networks with electrodes tips.

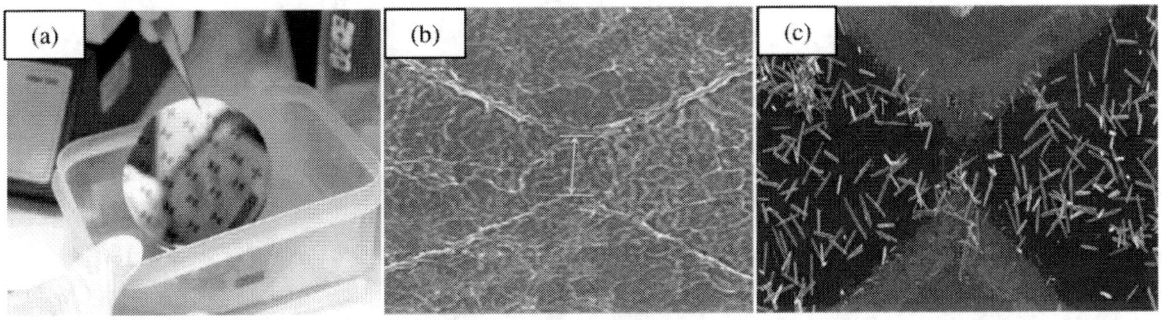

Figure 4. (a) Original sample at resist level. The deposited ZnO thin film and ZnO nanorods at the selected area of microgap electrodes can be seen in (a) and (b).

Both of the two different morphologies deposited sensors were tested electrically under the exposure of ultraviolet (UV) light and studied the effect of photrespose comparatively as can seen in Figure 5 (a) and (b). From the I-V curves we can easily visualize that the ZnO nanorods decorated microgap electrodes has a fast photorespose as compared with the ZnO thin film probably this might be because of nanorods bridging networks at the electrodes tips.

IOP Conf. Series: Materials Science and Engineering **99** (2015) 012017 doi:10.1088/1757-899X/99/1/012017

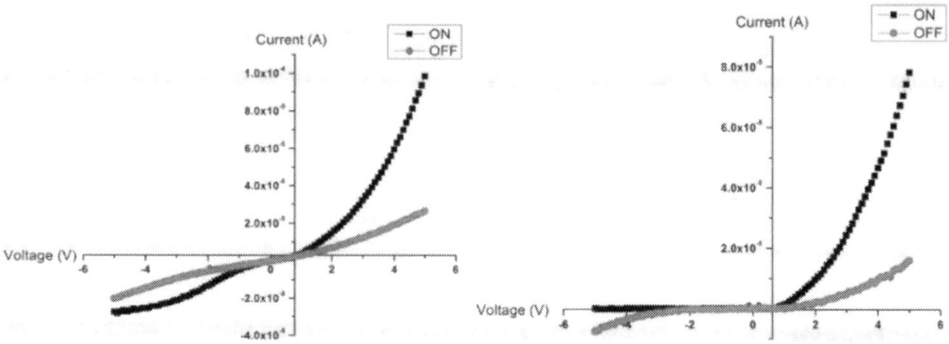

(a) (b)

Figure 5. Current to voltage measurement at UV ON/OFF states of (a) ZnO nanorods and (b) ZnO thin film.

The get more precise measurement for ultraviolet (UV) light photoresponse, current to time (i-t) measurement under ON/OFF states was conducted. From the two different (i-t) curves as shown in Figure 6, we reveal that thin film decorated microgap electrodes have 3 fold higher gain compared with ZnO nanorods by exhibiting rapid response and fast reproducibility which can be verified from the calculated value of repose and recovery timing (1.345 s, 0.132 s), and (2.343 s, 0.234 s), respectively. Finally we can conclude that the ZnO nanostrcutres deposited samples can be further used for low voltage operating sensors, having rapid response and reproducibility.

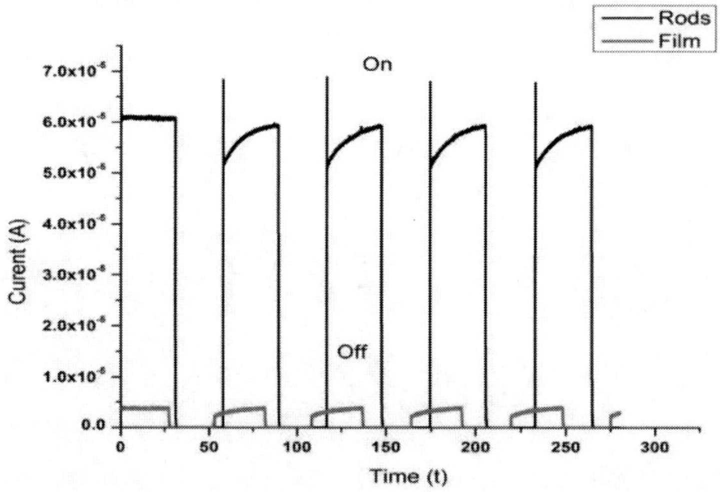

Figure 6. Thin film and nanorods current to time (i-t) response curves under ultraviolet (UV) On/off states.

4. Conclusion

The morphology of ZnO nanostructures as thin film and ZnO as nanorods were controlled at selective area of gold coated microgap electrodes by low cost conventional photolithography coupled with wet chemical etching and followed by low cost solution based spin coating technique. All the synthesize ZnO nanostructures measurement was conducted by using XRD and SEM data. Finally the devices were electrically tested for sensing application by conducting a comparative study with respect to ZnO thin film and ZnO nanorods under the exposure of ultraviolet (UV). The devises were also passed through a current to time (i-t) measurement under ultraviolet (UV) light ON/OFF. Both of the sensors exhibited a fast reproducibility and rapid response at 5 V bias moreover the ZnO nanorods exhibited a fast response probably might be because of bridging ZnO nanorods with electrodes tips. In our upcoming research the microgap size will be improved to nanogap size by implementing a thermal oxidation technique. Then the nano size gap coupled with nanostructure ZnO will be used for biosensing application, by inserting a biomolecules to produce a connecting network with nanostructures, which will definitely improve the sensing of single biomolecule.

5. Acknowledgment

The authors acknowledge the financial support from Univerisiti Malysia Perlis (UniMAP). The authors also would like to thank all of the team members and technical staff in the Institute of Nano Electronic Engineering and Faculty of Engineering and Technology Universiti Malaysia Perlis (UniMAP) for their kind support to smoothly perform the research.

6. References

[1] S. Ilican, Y. Caglar and M. Caglar, 2008 *Journal of Optoelectronics and Advanced Materials*, **10** 2578-2583
[2] H. Benelmadjat, N. Touka, B. Harieche, B. Boudine, O. Halimi and M. Sebais 2010 *Optical Materials* **32** 764-767
[3] R. Ghosh, D. Basak and S. Fujihara 2004 *Journal of Applied Physics* **96** 2689-2692
[4] Y. Li, F. Della Valle, M. Simonnet, I. Yamada and J.J. Delaunay 2009 *Applied Physics Letters*, **94**
[5] Humayun Q and Hashim U 2012 *Research and Development (SCOReD), 2012 IEEE Student Conference on: IEEE)* pp 63-6
[6] F. Guo, B. Yang, Y. Yuan, Z. Xiao, Q. Dong, Y. Bi and J. Huang 2012 *Nature Nanotechnology*
[7] J. Zhou, Y. Gu, Y. Hu, W. Mai, P.H. Yeh, G. Bao, A.K. Sood, D.L. Polla and Z.L. Wang 2009 *Applied Physics Letters*, **94**
[8] Q. Humayun and U. Hashim 2012 *Biomedical Engineering and Sciences (IECBES), 2012 IEEE EMBS Conference on: IEEE)* pp 383-6
[9] K.W. Liu, J.G. Ma, J.Y. Zhang, Y.M. Lu, D.Y. Jiang, B.H. Li, D.X. Zhao, Z.Z. Zhang, B. Yao and D.Z. Shen 2007 *Solid-State Electronics*, **51** 757-761.11, 102-105
[10] Q. Humayun and U. Hashim 2012 *Biomedical Engineering and Sciences (IECBES), 2012 IEEE EMBS Conference on*, pp 499-502
[11] S.P. Chang, R.W. Chuang, S.J. Chang, C.Y. Lu, Y.Z. Chiou and S.F. Hsieh 2009 *Thin Solid Films*, **517** 5050-5053
[12] L. Feng, A. Liu, M. Liu, Y. Ma, J. Wei and B. Man 2010 *Materials Characterization* **61**128 133
[13] Q. Humayun and U. Hashim 2012 *Semiconductor Electronics (ICSE), 2012 10th IEEE International Conference on: IEEE)* pp 22-5
[14] Gupta, R. Menon and K. Sreenivas 2008 *IEEE*, **55**-58
[15] N. Tang, J. Wang, H. Xu, H. Peng and C. Fan 2009 *Sci. China Ser. E-Technol Sci* **52** 2200-2203
[16] Q. Humayun and U. Hashim 2015 *Advanced Materials Research*, pp 304-8
[17] Q. Humayun and U. Hashim 2015 *Advanced Materials Research*, pp 200-4
[18] Z. Xu, H. Deng, J. Xie, Y. Li, Y. Li, X. Zu and S. Xue, 2005 Journal of Sol-Gel Sci Technol **36** 223-226
[19] P. Biswas, S. Kundu, P. Banerji and S. Bhunia 2013 Sensors and Actuators B: Chemical, **178** 331- 338
[20] Q. Humayun and U. Hashim 2012 *Semiconductor Electronics (ICSE), 2012 10th IEEE International Conference on: IEEE)* pp 774-7
[21] Q. Humayun and U. Hashim 2013 *Advanced Materials Research*, pp 867-77
[22] M. Farag, M. Cavaş, F. Yakuphanoglu and F. M. Amanullah 2011 *Journal of Alloys and Compounds* **509** 7900-7908
[23] S.J. Young, L.W. Ji, S.J. Chang and Y.K. Su 2006 *Journal of Crystal Growth* **293** 43-47
[24] S.J. Young, L.W. Ji, S.J. Chang, S.H. Liang, K.T. Lam, T.H. Fang, K.J. Chen, X.L. Du and Q.K. Xue, 2008 *Sensors and Actuators A: Physical* **141** 225-229

IOP Conf. Series: Materials Science and Engineering **99** (2015) 012017 doi:10.1088/1757-899X/99/1/012017

[25] D. Jiang, J. Zhang, Y. Lu, K. Liu, D. Zhao, Z. Zhang, D. Shen and X. Fan 2008 *Solid-State Electronics* **52** 679-682
[26] S.S. Hullavarad, N.V. Hullavarad, P.C. Karulkar, A. Luykx and P. Valdivia *2007 Nanoscale Research Letters*, **2** 161-167
[27] Q. Humayun and U. Hashim 2013 *Australian Journal of Basic and Applied Sciences* **7** 278-85
[28] F. Guo, B. Yang, Y. Yuan, Z. Xiao, Q. Dong, Y. Bi and J. Huang 2012 *Nature Nanotechnology*

Study on the effect of PVDF-TrFE layer to the electrical properties of MIS devices

L. N. Ismail[1], A. D. Ahmad Fuad[1], M. H. M. Wahid[2], Z. Habibah[3], S. Norsabrina[1], M. K, Nazirah[1]
[1]Faculty of Electrical Engineering, Universiti Teknologi MARA Pulau Pinang, 13500 Pulau Pinang, Malaysia.
[2]Faculty of Applied Sciences, Universiti Teknologi MARA, 40450 Shah Alam Malaysia
[3]Faculty of Electrical Engineering, Universiti Teknologi MARA, 40450 Shah Alam, Malaysia.

E-mail: lyly2909@gmail.com

Abstract. This paper investigates the effect of PVDF-TrFE layer to the electrical properties of MIS device. The MIS (Al/PMMA:TiO_2/PVDF-TrFE/Si) structures were fabricated on n-type Si substrate. The results indicate that the properties of MIS improved drastically with PVDF-TrFE layer. I-V and C-V characteristics shows that the MIS has a fast operating voltage approximately 1 V, low leakage current, large capacitance. No hysteresis was observed compared to MIS without PVDF-TrFE layer. The improvement in the electrical properties of MIS (Al/PMMA:TiO_2/PVDF-TrFE /Si) is due to the increment in the real permittivity, ε' value of the insulator.

1. Introduction

Polymer material has been applied in various electronic applications such, light emitting diode (LED) and solar cells [1]. The properties of polymer material such as light weight, high flexibility and also can be process at low temperature has be advantage if compare to inorganic material [2]. There are many polymer material have been studied to be used as insulator or dielectric layer in organic thin film transistor (OTFT) and organic field effect transistor (OFET) because some polymer material has good insulating properties which is similar and comparable with silicon dioxide (SiO_2) [3].

Poly (methyl methacrylate) (PMMA) is an example of polymer material that has been used as dielectric or insulating layer in OFET and OTFT. It has high resistivity, (10^{15} Ωcm) along with thermal and mechanical stability [4]. Nevertheless, using PMMA as dielectric layer in OTFT leads to high operating voltage because it has low dielectric permittivity, ε_r (~2.6 at 1MHz) [5] and formation of pinholes at minimal film thickness [6-8]. This high operating voltage will degrade the performance of the OFET and OTFT. In order to overcome the PMMA problem, some researcher introduces compositing the PMMA with inorganic material [5-7]. There are several inorganic material such as titanium dioxide (TiO_2), zinc oxide (ZnO) and silicon dioxide (SiO_2) are an example of inorganic

[1] To whom any correspondence should be addressed.

material that being used as filler in the nanocomposite material. TiO_2 were selected due to several characteristics such as it has high ε_r (in the range of 40 to 86 [9]. Our previous work in 2012 proven that the ε_r of PMMA increased from 2.6 to 12 measure at 1 MHz [10-12]. Even though, there is remarkable improvement in the value of ε_r, in the point of the morphology view, the poly (methyl methacrylate): titanium dioxide (PMMA:TiO_2) nanocomposite film are poor. To overcome the morphology problem, buffer layer were added on top of the PMMA:TiO_2 film. Poly (vinylidenefluoride-co-trifluoroethylene) (PVDF-TrFE) film was selected because it has a smooth, uniform morphology and also high ε_r (in range 12 ~ 40) measured at 1 MHz. Further, it has been reported that PVDF-TrFE is a semi-crystalline copolymer is miscible with other polymer for various compositions and that this binary blend could stabilize the phase morphology and improve the interfacial adhesion [13].

Metal-insulator-semiconductor (MIS) was fabricated to investigate the effect of PVDF-TrFE layer on top of PMMA:TiO_2 layer. MIS is a device that is formed from a layer of metal, a layer of insulating material and a layer of semiconductor material. There are several characteristics of MIS such as the capacitance value and leakage current value influence the performance of MIS devices [14]. This paper is to investigate the electrical characteristics of metal insulator semiconductor (MIS) using multilayer dielectric film. There are two samples of MIS devices with multilayer dielectric were fabricate to investigate and analyze the performance of MIS in the point of the electrical properties. MIS device with single layer dielectric film were also fabricate for comparison purpose. All MIS devices were fabricated on n-type silicon (*n-Si*) acting as semiconductor layer.

2. Experimental Setup

The MIS devices was fabricated on n-type Si substrate. The n-Si was used as substrate and also as semiconductor layer. The *n*-Si substrate was cleaned in an ultrasonic bath with acetone, methanol, de-ionized water (DI) successively for 10 minutes and then dried by nitrogen (N_2) gas.

2.1 MIS Device Fabrication

There are two types of MIS devices were fabricated. The first types is known as sample A consist only PMMA:TiO_2 layer while the second MIS (sample B) consist of PVDF-TrFE layer on top of the PMMA:TiO_2 layer. The single layer was fabricated for comparison purpose meanwhile the multilayer (PVDF-TrFE/PMMA:TiO_2) is to investigate the performance of MIS. All MIS devices were fabricated on *n*-type silicon (*n*-Si) substrates. The *n-Si* was used as substrate and also as semiconductor layer. The *n*-Si substrate was cleaned in an ultrasonic bath with acetone, methanol, de-ionized water (DI) successively for 10 minutes and then dried by nitrogen (N_2) gas.

2.2 MIS with PMMA:TiO_2 dielectric film

After the drying process of *n-Si*, the nanocomposite PMMA: TiO_2 dielectric films were prepared at 3wt% by dissolving PMMA powder and self prepared TiO_2 nanowire in Toluene (Sigma Aldrich). Six drops of trimethoxymethylsilane (TMOMS) was added into the solution to stabilize the PMMA and TiO_2. The solution was sonicated for 30 minutes at 50°C before being stirred for 24 hours. The PMMA: TiO_2 were spin coated on *n-Si* at 4000 rpm for 60 seconds. The film is dried at 50°C for 5 minutes to evaporate the solvent and then annealed at 120°C for 30 minutes resulting a PMMA:TiO_2 dielectric film of 500 nm. Aluminum (Al) with 60 and 100 nm thick was deposited as bottom and top contact electrode using lift-off mask. The characterization of the electrical properties was performed by Keithley 4200 Semiconductor Characterization Unit for I-V and C-V measurement. All measurement was taken at room temperature. The physical thicknesses were determined using Dektak surface profiler.

2.3 MIS with PVDF-TrFE/PMMA:TiO_2 dielectric film

The MIS with sample B film were fabricated on *n-Si* substrate. The deposition methods PMMA:TiO_2 dielectric film is similar in subsection 2.1. The PVDF-TrFE 70/30 copolymer was obtained from (Piezotech France). PVDF-TrFE solution was prepared by dissolving it in methyl ethyl ketone (MEK) (Sigma Aldrich) with the concentration of 15g/l. The PVDF-TrFE solution stirred for 24 hours at room temperature. Then the solution was sonicated for 1 hour at 50°C with ultrasonic water glass. The spin speed is 3500 rpm for 90 seconds. The thickness of PVDF-TrFE thin films with is approximately 70 nm. Aluminum (Al) with 30 and 60 nm thick was deposited as bottom and top contact electrode using lift-off mask. Aluminum (Al) with 60 and 100 nm thick was deposited as bottom and top contact electrode using lift-off mask. The film thickness of PMMA:TiO_2 /PVDF-TrFE dielectric film were measured ~ 567 nm. The characterization of the electrical properties was performed by Keithley 4200 Semiconductor Characterization Unit for I-V and C-V measurement. All measurement was taken at room temperature. The physical thicknesses were determined using Dektak surface profiler.

3. Results and Discussion

3.1 Surface Morphology

The insulator/semiconductor interface is equally important to control because this is where the conducting channel is formed. The surface morphologies of sample A and B are shown in Figure 2 and 3, respectively. It can be seen that sample A have rough surface than sample B. TiO_2 is the cause for sample a have rough surface. However, sample A have a better performance for current to flow in the insulator to semiconductor due to the agglomeration of TiO_2. An agglomeration of TiO_2 is a volcanic rock consisting of rounded and angular fragments fused together. The agglomeration of TiO_2 can create a path from metal contact to the semiconductor layer and lead to high leakage current.

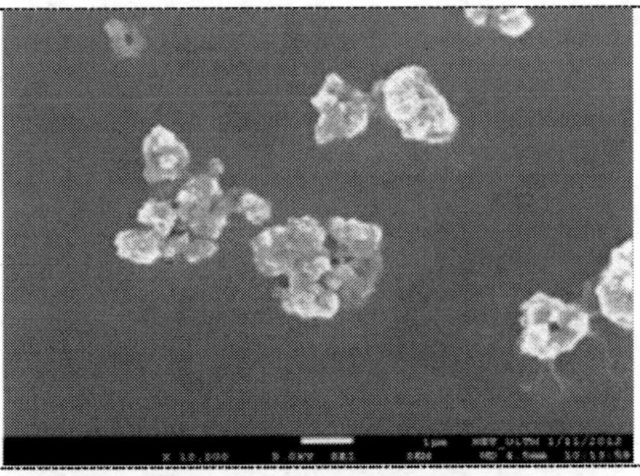

Figure 2. FESEM image of PMMA:TiO_2 dielectric film.

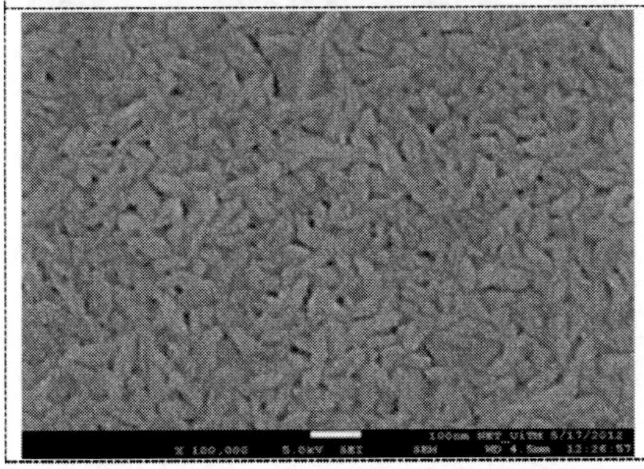

Figure 3. FESEM image of PVDF-TrFE layer on top of PMMA:TiO_2 dielectric film.

3.2 Current-voltage characteristics of MIS

I-V characteristics are important characteristics in most electronic devices. It basically to determine and understand the behavior of electronic device. In MIS device characteristics, the *I-V* curves determine the threshold voltage, V_{TH} in both its forward and reverse regions. Figure 4 show the *I-V* curves of MIS device with PMMA:TiO_2 and PVDF-TrFE/PMMA:TiO_2 layer. The voltage bias was sweep from 0V to +10 V. It can be seen that *I-V* for sample A are better than compared to sample B. The V_{TH} for sample A is 1.4 V and for sample B is 4.13 V. The V_{TH} values show sample A operate at low operating voltage compared to sample B. The increment in the V_{TH} for sample B are due to the increment in the film thickness of the insulator layer from 500 nm to 567 nm. Different work function (eV) different between Al-PVDF-TrFE-PMMA:TiO_2-Si-Al might also contribute to the higher V_{TH}. Another characteristic to evaluate the electrical properties of MIS is to examine the leakage current densities, *J* properties. The leakage current density is used to determine the amount of carrier tunneling through the insulator (dielectric) region in the MIS device. As addition, it also used to observe the suitability of the dielectric layer with semiconductor layer. Figure 5 show the leakage current density, *J* characteristics for all samples. It can be seen that there is slightly difference between in the *J* value for all samples. The *J* value for sample A and B is $1e^{-4}$ and $1e^{-5}$ A/cm^2, respectively measured at 5 V. Other characteristics that can be obtained from *J* curve is the ratio between current off, I_{off} and current on, I_{on}. I_{off} for all samples are in the range of $1e^{-8}$ to $1e^{-6}$ A/cm^2. It is important to keep I_{off} very small in order to minimize the static power that a circuit consumes. The I_{on} value is in the range of $1e^{-5}$ to $1e^{-4}$ A/cm^2.

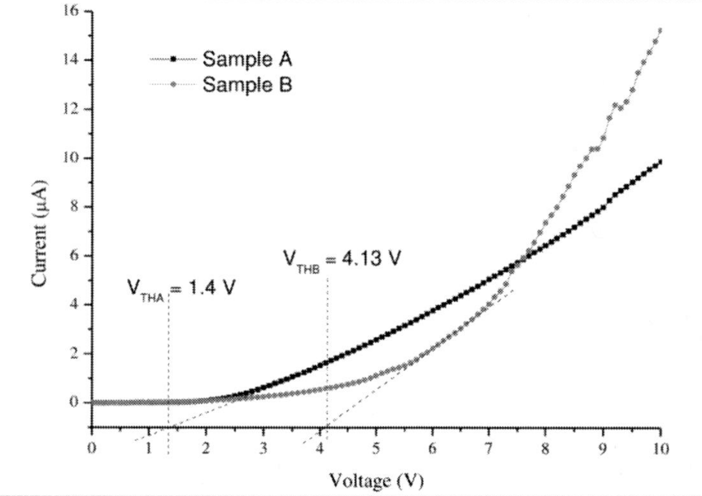

Figure 4. I-V curve for MIS for sample A (PMMA:TiO_2 layer) and sample B (PVDF-TrFE/PMMA:TiO_2 layer)

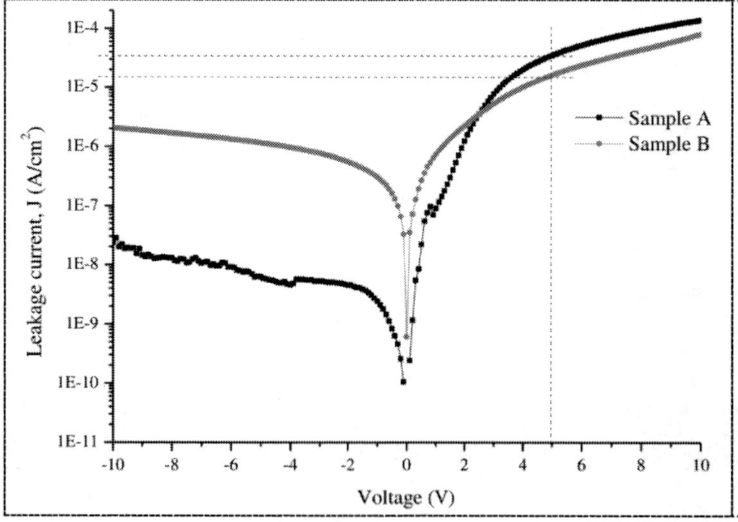

Figure 5. Leakage current for MIS for sample A (PMMA:TiO_2 layer) and sample B (PVDF-TrFE/PMMA:TiO_2 layer)

IOP Conf. Series: Materials Science and Engineering **99** (2015) 012018 doi:10.1088/1757-899X/99/1/012018

3.3 Capacitance-voltage characteristics of MIS

C-V characteristics are other important characteristics in most MIS devices. It basically to determine the capacitance value which reflected to the amount of charger were stored in the MIS device. On top of that *C-V* characteristics also indicate the accumulation, depletion and inversion region occur by applying different frequency. The *C-V* characteristics of MIS with PMMA:TiO_2 and PVDF-TrFE/PMMA:TiO_2 dielectric were plotted as shown in Figure 6 measured 1 MHz. It can be seen that *C-V* curve for sample A and B shows show a typical *n*-type behaviour with accumulation, depletion and inversion regions. Accumulation region is define state the trap charge in the MIS device when the voltage been applied below than flat band voltage, V_{FB}. V_{FB} indicate the separation region of accumulation and depletion. From the *C-V* curve, it also can be observed that both samples show the inversion strongly occur during high frequency. The capacitance for MIS at accumulation region with sample A is 24 nF while sample B just stored charge about 25 nF. Even though there is only a slight changes in the capacitance value for between MIS with sample A and B, it indicate that MIS with sample B has the capability to stored more charge compare MIS with sample A. The flatband voltage, V_{FB} for both MIS can be obtained from Figure 6. There is not so much different from the V_{FB} for MIS with sample A and sample B. V_{FB} for MIS with sample A is -4.4 V while MIS with sample B is -2.6 V.

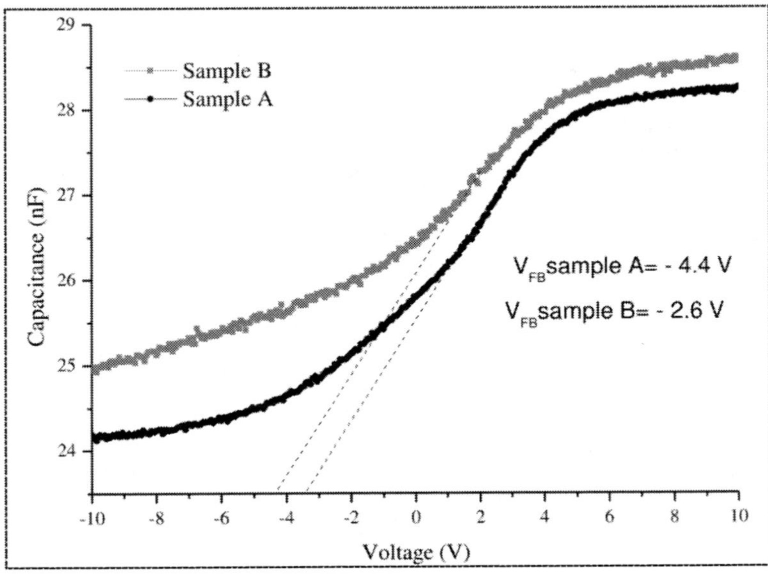

Figure 6. C-V curve for MIS for sample A (PMMA:TiO_2 layer) and sample B (PVDF-TrFE/PMMA:TiO_2 layer)

Table 1 show the summarization of I-V and C-V characteristics of MIS.

Table 1. Summarization of C-V characteristics of MIS

MIS Device	Film Thickness (nm)	Threshold Voltage, V_{TH} (V)	Leakage current, J (A/cm^2)	Capacitance, C (pF)	Flatband voltage, V_{FB} (V)
Sample A	500	1.4	$1e^{-4}$	24	- 4.4
Sample B	567	4.13	$1e^{-5}$	25	- 2.6

4. Conclusion

In this paper, MIS device with PVDF-TrFE/PMMA:TiO_2 thin films were successfully fabricated. Result from the surface morphology indicate by adding PVDF-TrFE layer on top of PMMA:TiO_2 layer, the agglomeration problem are overcome. Result from I-V show that increment in the V_{TH} value from 1.75 V using PMMA:TiO_2 to 5 V using PVDF-TrFE/PMMA:TiO_2. Nevertheless, in term of the leakage current properties there is significant improvement was observed. Result for C-V indicates that by adding PVDF-TrFE layer on top PMMA:TiO_2 improvement in the storage capability of MIS.

5. Acknowledgement

This work was supported by the RAG grant funded by Research Management Unit (RMU), Universiti Teknologi MARA (600-RMI/RAGS 5/3 (205/2014).

References

[1] K. Gurunathan, *1999 Materials Chemistry and Physics,* **61**, 173-191.
[2] M. R. Aguilar and J. San Román, Woodhead Publishing, 2014, 1-11.
[3] M. M. Ling and Z. Bao, 2004 *Chemistry of materials,* **16**, 4824-4840.
[4] M. Na and S.-W. Rhee, 2006 *Organic electronics,* **7**, 205-212.
[5] M. Estrada, *et al.*, 2008 *Solid-State Electronics,* **52**, 53-59.
[6] S. Gross, *et al, 2007 European polymer journal,* **43**, 673-696.
[7] M. Estrada, *et al.*, "Mobility model for compact device modeling of OTFTs made with different materials," *Solid-State Electronics,* vol. 52, pp. 787-794, 2008.
[8] N. Semaltianos, "Spin-coated PMMA films," *Microelectronics journal,* vol. 38, pp. 754-761, 2007.
[9] S. A. Campbell, *et al.*, "TiO_2-based gate insulators," *IBM Journal of Research and Development,* vol. 43, pp. 383-392, 1999.
[10] L. N. Ismail, *et al.*, "Dielectric and physical properties of PMMA:TiO thin films by varying TiO_2 concentration," in *Humanities, Science and Engineering Research (SHUSER), 2012 IEEE Symposium on*, 2012, pp. 259-262.
[11] L. N. Ismail, *et al.*, "Effect of annealing temperature on electrical properties of poly (methyl methacrylate): titanium dioxide nanocomposite films using spin coating deposition technique," *IOP Conference Series: Materials Science and Engineering,* vol. 64, p. 012051, 2014.
[12] I. Lyly Nyl, *et al.*, "Dielectric Properties of PVDF-TrFE/PMMA: TiO2 Multilayer Dielectric Thin Films," *Advanced Materials Research,* vol. 576, pp. 582-585, 2012.
[13] S. Wi, *et al.*, "Characterization of poly (vinylidene fluoride-trifluoroethylene) 50/50 copolymer films as a gate dielectric," *Journal of Materials Science: Materials in Electronics,* vol. 19, pp. 45-50, 2008.
[14] S. M. Sze and K. K. Ng, *Physics of semiconductor devices*: John Wiley & Sons, 2006.

Effect of Annealing Time Process on the pH Sensitivity of Spin-coated TiO_2/ ZnO Bilayer Film

R A Rahman*[1], M A Zulkefle[1], K A Yusof[1], W F H Abdullah[2], M. Rusop[1,2,3], S H Herman[2,4]

[1]NANO-Technology Centre (NET), Universiti Teknologi MARA (UiTM), 40450 Shah Alam, Selangor, Malaysia
[2]Faculty of Electrical Engineering, Universiti Teknologi MARA (UiTM), 40450 Shah Alam, Selangor, Malaysia
[3]NANO-Science Technology (NST), Institute of Science (IOS), Faculty of Applied Sciences, Universiti Teknologi MARA (UiTM), 40450 Shah Alam, Selangor, Malaysia
[4]Core of Frontier Materials & Industry Applications, University Teknologi MARA (UiTM), 40450 Shah Alam, Selangor, Malaysia

E-mail : rohanieza.abdrahman@gmail.com

Abstract. This paper presents an investigation on titanium dioxide (TiO_2) and zinc oxide (ZnO) bilayer film, which is used as sensing membrane for extended-gate field effect transistor (EGFET) for pH sensing application. TiO_2/ZnO thin films were deposited using sol-gel spin coating method on indium tin oxide (ITO) substrates. After the deposition, the bilayer films were annealed at constant temperatures which is 400 °C for 15, 30, 40 and 60 minutes. The sensitivity of the TiO_2 thin film towards pH buffer solution was measured by dipping the sensing membrane in pH4, pH7 and pH10 buffer solution. By varying the annealing time, we found that the TiO_2/ZnO thin film annealed at 400°C for 15 minutes gave the highest sensitivity compared to other annealing conditions, with the value of 64.87 mV/pH.

1. Introduction

Chemical sensors with metal oxides as sensing materials play important roles in industrial, medical, domestic and many more applications. Among the chemical sensor existed, pH and gas sensor have been widely investigated and applied due to their importance. For instance, the used of pH sensor was implemented in biomedical [1]. In biomedical field, pH sensor was utilized as blood monitoring, clinical detection and many more [2].

Researchers nowadays are focusing on EGFET [3]–[6], which is a type of pH sensor. This EGFET was designed as an alternative way to overcome the drawbacks of Ion Sensitive Field effect transistor (ISFET). EGFET was chosen because it can be characterize using simple connection and also flexibility of the extended gate design [7]. Other than that, EGFET has few advantages which is low cost, simple in packaging, and better-long stability [3].

In order to fabricate and prepare the EGFET sensing membrane, researchers usually choose metal oxides to be deposited. According to previous researches, vanadium oxide (V_2O_5) [8], tin dioxide (SnO_2) [9], titanium dioxide (TiO_2) [10] and zinc oxide (ZnO) [11] were used to be fabricated as sensing membrane. In comparison, between all the chosen metal oxides, ZnO and TiO_2 are most

IOP Conf. Series: Materials Science and Engineering **99** (2015) 012019 doi:10.1088/1757-899X/99/1/012019

favorable among researchers because of their performance. These two metal oxides have similar characteristics. Both of these metal oxides are chemically stable and have wide band gap. According to these characteristics, we choose ZnO and TiO_2 as the metal oxides as our sensing membrane for EGFET.

In this investigation, we fabricate bilayer film as the sensing membrane for our EGFET by using spin-coating technique. TiO_2 act as the upper layer while ZnO act as the lower layer. The characteristics of this bilayer film towards drying temperature was explored and characterized by EGFET measurement, and current-voltage (I-V).

2. Methodology

In this experiment, the glass substrates, which is ITO/glass was first cut into 1 x 2 cm. Then, the substrates were cleaned using standard organic cleaning method with methanol and distilled water, followed by drying with argon gas. After that, the substrates were sonicated (sonic temperature is 40˚C) for 10 minutes with the temperature was set to 50˚C.

2.1. ZnO sol-gel preparation

The solution were prepared by using zinc acetate dehydrate ($Zn(CH_3COO)_22H_20$) as starting material, 2-methoxyethanol ($C_3H_8O_2$) as a solvent and monomethanolamine (MEA, $C_2H_7N_{14}$) as a stabilizer. Zinc acetate was dissolved in a mixture of 2-methoxyethanol and MEA solution to produce 0.4M solution concentration. This solution was stirred at 300 rpm and heated at 80°C for 3 hours. In order to obtain a clear and homogenous solution, this solution was kept aging for 24 hours at room temperature.

2.2. TiO_2 sol-gel preparation

The 0.4M TiO_2 sol-gel solution was prepared by mixing two separate solutions (Solution A and B). For solution A, absolute ethanol (C_2H_5OH), glacial acetic acid, GAA (CH_3COOH) and titanium isoproxide, TTIP ($Ti(OCH_2CH_2CH_2CH_3)_4$) were used in the solution preparation process. For solution B, the chemicals used were absolute ethanol (C_2H_5OH), triton X-100 ($C_{14}H_{22}O(C_2H_4)n$ n=9-10) and deionized water. Both of these solutions were stirred separately for 1 hour at 2000 rpm. Then the stirring process for 1 hour was continued after the solutions were mixed up together.

2.3. TiO_2/ZnO bilayer deposition

The fabrication of these bilayer films was conducted by using spin-coating technique at room temperature. The deposition process was conducted by dropping ten drops of TiO_2 solution onto ITO/glass substrate at 3000 rpm of spin speed for 60s. After that, the deposition process continued by dropping the ZnO solution onto the TiO_2 layer with the same drops, spin speed and duration of speed. Then, the bilayer film was annealed at 400°C for 15 minutes. The steps were repeated and the annealing time were varied for 30, 45 and 60 minutes.

2.4. EGFET measurement and characteriztaion

The measurement process of TiO_2/ZnO bilayer film was done by connecting the bilayer film to the gate pin of commercialized MOSFET NDP6060L by using copper wire to form the extended gate of EGFET setup. Then, the MOSFET was connected to a readout interfacing circuit (ROIC). Kapton tape and silver paste were used as bonding agent between wire and substrate. The sensitivity of the bilayer film towards pH was done by dipping the fabricated sensing membrane with reference electrode (Ag/AgCl) into three different pH value, pH4, pH7 and pH10. The value of the sensitivity was obtained from the slope of the graph plotted. All of the fabricated TiO_2/ZnO bilayer films were characterized by using I-V characterization.

IOP Conf. Series: Materials Science and Engineering **99** (2015) 012019 doi:10.1088/1757-899X/99/1/012019

3. Result and discussion

3.1. Sensitivity of the TiO_2/ZnO bilayer film towards pH buffer solution

Sensitivity is defined as the slope of the output characteristic curve or, more generally, the minimum input of the value of V_{out} and sensitivity of the thin film physical parameter that will create a detectable output change. Table 1 below shows the sensitivity and linearity values for each of the bilayer films, annealed at different annealing time.

According to the result obtained, TiO_2/ZnO bilayer film annealed for 15 minutes produced highest sensitivity and linearity value towards pH buffer solution which is 64.87 mV/pH and 0.9776. Comparing with all of the results, sensitivity and linearity value for the bilayer film shows the decreasing trend with the increasing annealing time. TiO_2/ZnO bilayer film annealed at longest annealing duration exhibit lowest sensitivity and linearity which is 47.56 mV/pH and 0.9100. We plotted the graph in Figure 1 in order to define the sensitivity and linearity differences between all the bilayer films.

Based on this comparison, we assume that 15 minutes is the optimum annealing time for this bilayer film. According to N.S Kamarozaman *et.al*, longer annealing duration will produced more oxygen vacancies [12]. Commonly, pH buffer solution contained of H^+ and OH^-, depends on its value. Due to the positively charged of oxygen vacancies, the TiO_2/ZnO bilayer film will attract the OH^- ion that contained in the pH buffer solution. So that, this bilayer film becomes less sensitive towards H^+ ion that contained in the pH buffer solution. We can conclude that bilayer film annealed at less time produced highest sensitivity due to the less oxygen vacancies. So that it can attract more H^+ ions in the pH buffer solution.

Table 1. Sensitivity and linearity values for each bilayer film annealed at different duration

Annealing Time (min)	Sensitivity (mV/pH)	Linearity
15	64.87	0.9776
30	56.60	0.9750
45	49.63	0.9400
60	47.56	0.9100

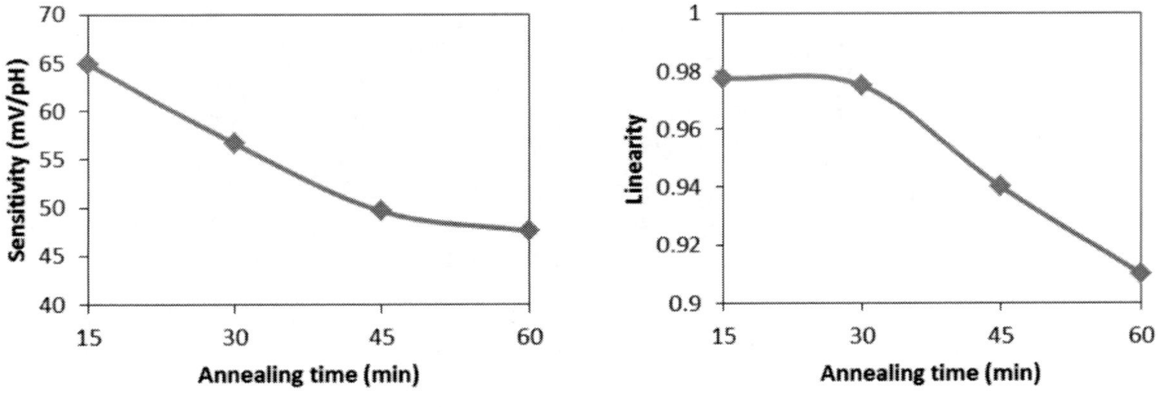

Figure 1 : Sensitivity and linearity comparison each of the TiO_2/ZnO bilayer films

3.2. Characterization of TiO_2/ZnO bilayer film

3.2.1. Atomic Force Microscopy (AFM) characterization

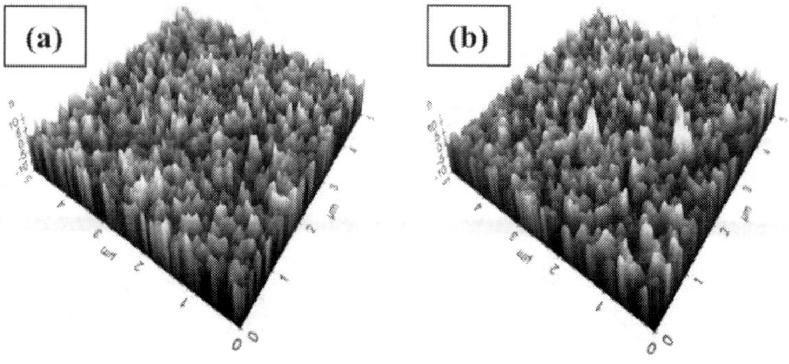

Figure 2 : AFM result for TiO_2/ZnO bilayer film annealed for (a) 15 minutes and (b) 30 minutes.

Figure 2 above shows the AFM result for TiO_2/ZnO bilayer film at two different annealing times. According to this figure, we can see the different topology when the annealing time increased. In comparison, TiO_2/ZnO bilayer film annealed for 15 minutes produces a uniform and high porosity surface. When annealing time was increased to 30 minutes, the topology and surface changed into less uniform and lower porosity. From this, we assume that when the annealing time increased, the surface and topology of the TiO_2/ZnO bilayer will have less uniformity and decreased in porosity. This conditions have support the sensitivity result that obtained in the EGFET measurement. We can conclude that bilayer film that has a uniform and high porosity produced the highest sensitivity and linearity value.

3.2.2. Field Emission Scanning Electron Microscope (FESEM) characterization

Figure 3 : Surface morphology of TiO_2/ZnO bilayer film annealed for 15 minutes at different magnification (a) 50,000x and (b) 100,000x

The surface morphology of TiO_2/ZnO bilayer film annealed for 15 minutes was shown in Figure 3. This figure shows the different magnification of the TiO_2/ZnO surface morphology. From this figure, TiO_2 has been observed on the surface of the bilayr film. The agglomeration of the TiO_2 also was observed. However, this agglomeration doesn't give a huge effect towards sensitivity value. Thus, it is proved that TiO_2 had been successfully deposited onto the ITO substrate and become as sensing membrane for the EGFET.

4. Conclusion

In conclusion, TiO_2 and ZnO were successfully deposited onto the ITO substrate as the bilayer film. It was proved that annealing time process will influenced the sensitivity of the bilayer film towards pH buffer solution. In this investigation, we successfully fabricate TiO_2/ZnO bilayer film that produces highest sensitivity when annealed at less annealing time, which is 15 minutes. The result obtained is 64.87 mV/pH and the linearity is 0.9776. This result was supported with the AFM result, which is a uniform surface topology was observed for TiO_2/ZnO bilayer annealed for 15 minutes.

Acknowledgement

The authors would like to thank all members of NANO-ElecTronic Center (NET) Universiti Teknologi MARA, UiTM for all the research facilities. The work is partially supported by MOSTI Malaysia under the Science Fund grant (Project code : 06-01-01 SF0662)

References

[1] J. Li, S. Chang, S. Chang, and T. Tsai, "Sensitivity of EGFET pH Sensors with TiO_2 Nanowires," *Solid State Letter*, vol. 3, no. 10, pp. 123–126, 2014.

[2] S. Chang and T. Yang, "Sensing Performance of EGFET pH Sensors with CuO Nanowires Fabricated on Glass Substrate," *Electrochemical Sciences*, vol. 7, pp. 5020–5027, 2012.

[3] T. Kieu, T. N. T. Nguyen, T. Quang, I. Yung, D. Kim, J. Jung, and N. Lee, "Reduced graphene oxide field-effect transistor with indium tin oxide extended gate for proton sensing," *Curr. Appl. Phys.*, vol. 14, no. 5, pp. 738–743, 2014.

[4] W. Tsai, B. Huang, K. Wang, Y. Huang, P. Yang, and H. Cheng, "Functionalized Carbon Nanotube Thin Films as the pH Sensing Membranes of Extended-Gate Field-Effect Transistors on the Flexible Substrates," *Transactions on Nanotechnology*, vol. 13, no. 4, pp. 760–766, 2014.

[5] E. M. Guerra and M. Mulato, "Titanium Oxide Nanorods pH Sensors : Comparison between Voltammetry and Extended Gate Field Effect Transistor Measurements," *Scientific Research*. no. May, pp. 459–466, 2014.

[6] L. Manjakkal, E. Djurdjic, K. Cvejin, J. Kulawik, K. Zaraska, and D. Szwagierczak, "Electrochemical Impedance Spectroscopic Analysis of RuO_2 Based Thick Film pH Sensors," *Electrochim. Acta*, vol. 168, pp. 246–255, 2015.

[7] C. E. Lue, I. S. Wang, C. H. Huang, Y. T. Shiao, H. C. Wang, C. M. Yang, S. H. Hsu, C. Y. Chang, W. Wang, and C. S. Lai, "pH sensing reliability of flexible ITO/PET electrodes on EGFETs prepared by a roll-to-roll process," *Microelectron. Reliab.*, vol. 52, no. 8, pp. 1651–1654, 2012.

[8] Z. Zhang, Y. V. Kaneti, X. Jiang, and A. Yu, "Hydrothermal synthesis of sodium vanadium oxide nanorods for gas sensing application," *Sensors Actuators B Chem.*, vol. 202, pp. 803–809, 2014.

[9] D. Wang, X. Chu, and M. Gong, "Gas-sensing properties of sensors based on single-crystalline SnO_2 nanorods prepared by a simple molten-salt method," *Sensors Actuators, B Chem.*, vol. 117, no. 1, pp. 183–187, 2006.

[10] M. A. Rosdan, S. H. Herman, W. F. H. Abdullah, N. S. Kamarozaman, and M. I. Syono, "Sputtered Titanium Dioxide Thin Film for Extended-Gate FET Sensor

Application," *Micro and Nanoelectronics (RSM), 2013 IEEE Regional Symposium on* vol. no. pp. 2–5, 2013.

[11] A. B. Rosli, N. D. H. A. Patah, M. a. Rosdan, M. M. Marbie, M. H. Juhari, S. S. Shariffudin, S. H. Herman, and M. Rusop, "Influence of metal catalyst for zinc oxide nanostructures grown by TCVD method for extended-gate FET sensor application," *RSM 2013 IEEE Reg. Symp. Micro Nanoelectron.*, pp. 171–174, Sep. 2013.

[12] N. S. Kamarozaman, Z. Aznilinda, S. H. Herman, R. A. Bakar, and M. Rusop, "Effect of Annealing Duration on the Memristive Behavior of Pt/ TiO_2/ITO Memristive Device," *Semiconductor Electronics (ICSE), 2012 10th IEEE International Conference on* vol. no. pp. 703–706, 2012.

Influence of Different Sol-gel Spin Coating Speed on Memristive Behaviour of Pt/TiO_2/ZnO/ITO Device

S. M. M. Kasim[1], N. A. A. Shaari[1], R. A. Bakar[1], Z. Aznilinda[1], Zulfakri Mohamad[3], S. H. Herman[1,2,3]

[1]NANO-Electronic Center, Faculty of Electrical Engineering, Universiti Teknologi MARA, 40450 Shah Alam, Selangor, Malaysia
[2]CoRe of Frontier Materials & Industry Applications, Universiti Teknologi MARA, 40450 Shah Alam, Selangor Darul Ehsan, Malaysia
[3]Faculty of Electrical Engineering, Universiti Teknologi MARA, 40450 Shah Alam, Selangor, Malaysia

Email: mardhiyana@gmail.com[1], hana1617@salam.uitm.edu.my[2]

Abstract. Composite titanium dioxide (TiO_2) and zinc oxide (ZnO) thin films were deposited on indium tin oxide (ITO) substrates using sol-gel spin coating technique. The electrical and physical characterizations of three different sol-gel spin coating speed were investigated using two-probe current-voltage (I-V) measurement, field emission scanning electron microscopy (FESEM) and surface profiler (SP) respectively. The I-V measurement results showed the pinched hysteresis loop for every single of devices thus indicate that all the devices are memristive. R_{OFF}/R_{ON} ratio which was defined from the hysteresis loop of device with higher spin speed was slightly higher compared to others.

1. Introduction

Memristive device which was introduced mathematically in 1971 [1] and physically was fabricated in 2008 [2] is known as the one of fundamental passive circuit elements. This two terminal device is constructed from metal-insulator-metal (MIM) structure where metals are represent as top and bottom electrodes and insulator is the thin film as well as the active layer of the device. The thin film can be deposited using several methods such as RF magnetron sputtering [3], chemical vapor deposition (CVD), spray pyrolysis, electrochemical anodization [4], atomic layer deposition (ALD) and sol-gel process. In this work, sol-gel process was selected due to it inexpensive, ease of handling and require no complex step to coat small or large desired area [5]–[7].

In memristive device, the thickness of thin films and dopant mobility give significant effect on the performance of the device. The correlation between these variables is show in equations (1) which was found by HP Lab researchers [2].

$$M(q(t)) = \left[R_{ON}\frac{w(t)}{D} + R_{OFF}\left(1 - \frac{w(t)}{D}\right)\right] \quad (1)$$

and

$$w(t) = \mu_V \frac{R_{ON}}{D} q(t) \tag{2}$$

where R_{ON} is the lowest resistance in doped region, R_{OFF} is highest resistance in undoped region, D is the overall film thickness, $w(t)$ is state variable, $q(t)$ is charge and μ_V is the dopant mobility. By substituting equation (2) into equation (1) and consider on $R_{ON} << R_{OFF}$ for the higher R_{OFF}/R_{ON} ratio the memristance can be simplifies to

$$M(q) = R_{OFF}\left(1 - \frac{\mu_V R_{ON}}{D^2} q(t)\right) \tag{3}$$

Equation (3) shows that the thinner film thickness in nanoscale will gives the significant memristance by referring to the factor of $1/D^2$. In sol-gel spin coating method, the film thickness is inversely proportional to the spin speed and spin time. Therefore, higher speed of spin coating will give thinner film thickness.

Three different speeds of sol-gel spin coating technique were applied to deposit composite bilayer TiO_2 and ZnO thin films on ITO/glass substrate in order to perceive the effect of spin coating speed toward the performance of device. The higher spin speed should give the thinner thin film thus contribute to better memristive performance.

2. Experimental

2.1. Substrates Preparation

The substrates were sonicated in ethanol for 10 minutes, followed by using deionized (DI) water with the same step to rinse the substrates. Then substrates were blown-dried under nitrogen gas, N_2.

2.2. Solutions Preparation

2.2.1. ZnO solution. Zinc acetate dehydrate ($Zn(CH_3COO)_2 2H_2O$) and aluminum nitrate ($Al(NO_3)_3$) were dissolved into solvent 2-methoxyethanol ($C_3H_8O_2$). Zinc acetate and aluminium nitrate were used as precursor. Then, monoethanolamine (MEA, (C_2H_7NO)) which act as a stabilizer was added into the solution. At 80 °C, the solution was stir-heated for 3 hours and continued stirred for another 24 hours in room temperature before ready to be deposited.

2.2.2. TiO_2 solution. Two separated solutions which was labelled as Solution A and Solution B were prepared initially. Solution A consists of absolute ethanol, glacial acetic acid (GAA, CH_3CO_2H) and titanium (IV) isopropoxide (TTIP, $Ti(OCH(CH_3)_2)_4$) where absolute ethanol performed as solvent, GAA as stabilizer and TTIP as precursor. While Solution B is a mixture of absolute ethanol which act as solvent, DI water as a substance to add O_2 element and triton X-100 as the stabilizer. Both solutions were stirred for 200 rpm in ambient temperature for 1 hour then they had been mixed up and subsequently stirred at the same speed for another 1 hour.

2.3. Thin Films Deposition and Interconnection

The spin coater was set to have two levels of spin speed. The first spin speed was set to be low at 500 rpm for 10 seconds to spread the solution evenly over the substrate and the second speed was set higher in order to have the required final thickness. In this work, 10 drops of solution were deposited on the substrate during spinning at the low speed then continued to higher speed at 1000 rpm, 2000 rpm and 3000 rpm. These steps were implemented for both TiO_2 and ZnO solutions deposition. ZnO thin film is the first layer to be coated on substrate was annealed in air ambient at 450°C for 1 hour then followed by TiO_2 at 250°C for 20 minutes in air ambient as well.

For the metal contact, Pt was selected because it may give the smooth memristive behaviour and wider hysteresis loop compared to other metal [3]. RF sputtering technique was used to deposit 60 nm Pt on top of thin films as sputtering is the suitable method for high melting point metal (Melting point of Pt: 1769°C).

Figure 1 illustrates the final structure of memristive device with Pt on the top as top electrode, then TiO_2 and ZnO thin films as well as active layer of the device deposited on ITO for the bottom electrode which was coated on glass.

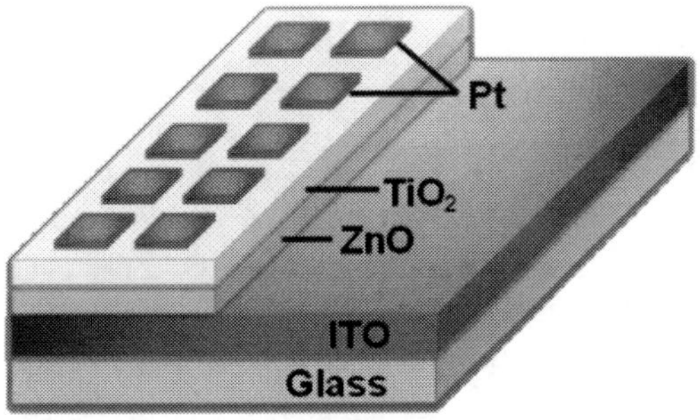

Figure 1. Final structure of Pt/TiO_2/ZnO/ITO/Glass memristive device

2.4. Measurement and Characterizations

Surface morphology and thicknesses of individual thin films were defined using the field effect scanning electron microscope (FESEM) and surface profiler respectively. Surface morphology was defined to prove the existence of thin films while for the thin films thickness itself, the thicknesses for individual layer with different speed of spin coating were measured in order to observe the changes of thin films thickness. For electrical characterization, I-V measurement was performed to get memristive behavior of the device. In order to get the pinched hysteresis loop, the bias voltage was swept from 0 V to 5 V, then from 5 V to -5 V and back to 0 V.

3. Result and Discussions

Single layer of TiO_2 and ZnO films thickness and composite TiO_2/ZnO which were deposited at 1000, 2000 and 3000 rpm were measured using surface profiler was plot in Figure 2. It can be seen that film thickness of both TiO_2 and ZnO as well as composite TiO_2/ZnO are inversely proportional to the spin coating speed [8] which mean the higher spin speed applied, the thinner film thickness will produce. Besides, both thin films were presented in exponential meet the spin coat theory, where thickness is inversely proportional to the square root of angular velocity, $t \propto (\sqrt{\omega})^{-1}$. By comparing of TiO_2 and ZnO thicknesses, it can be said that viscosity of TiO_2 solution is lower than ZnO solution since TiO_2 thin films is thinner even same spin speed were setup for both deposition [9]. Single layer of TiO_2 and ZnO films thickness were measured from different samples which were prepared separately with the composite TiO_2/ZnO but same parameter were applied. Sample with single TiO_2 and ZnO thin films was coated on ITO substrate and annealed in air ambient at 250°C for 20 minutes and 450°C for 1 hour respectively.

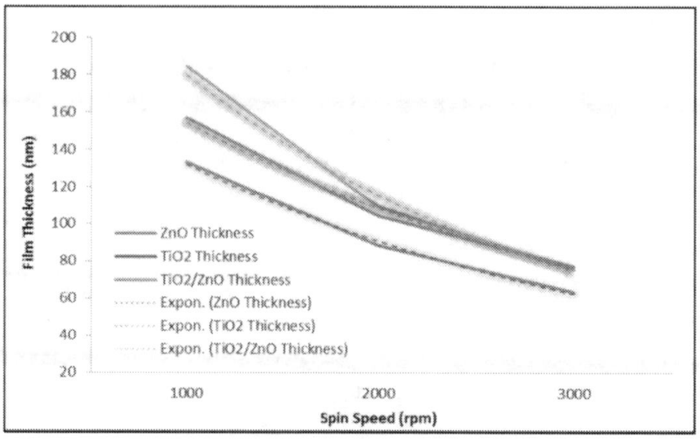

Figure 2. Film thicknesses of TiO_2 and ZnO for sample deposited at different spin speed.

Figure 3 shows the I-V measurement results for Pt/TiO_2/ZnO/ITO/Glass samples deposited with 1000 rpm, 2000 rpm and 3000 rpm spin speed in linear and semilog scale. The results were taken from 2nd measurement and above by considering the first measurement initially performs to create a path for ions movement [7]. This process can be considered as electroforming where high voltage bias will be applied to the virgin sample in order to conduct ions movement through the active layer. All samples successfully exhibited the pinched hysteresis. Sample with 1000 rpm spin speed was plot in blue triangle line has a wider loop at positive side compared to negative side while sample with 2000 and 3000 rpm in red circle and green square line respectively give wider loop at negative side.

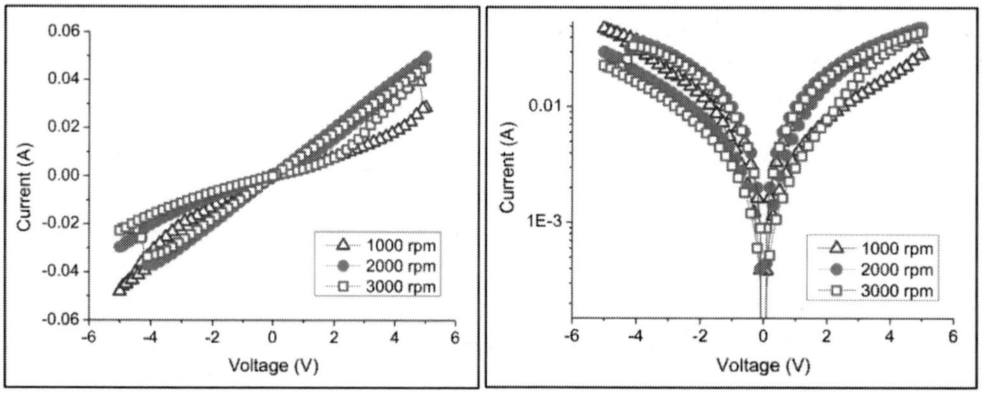

Figure 3. I-V characteristics of Pt/TiO_2/ZnO/ITO/Glass memristive device for samples deposited at different spin speed.

Table 1 was tabulated to show the R_{OFF}/R_{ON} ratio of the samples. The values of R_{OFF}/R_{ON} ratio was calculated based on

$$\text{Ohm's law:} \quad V = IR \tag{4}$$

Substitute in

$$\frac{R_{OFF}}{R_{ON}} = \frac{V/I_{OFF}}{V/I_{ON}} = \frac{I_{ON}}{I_{OFF}} \tag{5}$$

The I_{OFF} *and* I_{ON} values were taken after several times of measurement to ensure the I-V characteristics are wider and stable enough to be used after electroforming process. Table 1 shows that R_{OFF}/R_{ON} ratio of deposition at 3000 rpm is slightly higher 2.07 compared to 1000 rpm and 2000 rpm with ratio 1.84 and 1.95 respectively. These results seem to be consistent with other research [7] which found that device with thinner film thickness will give the higher R_{OFF}/R_{ON} ratio thus better memristive behavior as the loop is wider which also obeys Equation (3).

Table 1. R_{OFF}/R_{ON} ratio for samples deposited at different spin speed

Spin Speed	Thickness	Voltage at Maximum R_{OFF}/R_{ON} ratio	R_{OFF}/R_{ON} Ratio
1000	184.386	4.1 V	1.84
2000	109.417	-3.7 V	1.95
3000	75.8	-4.0 V	2.07

Several recent studies performing surface morphology characterization to define the effect of annealing temperature [6], [10], [11], doping [12], different type of metal oxide or substrate use [13] towards the structure, surface roughness and film thickness. Here, the surface morphology characterization was carried out to observe the existence of TiO_2 and ZnO thin film on top of the substrate regardless the effect on surface structure and roughness. Figure 4 shows the surface morphology of samples deposited at 2000 rpm and 3000 rpm with grain diameters between 80 nm and 110 nm. It can be seen that sample with 3000 rpm gave cleared agglomerate structure compared to sample with 2000 rpm spin speed but these changes of spin speed not really impact the size of the grain. Same results were reported by other researcher where the grain size was not so being affected compared to film thickness when the spin coating time was varied [14]. Previous studies show that it will be more significant if cross-section image could be performed as it will specify the film structure and thickness of the sample [15], [16].

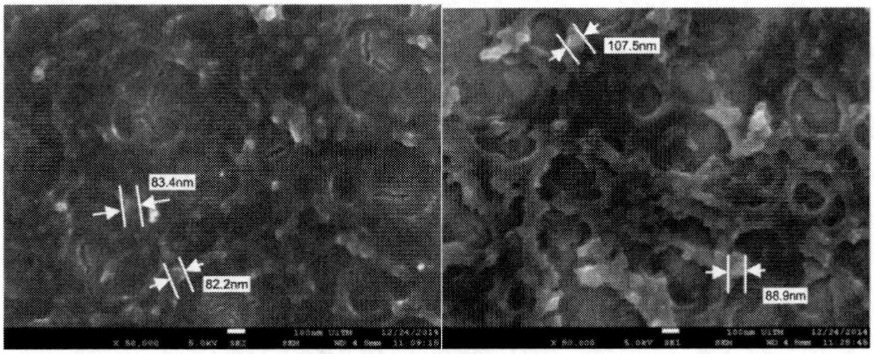

Figure 3. Surface morphology of sample deposited at 2000 rpm and 3000 rpm respectively.

4. Conclusion

Three samples which were deposited at different spin coating speed 1000, 2000 and 3000 rpm were fabricated on ITO coated glass to study on the influence of film thickness towards memristive behavior. Sample deposited with higher spin coating speed gave thinner film thickness as per proven by the thickness measurement on single TiO_2 and ZnO layer and composite TiO_2/ZnO. The results shown that all samples exhibit the characteristics of memristive device but sample with 3000 rpm spin speed gives a wider hysteresis loop with 2.07 R_{OFF}/R_{ON} ratio compared to samples deposited at 1000 and 2000 rpm which have 1.84 and 1.95 respectively. Thus, we can conclude that memristive device with thinner film thickness may give the better memristive behavior.

Acknowledgement
This work is partially supported by Malaysian Ministry of Education (MOE) under the Niche Research Grant Scheme (NRGS) (Project code: 600-RMI/NRGS 5/3 (7/2013)), Research Acculturation Grant Scheme (RAGS) (Project code: 600-RMI/RAGS 5/3 (196/2014)) and Universiti Teknologi MARA (UiTM).

References

[1] L. Chua, "Memristor-the missing circuit element," *Circuit Theory, IEEE Trans.*, vol. CT-18, no. 5, pp. 507–519, 1971.

[2] D. B. Strukov, G. S. Snider, D. R. Stewart, and R. S. Williams, "The missing memristor found.," *Nature*, vol. 453, no. 7191, pp. 80–3, May 2008.

[3] Z. Aznilinda, S. H. Herman, R. a. Bakar, and M. Rusop, "Effect of Electrode Types on the Resistive Switching Behavior of Titania Thin Films," *Appl. Mech. Mater.*, vol. 393, pp. 74–78, Sep. 2013.

[4] K. Miller, K. S. Nalwa, A. Bergerud, N. M. Neihart, and S. Chaudhary, "Memristive Behavior in Thin Anodic Titania," *IEEE Electron Device Lett.*, vol. 31, no. 7, pp. 737–739, Jul. 2010.

[5] Y. Li, S. Long, H. Lv, Q. Liu, Q. Wang, and Y. Wang, "A Low-cost Memristor Based on Titanium Oxide," *2010 10th IEEE Int. Conf. on. IEEE Solid-State Integr. Circuit Technol. (ICSICT)*, pp. 2–4, 2010.

[6] R. A. Bakar, 'Awatif Harun Mohd Nur Nazmi, N. A. A. S. Nur Syahirah Kamarozaman, and S. M. M. Kasim, "Annealing Temperature Dependence of Resistive Switching Behavior for Sol-gel Spin Coated Zinc Oxide Thin Films," *2014 IEEE Int. Conf. Semicond. Electron. (ICSE)*, pp. 452–455, 2014.

[7] N. S. Kamarozaman, M. A. R. Rashid, M. Z. Musa, S. H. Herman, R. A. Bakar, W. F. H. Abdullah, and M. Rusop, "Effect of Film Thickness on the Memristive Behavior of Spin Coated Titanium Dioxide Thin Films," *2013 IEEE Reg. Symp. Micro Nanoelectron. (RSM)*, pp. 155–158, 2013.

[8] G. A. Luurtsema, "Spin coating for rectangular substrates," University of California, Berkeley, 1997.

[9] H. Shi, F. Liu, E. Han, and Y. Wei, "Effects of Nano Pigments on the Corrosion Resistance of Alkyd Coating," *J. Mater. Sci. Technol*, vol. 23, no. 4, pp. 551–558, 2007.

[10] N. S. Kamarozaman, Z. Aznilinda, S. H. Herman, R. A. Bakar, and M. Rusop, "Effect of Annealing Duration on the Memristive Behavior of Pt / TiO_2 / ITO Memristive Device," *2012 10th IEEE Int. Conf. Semicond. Electron. (ICSE)*, pp. 703–706, 2012.

[11] N. R. Mathews, E. R. Morales, M. a. Cortés-Jacome, and J. a. Toledo Antonio, "TiO_2 thin films – Influence of annealing temperature on structural, optical and photocatalytic properties," *Sol. Energy*, vol. 83, no. 9, pp. 1499–1508, Sep. 2009.

[12] N. a. Vorobyeva, M. N. Rumyantseva, R. B. Vasiliev, V. F. Kozlovskiy, Y. M. Soshnikova, D. G. Filatova, V. B. Zaytsev, a. V. Zaytseva, and a. M. Gaskov, "Doping effects on electrical and optical properties of spin-coated ZnO thin films," *Vacuum*, vol. 114, pp. 198–204, Apr. 2015.

[13] C.-H. Wei and C.-M. Chang, "Polycrystalline TiO_2 Thin Films with Different Thicknesses Deposited on Unheated Substrates Using RF Magnetron Sputtering," *Mater. Trans.*, vol. 52, no. 3, pp. 554–559, 2011.

[14] M. Schneider, a. Bittner, and U. Schmid, "Impact of film thickness on the temperature-activated leakage current behavior of sputtered aluminum nitride thin films," *Sensors Actuators A Phys.*, vol. 224, pp. 177–184, Apr. 2015.

[15] D. Seo, J. Na, C. Kim, C. Jeong, and S. Lim, "Improvement of Cu_2ZnSnS_4 thin film properties by a modified sulfurization process," *Thin Solid Films*, pp. 6–11, May 2015.

[16] M. Wang, Y. Wang, and J. Li, "The fabrication of TiO_2 nanoparticle / ZnO nanowire arrays bi-filmed photoanode and their effect on dye-sensitized solar cells," *J. Sol-Gel Sci. Technol.*, vol. 61, no. 3, pp. 613–619, 2012.

IOP Conf. Series: Materials Science and Engineering **99** (2015) 012021 doi:10.1088/1757-899X/99/1/012021

Annealing time dependence of the physical, electrical and pH response characteristics of spin coated TiO_2 thin films

M A Zulkefle[1], R A Rahman[1], K A Yusoff[1], W F H Abdullah[2], M Rusop[3], S H Herman[1,2,4]

[1]NANO-Electronic Centre, Universiti Teknologi MARA (UiTM), 40450 Shah Alam, Selangor, Malaysia
[2]Faculty of Electrical Engineering, Universiti Teknologi MARA (UiTM), 40450 Shah Alam, Selangor, Malaysia
[3]NANO-Science Technology (NST), Institute of Science (IOS), Faculty of Applied Sciences, Universiti Teknologi MARA (UiTM), 40450 Shah Alam, Selangor, Malaysia
[4]Core of Frontier Materials & Industry Applications, Universiti Teknologi MARA (UiTM), 40450 Shah Alam, Selangor, Malaysia

E-mail: alhadizulkefle@gmail.com

Abstract. Titanium dioxide (TiO_2) thin film was deposited on indium tin oxide (ITO) substrate and used as sensing membrane of EGFET pH sensor. The thin film was fabricated using sol-gel spin coating method. All samples were annealed at 400 °C but the annealing time was varied. This is done to study the effects of annealing time on physical and electrical properties of titanium dioxide thin film. The sensitivity of each sample towards H^+ ion was measured and result shows that sample annealed for 45 minutes has the highest sensitivity (52.6 mV/pH). It is found that increasing annealing duration will increase the pH sensitivity but a limit will be reached at certain point. Longer annealing processes done beyond this point will results in lower pH sensitivity.

1. Introduction

The glass selective membrane electrode is the popular structure for pH sensor. Another well-known structure for pH sensor is ion sensitive field effect transistor (ISFET) [1]. Due to brittleness of the former and device instability of the latter, another pH sensor was proposed. The device is known as extended gate field effect transistor (EGFET) and was first reported by Spiegel in 1983 [2]. It has the same working principal as ISFET since it was developed from it. The main difference between ISFET and EGFET is that the sensing membrane of EGFET is extended from the sensing MOSFET [3]. This extended structure gives few advantages such as device stability, temperature insensitivity and flexibility in designing shape of the extended gate [4-6].

Metal oxides in form of thin films has been investigated to be used as sensing layers to detect presence of H^+ in solutions. Among them are palladium oxide [7], indium tin oxide [8], silicon dioxide [9], titanium dioxide [10] and zinc oxide [11]. TiO_2 is chosen to be used in this experiment since it has many interesting properties such as chemically stable and shows good electrical and optical properties [12-14].

There are several reported ways of depositing titanium dioxide thin films. Examples are chemical vapor deposition (CVD) [15], spin coating [16], chemical bath deposition (CBD) [17], atomic layer deposition (ALD) [18], and magnetron sputtering [19]. Due to its easy handling procedure and good uniformity of thin film fabricated, sol-gel spin coating is more favored compared to the other processes [20-21]. An important factor in controlling the quality of thin films produced is by controlling the post-deposition annealing temperature and time. In this paper, study on effects of increasing annealing time towards sensitivity of TiO_2-based EGFET pH sensor was done.

2. Experimental details

ITO substrate was cleaned using methanol and deionized (DI) water in ultrasonic cleaner. The substrate was then dried by blowing with nitrogen gas. TiO_2 solution was prepared from mixture of Solution A and Solution B. Solution A comprise of titanium (IV) isopropoxide (precursor), absolute ethanol (solvent) and glacial acetic acid (stabilizer). Solution B contains Triton X-100 (surfactants), deionized water and absolute ethanol (solvent). Solution A and B were stirred for 1 hour. Then both solutions were mixed and the mixture was continued to be stir for another hour.

The cleaned ITO coated glass substrate was spun with the speed of 3000 rpm for one minute. During the spinning process, ten drops of the prepared TiO_2 solutions were dropped onto the substrate. The TiO_2 solution will disperse evenly throughout the substrate surface as a result of the high speed spinning. The TiO_2 (now in form of a thin film) would then be dried at temperature of 200 °C for duration of 10 minutes.

Then the samples were annealed at a temperature of 400 °C. However the duration of the annealing process were varied for each particular sample. First, second, third and fourth sample were all annealed for 15, 30, 45 and 60 minutes respectively. Each sample was labeled based on their respective annealing time. Sample annealed for 15 minutes is named t_{15}, 30 minutes is t_{30}, 45 minutes is labeled t_{45} and t_{60} is sample annealed for 60 minutes.

The samples were measured for its sensitivity towards sensing H^+ ions in solutions. It is done by connecting the sample to the gate of MOSFET. The sample now acts as the gate of the MOSFET which is then attached with readout interface (ROIC) [22] which is linked to a power supply and multimeter. The physical characterization includes of determining thickness of TiO_2 thin films that had been annealed at different time. Surface studies of the samples were done by using atomic force microscopy (AFM). Electrical properties of each thin film were also found after analyzing graphs from the resulted I-V measurement.

3. Results and discussion

Surface profiler was used to determine thickness of each sample. The value was taken from average of three readings. Table 1 shows the data obtained. It was shown from the data that increasing annealing time of samples will decrease the thickness of thin film. This might be due to elimination of solvent and organic residues during annealing [23]. It is noteworthy to state that annealing process also would result in densification [24-25] and shrinkage of thin films.

Table 1. Thickness of thin films

Annealing time (minutes)	Thickness (nm)
15	30.27
30	29.65
45	21.57
60	15.57

AFM is done to measure and obtaining roughness of thin films. Depending on the application, rough surface might be unwanted or required. According to studies done by [26], annealing process may modify the surface of thin films where grain size will increase as well as the surface roughness. The changes in mentioned physical aspects will influence several properties such as electron movement and light scattering [27]. Figure 1 shows AFM images obtained from the samples.

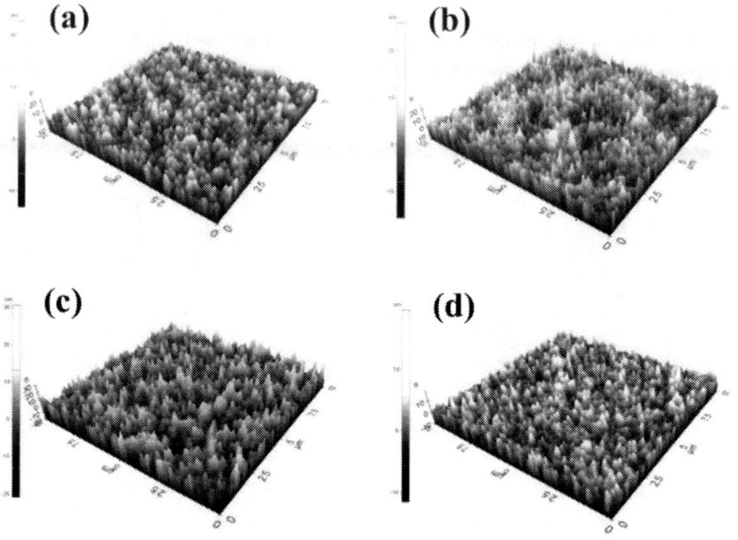

Figure 1. AFM images representing surface roughness of (a) t_{15}, (b) t_{30}, (c) t_{45}, and (d) t_{60}

The surface roughness is 2.745 nm, 3.174 nm and 5.359 nm for t_{15}, t_{30} and t_{45}, respectively. The annealing process conducted in this experiment was found to increase the surface roughness of the thin films. This result is comparable with works by [28-29] where they also reported an increase in the surface roughness of their thin films when annealing process was applied. The roughness however decreases when sample being annealed for 60 minutes. t_{60} has roughness of 2.265 nm.

Besides the physical alteration, annealing would also change the electrical properties of thin films. The TiO_2 thin films had been seen to be decreasing in their electrical performance when being annealed as can be seen in the I-V in characteristics in Figure 2. Ohmic characteristic was observed for all samples. Based on data from the graphs, the resistivity and conductivity of each sample was calculated.

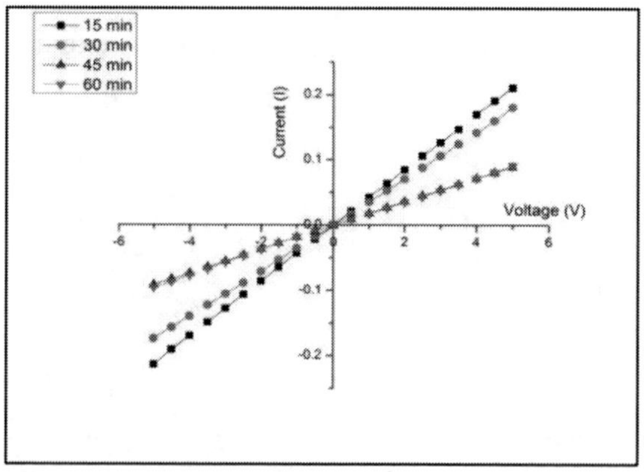

Figure 2. I-V measurement graph

The resistivity, ρ and conductivity, σ is shown in Table 2. It can be said that annealing process decreases the electrical conductivity of TiO_2 thin film. Further annealing done at a longer period of time will further decrease the conductivity value. Since the resistivity is inversely proportional to conductivity, increase in annealing time would mean an increase in resistivity of thin films.

Electron scattering would be reduced if smoother surface was obtained. The reduction in electron scattering in exchange would increase conductivity of thin films [30]. Considering the facts that annealing process done in this experiment increases the surface roughness of TiO_2 thin films, it explains why conductivity of thin films undergoes decline in values.

Table 2. Fabricated samples resistivity and conductivity

Annealing time (minutes)	**Resistivity, ρ (Ω.cm)**	**Conductivity, σ (S.cm^{-1})**
15	6.68×10^{5}	1.50×10^{-6}
30	8.20×10^{5}	1.22×10^{-6}
45	2.20×10^{6}	4.55×10^{-7}
60	3.04×10^{6}	3.29×10^{-7}

The sensitivity measurement resulted in data in form of voltage output. Each voltage output produced at corresponding pH buffer solutions were plotted in Figure 3(a). The graph was then analysed to determine the sensitivity and linearity of each sample. A trend is spotted from the tabulated data and plotted graph. Longer annealing time was seen to improve the sensitivity of TiO_2 sensing membrane towards H^+ ions in the pH buffer solution. All sensitivity and linearity values were tabulated in Table 3.

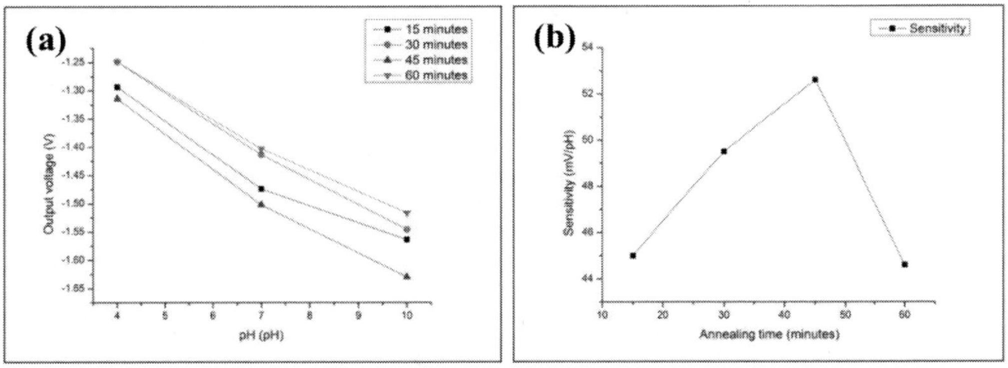

Figure 3. Graphs of (a) output voltage (b) sensitivity of samples fabricated

t_{15} has good sensitivity with value of 45.0 mV/pH. The t_{30} sample which was annealed 15 minutes longer than t_{15} sample has 49.5 mV/pH sensitivity. The sensitivity was further increased when TiO_2 thin film was annealed for 45 minutes. t_{45} sample was also the sensing membrane with the highest sensitivity. With sensitivity of 52.6 mV/pH, it has sensitivity approaching Nernstian theoretical value (59 mV/pH).

Table 3. Sensitivity and linearity value for each sample

Annealing time (minutes)	**Sensitivity (mV/pH)**	**Linearity (%)**
15	45.0	96.33
30	49.5	99.60
45	52.6	98.78
60	44.6	99.24

However, there was decrease in sensitivity of TiO_2 sensing membrane when the sample being annealed longer than 45 minutes. It is shown when t_{60} sample has lowest sensitivity compared to t_{45} sample and other previous samples. It is concluded that 45 minutes is the ideal annealing time for TiO_2 sensing membrane for EGFET pH sensor.

The pattern shown by sensitivity value were the same as pattern observed in roughness value of samples. Annealing time affects the roughness of the samples and these changes in surface roughness was most likely to be the reason behind the increment and decrement of sensitivity of samples. In term of linearity, all sample fabricated shows good sensing performance when being use as sensing membrane. With linearity value of 96.33%-99.60%, TiO_2 is proven to be a good material for detecting H^+ ions in liquid solutions.

4. Conclusion

Instead of using ISFET structure, TiO_2 was integrated as sensing membrane in EGFET structure for pH sensing. Studies on effects of applying different annealing time on fabricated TiO_2 thin films were effectively conducted. Each thin film was characterized for its physical and electrical performance. The annealing time was varied from 15 minutes, 30 minutes, 45 minutes and 60 minutes and was labeled as t_{15}, t_{30}, t_{45} and t_{60} respectively. Physical characterization was completed and first significant finding was that thickness of TiO_2 thin films is dependent on annealing time. Samples anneal for 15 minutes, 30 minutes, 45 minutes and 60 minutes has thickness of 30.27, 29.65, 21.57 and 15.57 nm correspondingly. It means that the longer annealing time, the thinner the thin film would be. The TiO_2 surface was noticed to gain roughness when being annealed for longer period of time. However it was found that the roughness decreases when the thin film was annealed for 60 minutes.

The second part of characterization focused on electrical characterization. I-V measurement was carried out and I-V characteristics graph was obtained. The graph indicates that all TiO_2 sample produced has ohmic characteristic. Data information from the graph was extracted to calculate resistivity, ρ and conductivity, σ. Resistivity of TiO_2 thin films increases when annealing time increases. Conductivity of the samples however decline when longer annealing period was applied. TiO_2 was recognized as suitable sensing material to be use in detecting H^+ ions in different pH solutions since it has good linearity value of 96.33%-99.60%. t_{15}, t_{30}, t_{45} and t_{60} have sensitivity of 45.0 mV/pH, 49.5 mV/pH, 52.6 mV/pH and 44.6 mV/pH respectively. Hence it was concluded that 45 minutes is the best annealing time to anneal TiO_2 thin films in order for the film to give optimized performance when being used as sensing membrane in EGFET pH sensor.

5. Acknowledgement

This study was partially supported by MOSTI Malaysia under the ScienceFund grant (Project code: 06-01-01-SF0662) and Universiti Teknologi MARA (UiTM).

References

[1] P. Bergveld, "Thirty years of ISFETOLOGY What happened in the past 30 years and what may happen in the next 30 years," *Sensors and Actuators B: Chem*, vol. 88, pp. 1–20, 2003.

[2] J. van der spiegel, I. Lauks, P. Chan, D. Babic, "The extended gate chemically sensitive field effect transistor as multi-species microprobe," *Sensors and Actuators*, vol. 4, pp. 291-298, 1983.

[3] T. K. Truong, T. N. T. Nguyen, T. Q. Trung, I. Y. Sohn, D.-J. Kim, J.-H. Jung, and N.-E. Lee, "Reduced graphene oxide field-effect transistor with indium tin oxide extended gate for proton sensing," *Curr. Appl. Phys.*, vol. 14, no. 5, pp. 738–743, May 2014.

[4] J.-C. Lin, B.-R. Huang, and Y.-K. Yang, "IGZO nanoparticle-modified silicon nanowires as extended-gate field-effect transistor pH sensors," *Sensors Actuators B Chem.*, vol. 184, pp. 27–32, Jul. 2013.

[5] H.-H. Li, C.-E. Yang, C.-C. Kei, C.-Y. Su, W.-S. Dai, J.-K. Tseng, P.-Y. Yang, J.-C. Chou, and H.-C. Cheng, "Coaxial-structured ZnO/silicon nanowires extended-gate field-effect transistor as pH sensor," *Thin Solid Films*, vol. 529, pp. 173–176, Feb. 2013.

[6] G. M. Ali, R. H. Dhaher, and A. a. Abdullateef, "pH sensing characteristics of EGFET based on Pd-doped ZnO thin films synthesized by sol-gel method," *2015 Third Int. Conf. Technol. Adv. Electr. Electron. Comput. Eng.*, vol. 4, no. 2, pp. 234–238, Apr. 2015.

[7] A. Das, D. Hsu, C. Chen, L. Chang, C. Lai, F. Chu, L. Chow, and R. Lin, "Sensors and Actuators B : Chemical Highly sensitive palladium oxide thin film extended gate FETs as pH sensor," *Sensors Actuators B. Chem.*, vol. 205, pp. 199–205, 2014.

[8] L. Yin, J. Chou, W. Chung, T. Sun, and S. Hsiung, "Study of indium tin oxide thin film for separative extended gate ISFET," *Mater. Chem. and Phy*, vol. 70, pp. 12–16, 2001.

[9] H.-K. Liao, J.-C. Chou, W.-Y. Chung, T.-P. Sun, and S.-K. Hsiung, "Study of amorphous tin oxide thin films for ISFET applications," *Sensors Actuators B Chem.*, vol. 50, no. 2, pp. 104–109, Jul. 1998.

[10] M. a. Rosdan, S. H. Herman, W. F. H. Abdullah, N. S. Kamarozaman, and M. I. Syono, "Sputtered titanium dioxide thin film for Extended-Gate FET sensor application," *RSM 2013 IEEE Reg. Symp. Micro Nanoelectron.*, pp. 219–222, Sep. 2013.

[11] A. Fulati, S. M. U. Ali, M. Riaz, G. Amin, O. Nur, and M. Willander, "Miniaturized pH Sensors Based on Zinc Oxide Nanotubes/Nanorods.," *Sensors (Basel).*, vol. 9, no. 11, pp. 8911–23, Jan. 2009.

[12] J.-C. Chou and L. P. Liao, "Study on pH at the point of zero charge of TiO_2 pH ion-sensitive field effect transistor made by the sputtering method," *Thin Solid Films*, vol. 476, no. 1, pp. 157–161, Apr. 2005.

[13] L. H. Chong, K. Mallik, C. H. De Groot, and R. Kersting, "The structural and electrical properties of thermally grown TiO_2 thin films," *J. Phys. Condens. Matter*, vol. 18, no. 2, pp. 645–657, Jan. 2006.

[14] K. N. Rao, "Influence of deposition parameters on optical properties of TiO_2 films," *Opt. Eng.*, vol. 41, no. 9, p. 2357, Sep. 2002.

[15] A. Ito, T. Sato, and T. Goto, "Transparent anatase and rutile TiO_2 films grown by laser chemical vapor deposition," *Thin Solid Films*, vol. 551, pp. 37–41, Jan. 2014.

[16] Y.-H. Liao and J.-C. Chou, "Preparation and characterization of the titanium dioxide thin films used for pH electrode and procaine drug sensor by sol–gel method," *Mater. Chem. Phys.*, vol. 114, no. 2–3, pp. 542–548, Apr. 2009.

[17] U. M. Patil, S. B. Kulkarni, P. R. Deshmukh, R. R. Salunkhe, and C. D. Lokhande, "Photosensitive nanostructured TiO_2 grown at room temperature by novel 'bottom-up' approached CBD method," *J. Alloys Compd.*, vol. 509, no. 21, pp. 6196–6199, May 2011.

[18] a. P. Alekhin, S. a. Gudkova, a. M. Markeev, a. S. Mitiaev, a. a. Sigarev, and V. F. Toknova, "Structural properties of the titanium dioxide thin films grown by atomic layer deposition at various numbers of reaction cycles," *Appl. Surf. Sci.*, vol. 257, no. 1, pp. 186–191, Oct. 2010.

[19] D. Pjevi, M. Obradovi, T. Marinkovi, A. Grce, M. Milosavljevi, R. Grieseler, T. Kups, M. Wilke, and P. Schaaf, "Properties of sputtered TiO_2 thin films as a function of deposition and annealing parameters," *Phy. B: Condens. Matter*, vol. 463, pp. 20–25, 2015.

[20] J. Tian, L. Chen, J. Dai, X. Wang, Y. Yin, and P. Wu, "Preparation and characterization of TiO_2, ZnO, and TiO_2/ZnO nanofilms via sol–gel process," *Ceram. Int.*, vol. 35, no. 6, pp. 2261–2270, Aug. 2009.

[21] D. Raoufi and T. Raoufi, "The effect of heat treatment on the physical properties of sol–gel derived ZnO thin films," *Appl. Surf. Sci.*, vol. 255, no. 11, pp. 5812–5817, Mar. 2009.

[22] W.F.H. Abdullah, M. Othman, M.A.M. Ali, "Chemical field-effect transistor with constant-voltage constant-current drain-source readout circuit," *Research and Development (SCOReD), 2009 IEEE Student Conference* on , vol., no., pp.219,221, 16-18 Nov 2009.

[23] A. Lewkowicz, A. Synak, B. Grobelna, P. Bojarski, R. Bogdanowicz, J. Karczewski, K. Szczodrowski, and M. Behrendt, "Thickness and structure change of titanium (IV) oxide thin films synthesized by the sol – gel spin coating method," *Opt. Mater. (Amst).*, vol. 36, no. 10, pp. 1739–1744, 2014.

[24] N.R. Mathews, Erik R. Morales, M.A. Cortés-Jacome, J. A. T. Antonio, "TiO_2 thin films – Influence of annealing temperature on structural , optical and photocatalytic properties," *Solar Energy*, vol. 83, pp. 1499–1508, 2009.

[25] M. H. Habibi and M. Nasr-Esfahani, "Preparation, characterization and photocatalytic activity of a novel nanostructure composite film derived from nanopowder TiO_2 and sol–gel process using organic dispersant," *Dye. Pigment.*, vol. 75, no. 3, pp. 714–722, 2007.

[26] C. Lin, H. Chen, A. Nakaruk, P. Koshy, and C. C. Sorrell, "Effect of Annealing Temperature on the Photocatalytic Activity of TiO_2 Thin Films," *Energy Procedia*, vol. 34, pp. 627–636, 2013.

[27] A. López, D. Acosta, A. I. Martínez, and J. Santiago, "Nanostructured low crystallized titanium dioxide thin films with good photocatalytic activity," *Powder Technol.*, vol. 202, no. 1–3, pp. 111–117, Aug. 2010.

[28] J. Ben Naceur, M. Gaidi, F. Bousbih, R. Mechiakh, and R. Chtourou, "Annealing effects on microstructural and optical properties of Nanostructured-TiO_2 thin films prepared by sol-gel technique," *Curr. Appl. Phys.*, vol. 12, no. 2, pp. 422–428, 2012.

[29] S. K. Sharma and S. Mohan, "Applied Surface Science Influence of annealing on structural , morphological , compositional and surface properties of magnetron sputtered nickel – titanium thin films," *Appl. Surf. Sci.*, vol. 282, pp. 492–498, 2013.

[30] N. K. Ponon, D. J. R. Appleby, E. Arac, P. J. King, S. Ganti, K. S. K. Kwa, and A. O'Neill, "Effect of deposition conditions and post deposition anneal on reactively sputtered titanium nitride thin films," *Thin Solid Films*, vol. 578, pp. 31–37, Mar. 2015.

The Effect of the Sol-gel Spincoating Deposition Technique on the Memristive Behaviour of ZnO-based Memristive Device

N.A.A.Shaari[1*]. S.M.M. Kasim[1], N.S.M.Sauki[1], and S.H. Herman[1,2]

[1]NANO-Electronic Centre (NET) Faculty of Electrical Engineering UiTM Shah Alam.
[2]CoRe of Frontier Materials & Industry Applications,UiTM Shah Alam.

*noraziraakma.shaari@yahoo.com.my

Abstract. This paper presents the memristive behavior of zinc oxide thin films deposited on ITO substrate by sol-gel spin coating technique. The spin coating speed was varied from 1000 rpm to 5000 rpm to study the effect it has on the memristive device fabricated. The electrical properties were characterized by using a two-point probe IV (current-voltage) measurement system (Keithley 2400). The thicknesses of the thin films were measured by Veeco Dektak 150 Surface Profiler and it shows that the thickness decreased with the spin coat speed. The lowest thickness was obtained from thin film deposited at 5000rpm which is 17.47 nm. The highest resistance Roff/Ron ratio was obtained from thin film spin coated at 3000 rpm which is 1.346 with visible ZnO nanoparticle characterized by FESEM (JEOL JSM 6701F). This indicated that the optimum spin coat speed for the zinc oxide-based memristive device is 3000 rpm as it exhibited the best switching behavior.
Keyword: spin coat, speed, memristive, ZnO, resistance ratio

1. Introduction

Memristor is a two terminal device that is memory dependent and non-volatile, which means it has the ability to store resistance value indefinitely. It was proposed by Professor Leon Chua back in 1971[1]. Since the thickness of the active layer needs to be in nanometer scale, its physical state only been discovered in HP lab in 2008 with advancing nanotechnology that enabled the fabrication of nanoscale devices [2]. It is also reported to be viable to use in microscale as well [3]. Memristor is considered as the fourth fundamental circuit element. It extends the fundamental passive circuit elements which are revolved around capacitors, resistors and inductors. The memristive device is defined by its IV characteristic that exhibits a hysteresis loop in a shape of bow tie and that loop is necessarily crossing the origin point [4].

The most commonly used material as the active layer of the memristive device is titanium dioxide (TiO_2) since it is compatible with standard complementary (CMOS) technology. For this study, zinc oxide is chosen instead as an active layer for the memristive device. It has similar bandgap compared to TiO_2 but it is a direct semiconductor. ZnO has shown to be a promising material in memristive device as it has been widely used in other applications such as blue and ultraviolet light emitter, solar cell windows, photovoltaic device [5], flexible memory application (RRAM) [6] and surface acoustic wave device [7].

In this paper, ZnO solution was prepared by using sol-gel spin coating technique to deposit the ZnO solution onto the ITO substrate. The spin coating speed is varied from 1000 rpm, 3000 rpm and 5000 rpm to study its effect on the memristive behavior. Sol-gel is chosen as it is cheap and easy to coat the substrate in whichever shape and area[8].

2. Methodology

2.1. Memristor Fabrication

The ITO substrate as a bottom electrode would first go through the cleaning process. Acetone, methanol and deionized water are used to clean the substrate, followed by drying it off with nitrogen gas. This cleaning process is crucial to get rid of all the contamination that could affect the properties of the thin films. The sol-gel was prepared by mixing together zinc acetate dehydrate ($Zn(CH_3COO)_2H_2O$) as a precursor material, 2-methoxyethanol ($C_3H_8O_2$) as solvent and monoethanolamine (C_2H_7NO, MEA) as stabilizer. The solution was then sonicated for 30 minutes under 50°C temperature. After that it will be stirred on the hot plate stirrer for 3 hours at 80°C. It was then left to age for 24 hours to produce the clear homogeneous solution.
The thin film deposition is then carried out by spin coating technique. This is where the speed of spin coater is varied. The thin films are spin coated at 1000rpm, 3000rpm, and 5000rpm in 1 minute each while 10 drops of ZnO solution is dropped. After the spin coating process, the thin films are left to dry for 10 minutes at 150°C to evaporate the solvent and remove the residuals. Finally the thin films were then annealed in the furnace for 1 hour at 350°C temperature

2.2 Memristor Characterization

The memristive device was characterized for both electrical and physical properties. For the physical properties, the thin films thickness was characterized by using Surface Profiler, the surface morphology was characterized by FESEM. The electrical characterization for current-voltage measurement was done by using two-point probe Keithley 4200 semiconductor characterization system connected to the probe station. The main purpose of this characterization was to obtain the memristive behavior of ZnO thin films by sweeping the voltage from 0V to -5V, -5V to 5V, 5V to 0V. The Figure 1 shows the memristive device structure for its IV measurement technique used in this work.

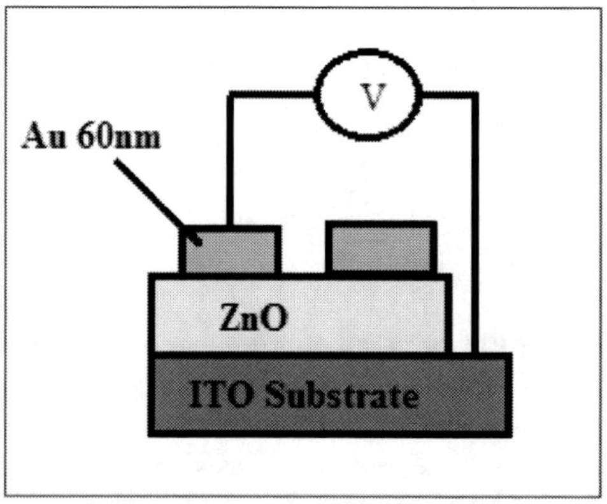

Figure 1: The cross section diagram of IV measurement for memristive device structure of Au/ZnO/ITO

3. Result and Discussion

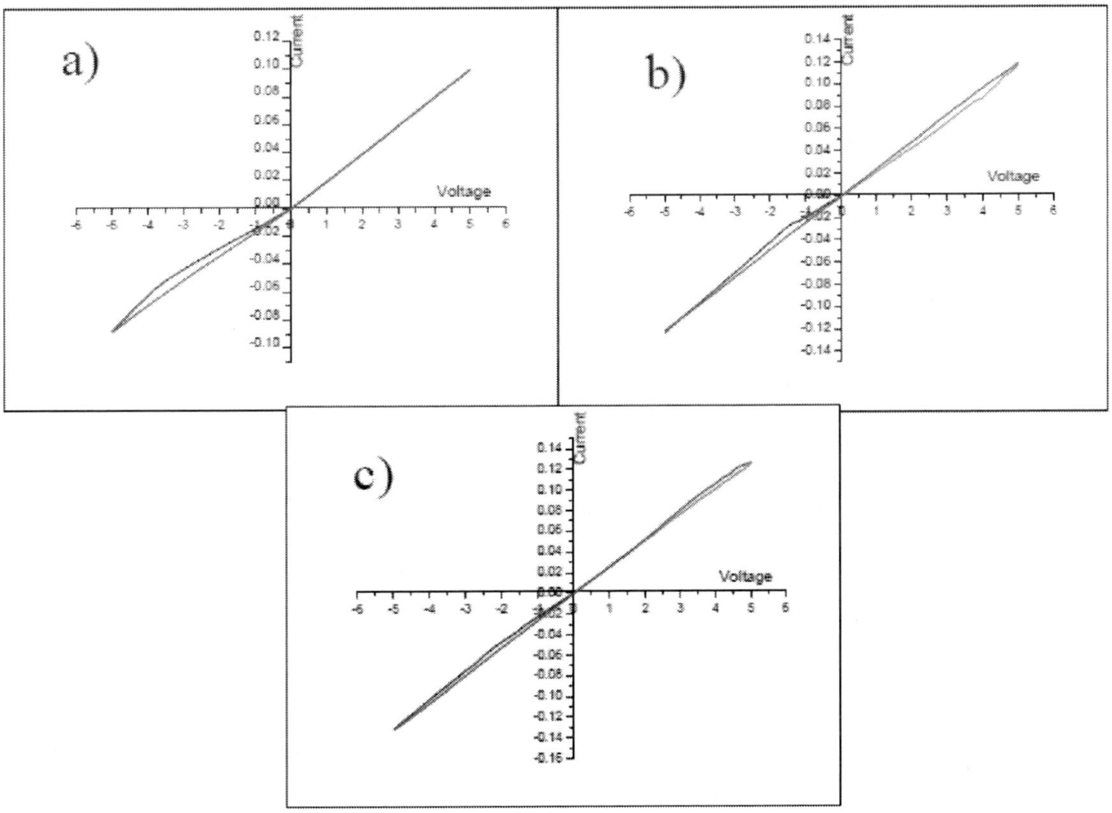

Figure 2: *IV Characteristics of Au/ZnO/ITO memristive device spin coated at a) 1000rpm, b) 3000rpm, and c)5000rpm*

From Figure 2(a) the thin film annealed at 1000rpm, the IV characteristic graph resulted in asymmetrical IV characteristic with only hysteresis loop exists in negative biased. This can be observed clearly from the semilog asymmetrical IV characteristic graph as shown in Figure 3(a).

The result shows that the ZnO spin coated at 3000rpm and 5000rpm show memristive behavior with resistance ratio of 1.346 and 1.125 respectively as tabulated in Table 1. This switching behavior may be resulted from the filament formation within the oxide layer that contains movement of oxygen vacancies that allows the electrons to drift towards the cathode [9]. In Figure 2(b) it can be said that the thin film's conductivity is increasing compared to thin film in Figure 2(a) as a result of better oxygen vacancies created in the ZnO active layer [10]. Further increment of spin coat speed to 5000rpm as shown in Figure 2(c) resulted in thinning loops of memristive behavior and decreasing resistance ratio to 1.125. The symmetrical semilog IV characteristic graph as shown in Figure 3(b) and 3(c) shown the existent of hysteresis loops in both negative and positive biased. Spin speed is somewhat contributed to the change in film thickness. It was proven that the forming of metallic gold filaments passing through the thin ZnO layer contributed to an ohmic resistance switches for sample spincoated at 5000 rpm as shown in Figure 3(c). These filaments could be disrupted by Joule heating. The gold was also reported to give a fuse-like behaviour, which contributed to poor memristive behaviour [11].

The higher the spincoating speed, the thinner thin film gets, causing the ZnO particles to be scattered in the substrate. This is proven in Table 1. Thinnest film makes contributes better resistance ratio with memristive behavior. As seen from Figure 2, although 5000 rpm has lowest resistance ratio compared to samples spincoated at 1000rpm and 3000rpm, it shows symmetrical memristive behavior with loops on negative and positive sides. The sample spin coated at 1000rpm is not symmetry, Figure 3(a) only shows loop on the negative side, since it has the thickest film. It was reported that the increase in film thickness contributed to defects in oxygen-related [12]. It is suspected that the fall in resistance is due to disruption in gold bridge [11].

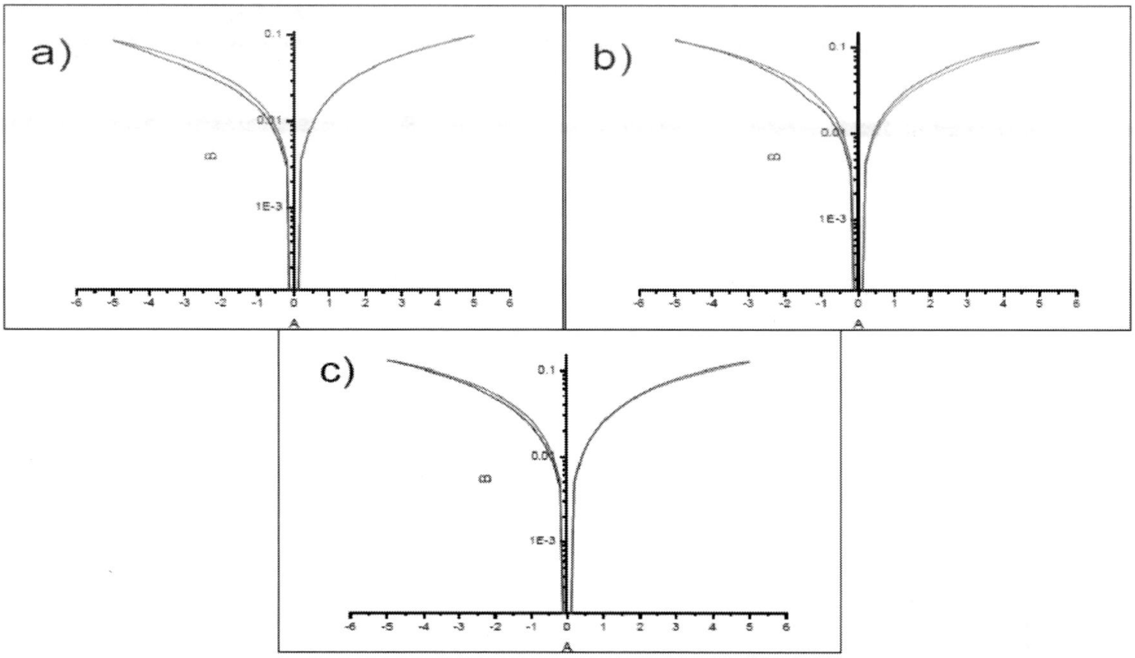

Figure 3: *Semilog IV Characteristic for thin films spincoated for a) 1000rpm, b) 3000rpm and c) 5000rpm*

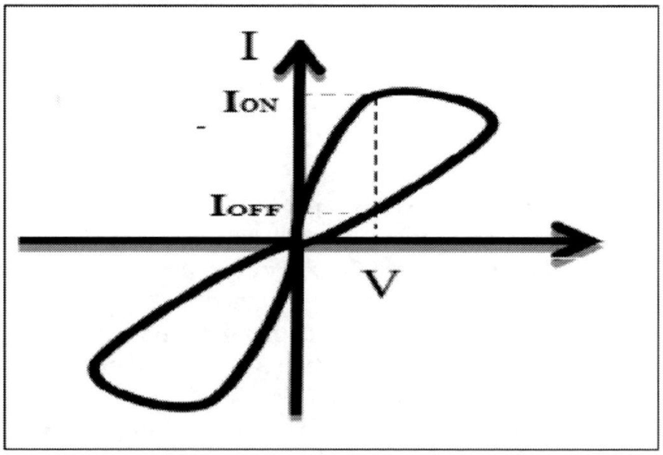

Figure 4: *The resistance ratio measurement for particular voltage*

The resistance ratios were calculated from the IV characteristic graph as shown in Figure 4 by using the formula (1) and the results are tabulated in Table 1 along with thin film thicknesses measured by Veeco Dektak Surface Profiler.

$R_{ON} = V/I_{ON}$, $R_{OFF} = V/I_{OFF}$ ……………………………………………………………………(1)

TABLE 1. *Thin film thicknesses and resistance ratio for different spincoat speed*

Spincoat Speed (rpm)	Thin Film Thickness (nm)	Resistance Ratio
1000	76.52	1.178
3000	34.48	1.346
5000	17.47	1.125

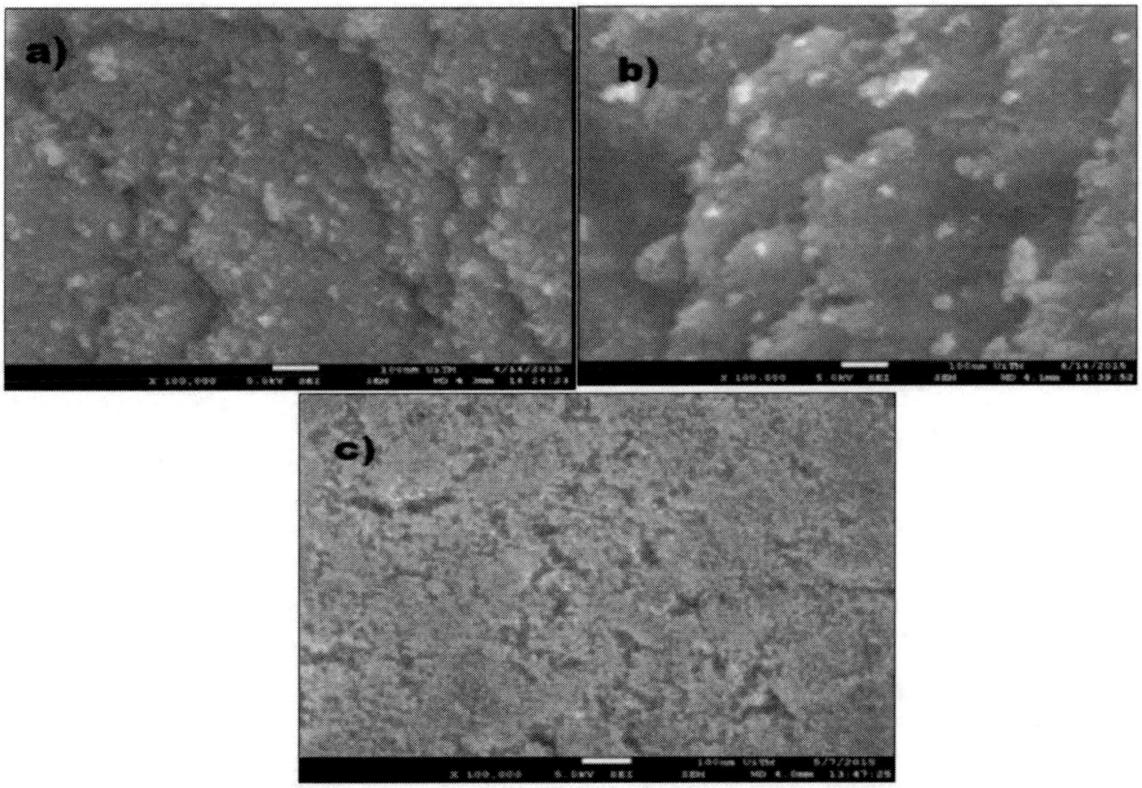

Figure 5: *FESEM image for ZnO thin films initially spincoated at a) 1000rpm, b) 3000rpm and c) 5000rpm*

Based on the thin film thicknesses tabulated in Table 1, it can be seen that the thicknesses decreased as the spin coating speed increasing [13]. As the spin coating speed increased, the more ZnO particles will be scattered away from the ITO substrate, thus leaving the thinner films [14].

Figure 5(a)-(c) shows the FESEM image of surface of ZnO thin films spin coated at 1000 rpm, 3000 rpm and 5000 rpm respectively. It can be seen that the nanoparticles were formed on the thin films. The images show the height difference possibility in the structures. From the figure 5(b), the ZnO thin film spin coated at 3000 rpm shows less dense structure compared to ZnO thin film spin coated at 1000rpm. However, the grain becomes more uniform, denser and reduced grain size for ZnO thin film annealed at 5000 rpm as it has the thinnest layer as proven in Figure 5(c).

4. Conclusion

The work was carried out to study the effect of spin coating speed on the resistive switching behavior of ZnO thin films. It is known that the film thickness decreases as the spin coating speed increasing. The higher spin coat speed is what improving the grain size of the ZnO nanoparticles by reducing it and making it uniform. The spin coat speed of 3000 rpm has the memristive behavior with the highest resistance ratio of 1.346 and also the optimum spin coating speed for this study. For thin film spin coated at 5000 rpm speed was seen to drop slightly in term of the resistance ratio by 1.125 due to the metallic gold filaments forming passing through the thin ZnO layer as it is proven to have the thinnest, densest structure of ZnO nanoparticles for the surface morphology characteristic.

5. Acknowledgment

This work is partially supported by the Ministry of Education (MOE) Malaysia under the Niche Research Grant Scheme NRGS (Project Code: 600-RMI//NRGS 5/3 (7/2013)), Research Acculturation Grant Scheme (Project Code: 600-RMI/RAGS 5/3 (198/2014)) and Universiti Teknologi MARA Shah Alam.

References

[1] L. O. Chua, "Memristor The Missing Circuit Element," *IEEE Trans.Circuit Theory*, vol. 5, no. CT-18, pp. 507–519, 1971.

[2] D. B. Strukov, G. S. Snider, D. R. Stewart, and R. S. Williams, "The missing memristor found.," *Nature*, vol. 453, no. 7191, pp. 80–3, May 2008.

[3] T. Prodromakis, K. Michelakis, and C. Toumazou, "Switching mechanisms in microscale memristors," *Electron. Lett.*, vol. 46, no. 1, p. 63, 2010.

[4] R. S. Williams, ""How We Found the Missing Memristor," *IEEE Spectr.*, vol. 45, no. 12, pp. 28–35, 2008.

[5] M. H. Mamat and M. Z. Sahdan, "Electrical Characteristics of Sol-Gel Derived Aluminum Doped Zinc Oxide Thin Films at Different Annealing Temperatures," *Electron. Devices, Syst. Appl.*, no. 100, pp. 408–411, 2010.

[6] S. Kim, H. Moon, D. Gupta, S. Yoo, and Y.-K. Choi, "Resistive Switching Characteristics of Sol–Gel Zinc Oxide Films for Flexible Memory Applications," *IEEE Trans. Electron Devices*, vol. 56, no. 4, pp. 696–699, Apr. 2009.

[7] Z. Sin, N.D.M, Mamat, M.H. ; Rusop, M. ; Zulkifli, "Electrically Conductive Zinc Oxide (ZnO) Nanostructures Prepared By Sol gel Spin-coating," *Enabling Sci. Nanotechnol. (ESciNano), 2010 Int. Conf.*, vol. 3, no. 9, pp. 9–10, 2010.

[8] A. A. A. Halim, H. Hashim, M. Rusop, M. H. Mamat, and A. S. Zoolfakar, "Study on Electrical Properties of Zinc Oxide Thin Film," *Innov. Technol. Intell. Syst. Ind. Appl. 2008. CITISIA 2008. IEEE Conf.*, no. July, pp. 123–126, 2008.

[9] R. Dong, X. Yan, Z. Zhang, and M. Wang, "Memristive behavior and forming mechanism of homogeneous TiOx device," *Mater. Res. Bull.*, vol. 61, pp. 101–104, 2015.

[10] R. A. Bakar, A. Faiz, M. Zohaimi, U. T. Mara, and S. H. Herman, "Annealing Temperature Dependence of Resistive Switching Behavior for Sol-gel Spin Coated Zinc Oxide Thin Films," *ICSE 2014 IEEE Int. Conf. Semicond. Electron.*, pp. 0–3, 2014.

[11] E. Gale, D. Pearson, S. Kitson, A. Adamatzky, and B. D. L. Costello, "The effect of changing electrode metal on solution-processed flexible titanium dioxide memristors," *Mater. Chem. Phys.*, vol. 162, p. 29, 2011.

[12] Q. Mao, Z. Ji, and J. Xi, "Realization of forming-free ZnO-based resistive switching memory by controlling film thickness," *J. Phys. D. Appl. Phys.*, vol. 43, no. 39, p. 395104, Oct. 2010.

[13] W. Wang, R. Dong, X. Yan, B. Yang, and X. An, "Memristive Behavior of ZnO/Au Film Investigated by a TiN CAFM Tip and Its Model Based on the Experiments," *IEEE Trans. Nanotechnol.*, vol. 11, no. 6, pp. 1135–1139, Nov. 2012.

[14] N. S. Kamarozaman, M. a. R. M. Rashid, M. Z. Musa, S. H. Herman, R. a Bakar, W. F. H. Abdullah, and M. Rusop, "Effect of film thickness on the memristive behavior of spin coated titanium dioxide thin films," *RSM 2013 IEEE Reg. Symp. Micro Nanoelectron.*, pp. 155–158, Sep. 2013.

Discriminant Analysis with Visible Lighting Properties for White Root Disease Infected Rubber Tree

Hadzli Hashim[1], Ahmad Faiz Mohd Sampian[1], Noor Aishah Khairuzzaman[1], Noor Ezan Abdullah[1], Mohd Suhaimi Sulaiman[2], Mohd Nasaruddin Mohd Aris[3], Nurmi Rohayu Abd Hamid[3]
[1]Faculty of Electrical Engineering, Universiti Teknologi MARA, 40450 Shah Alam, Selangor, Malaysia.
[2]Faculty of Electrical Engineering, Universiti Teknologi MARA, 23000 Dungun, Terengganu, Malaysia.
[3]Rubber Research Institute Malaysia (RRIM), Malaysian Rubber Board, 47000 Sungai Buloh, Selangor Malaysia.

Emails:hadzli66@salam.uitm.edu.my

ABSTRACT
The most serious disease known in rubber industry is root disease and among the major root diseases, white root is the most destructive agent of trees and agricultural crops. Since, it is too difficult and expensive to treat root disease infection on trees; prevention is important whereby one must rely on symptoms appearing on roots themselves in order to recognize the disease. Infection symptom on the tree is detected when leaves became discolored yellowish and dying. The main objective of this work is to investigate empirically an infected rubber trees infected by white root disease where its symptom could be detected visually from leaves gradual discolouration. Visible spectrum of optical measurements is taken on four different regions of interest (ROI) locations of the top side leaf sample features such as petiolule, main vein/midrib, vein and leaf cell of rubber trees. Statistical techniques is used to analyse for conclusive scientific findings of which ROIs above has shown clear discrimination between the healthy, medium and worst condition. The scope of work involves raw data inclusion of leaf samples belong only to 2025 rubber tree clone. This clone is recommended by Rubber Research Institute of Malaysia (RRIM) management due to its popularity and commercially used by small scale planters. Outcomes of this work has suggested that only main vein/midrib and leaf cell ROIs produced convincing significant discrimination between healthy, medium and worst case. Thus, their measurements can be recommended for developing on the shelf technology engineering sensor instrument using non-invasive advanced signal processing techniques and intelligent system for early detection of white root disease.

Keywords: rubber tree, white root disease

1. Introduction

Rubber is one of the most important products that have improved Malaysian economy for the past decades [1]. Every year, large amount of latex from rubber trees are being produced and due to the increasing of the rubber products in the market, it is a necessity to maintain the quality and quantity of this latex. Therefore, health of the rubber trees must be taken seriously so that they are free from any diseases. Not only that, issues pertaining yield should be a significant topic in the upstream production. This would lead to proper management on crop improvement and protection in producing new rubber clones that not only would improve latex and timber yield, but also have vigorous growth, good tree morphology, and resistant to wind, diseases and brown-bast [2]. It is known in this industry that diseases of rubber are divided into four categories; root, panel, stem and branch and leaf diseases [3]. Root disease is the most serious of all diseases in rubber plantation because the fungi kill the tree slowly [4].

Not only that, this disease could spread from tree to tree by root contact. Among the major root diseases, white root is a well-known and most destructive agent of trees and agricultural crops especially hevea brasiliensis. It was also reported to aggressively kill several other agriculture crops and fruit trees [4]. It is too difficult and expensive to treat trees that have been infected by root disease. So, prevention is important compared to treatment because of the constraint of money and time [5]. One must rely on symptoms appearing on the roots themselves in order to recognize the disease. Other symptom on the tree is when leaves go through discolouration and after they turn yellow, the tree will die .Conventionally, early detection of diseases is carried out via visual inspection by an expert person regularly where sometimes decision is made after matching the above infected features with the closest appearance photos from a library text. These evaluation processes however, are time consuming, have low percentage accuracy, as well as costly [4]. From the published work described in these literatures [6-8] digital imaging techniques and optical measurements could be employed to examine the leaves of infected rubber trees.

Normally, infected rubber trees would show symptoms of leaf discolouration and a previous study on empirical RGB (Red Green Blue) color extraction of these leaves through captured digital images has shown that various rubber tree diseases can be discriminated from one another [9]. Further analysis was also done on reflectance light properties using spectrometer [7]. The statistical results obtained show that there is strong evidence that these diseases can be discriminated. However, detection on color variegation and variation of the leaves at this stage is considered to be quite late for any remedial action [5]. Therefore, it is a novel idea to develop on the shelf technology engineering sensor using advanced signal processing techniques and intelligent system for non-invasive and early detection of white root disease. Such development needs fundamental studies for significant and appropriate feature extractions from the pathological perspectives of leaf discolouration [2].

2. Methodology

2.1. Introduction

This section discusses the data collection processes, the instruments used and the measurement procedures in this research. This is followed by the descriptions of the experimental set-up for quantification of the data images and elaboration of the proposed statistical methods for discriminations. The spectrum responses after the respective ROIs' were being measured with respect to healthy, medium and worst case would also be shown. Before the analysis of optical measurements can be done, it is important to discuss the methodology which basically involves the data collection steps, type of instrument used and measuring procedures. White root disease is selected as this infection is generally common for all rubber trees regardless of which clone they came from. The next step would be the collection of rubber tree leaf samples which at a later stage were divided into three groups, categorized as healthy, medium and worst case with respect to infected white root disease. These leaves were then being measured using optical instrumentation for their reflectance quantifications. Prior to that, this instrument was calibrated in order to give a controlled specification about the surrounding condition so that any parameters being measured inevitably would produce good and reliable measurements. Finally, these measured data were numerically analysed using statistical tool to highlight significant findings and conclusion.

2.2 Instruments

MCS 600 series is the latest MicroSpectrosCopy innovated with the finest photodiode array technology introduced by Carl Zeiss [11]. With the process-tested spectrometers, it performs even the most challenging measurements in UV (Ultraviolet), VIS (Visible), NIR (Near-Infrared) spectral regions quickly, efficiently and precisely. The cost-effective MCS 600 systems as shown in Figure 1, featuring the time-tested, modular principle of the MCS 500 series with a number of major enhancements.

Highlights of the new design include an increased signal-to-noise ratio, improved wavelength accuracy and an intelligent housing system, allowing different UV/VIS/NIR spectrometers to be directly combined for the very first time – covering the spectral range from 190 to 2150 nm. In addition, the MCS 600 sets a new standard for data processing through robust communications with peripheral systems and databases employing protocols [11]. For space, application or energy reasons, some specific applications require reflectance measuring heads with an integration sphere, the maximum diameter of which measures 30 mm. The model OFK 30 measuring head shown in Figure 2, has therefore been added to the MCS 600 accessory program [23]. It has the following specifications: Meas. geometry: d/8°; Sphere diam.: 30 mm; Range: 220 – 2200 nm and Illumination: ext. light guide, d=4 mm.

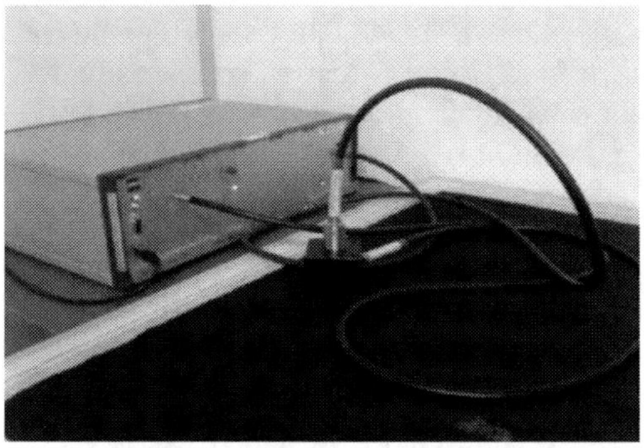

Figure 1. MCS 600 spectrometer

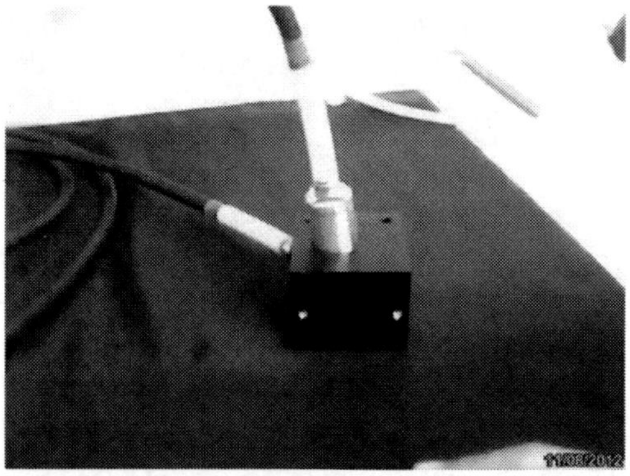

Figure 2. OFK 30 measuring head

For the purpose of this research, some modification was being made to the OFK head where a small hard cardboard of the same surface area size at its bottom was pasted to it. In the cardboard, a small rectangular window of 1cm in length and 3mm in width was being cut right at the position of the lens window of the head as depicted in Figure 3. The background color of the cardboard is black so that it could absorb any reflected white light during the measurement process. In this case, the transmitted and reflected light energy generated from the MCS 600 would only travel through the available rectangular window. The small window size is enough and suitable to be placed at the respective region of interest (ROI) locations of each leaf sample.

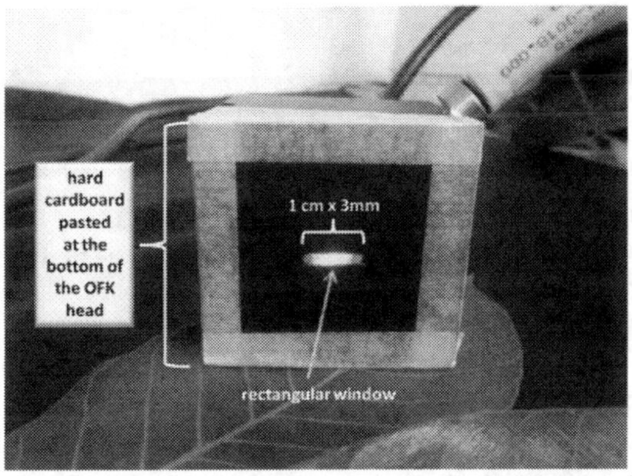

Figure 3. Hard cardboard (1cm x 3mm) rectangular window attached to OFK head

2.3. Software: Aspect Plus

Aspect Plus is a flexible, modular spectral analysis software package that runs under MS Windows and is the universal spectroscopy program from Carl Zeiss. Aspect Plus guarantees extensive functions as pictured in Figure 4, combined with outstanding ease of use. The File menu contains operations that are used for managing files that have been created using Aspect Plus or other compatible programs. Aspect Plus supports the import/ export of GRAMS (spc) ASCII (csv) and JCAMP format files. The Edit menu contains operations that are useful for inter-program transfer of data. A Copy command allows a spectrum to be easily inserted into Windows-based programs such as word processors. The View menu contains operations that can be used to alter the visual display of spectra and the set-up of the screen.

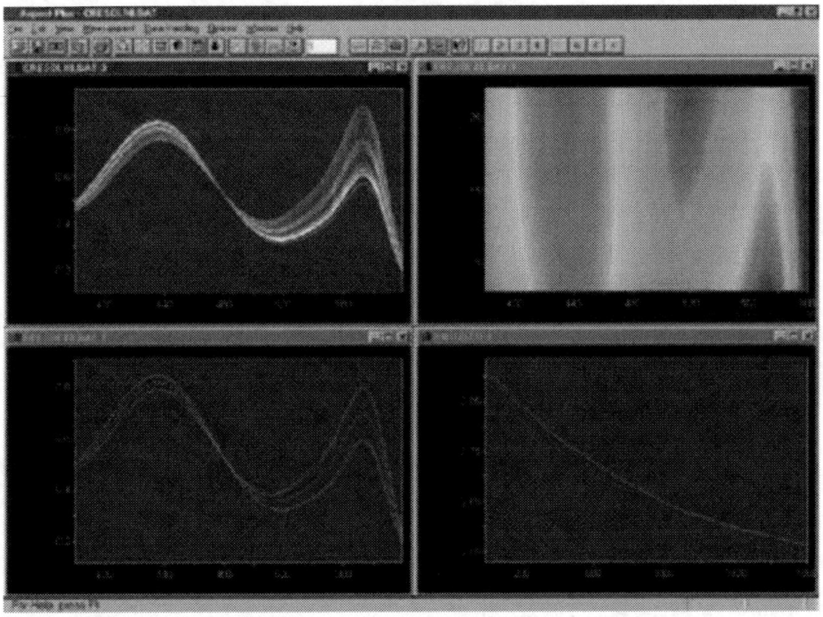

Figure 4. Aspect Plus display features

2.4. Calibration

Calibration is a mandatory procedure for any measuring equipment particularly spectrometer [15].. Before proceeding with any new measurements, calibration of the OFK head must be done depending on the spectrum usage during measurement i.e. whether for the VIS and NIR spectrum. Description about calibration in this section is on VIS spectrum where dark reference has to be measured firstly by putting the OFK head on a dark or black material as demonstrated in Figure 5. After that, the measure button was activated producing a reference spectrum representing dark color. As expected, all recorded percentage reflectance for the VIS spectrum are confined to 0% as depicted in the figure. The result indicates that all light components were not being reflected from the material. These data were then saved in the system database.

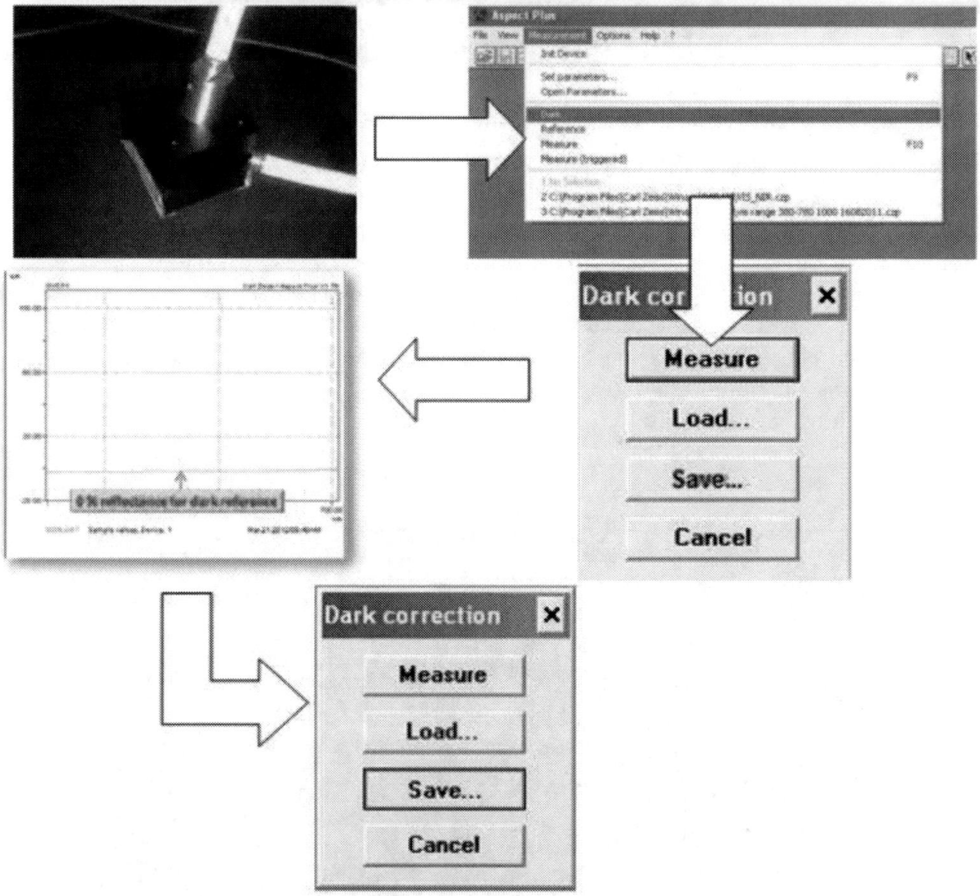

Figure 5. OFK head calibration with dark reference

The whole procedure was then being repeated for measuring white reference by replacing the black material with white material as shown also in Figure 6. However this time as observed in the figure, all recorded percentage reflectance are confined to 100%, suggesting that all light components were reflected from the material. Again, these data were also saved in the system database. The final stage of calibration process was on testing the OFK head onwards a green based color material as shown in Figure 7. It is observed that the peak recorded percentage reflectance is between 510-550nm, which is actually the range for green component. Finally, these measured data were also being saved.

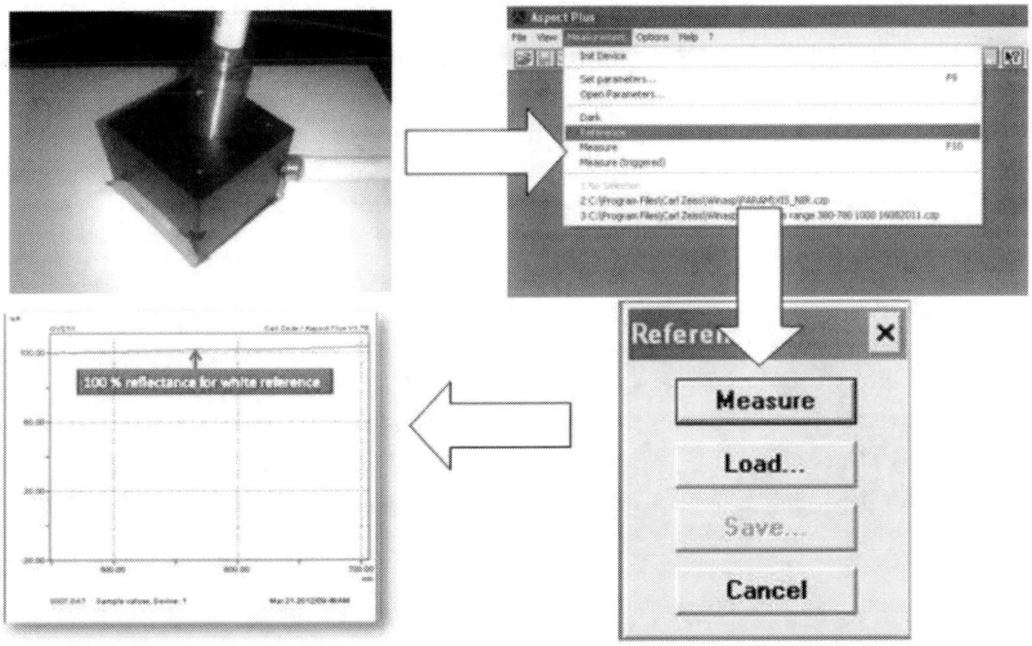

Figure 6. OFK head calibration white reference

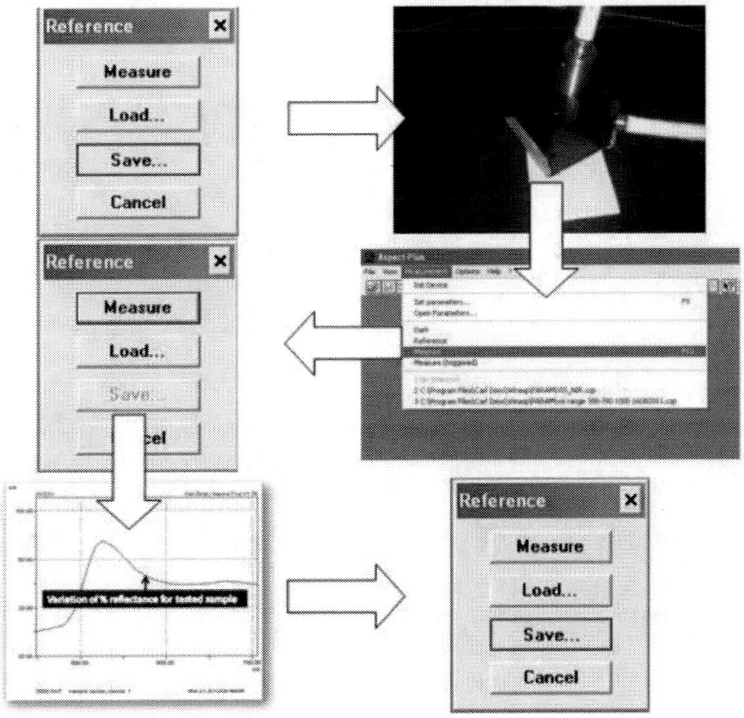

Figure 7. OFK head calibration green reference

2.5. Data Collection

Samples of rubber tree leaf were collected from nursery fields at the Research Station, Rubber Research Institute of Malaysia (RRIM), Sungai Buloh, Plucking of the leaves was done by senior workers from the Crop Improvement and Protection Unit (CIPU) as soon as the infected trees with white root disease were first identified by their research officers. There are two conditions of infected trees; medium and worst conditions which are reflected by the degree of leaf discolouration while the other condition is the healthy ones. Figure. 8 shows clearly examples of leaf green discolouration or degrading for the healthy, medium and worst conditions respectively.

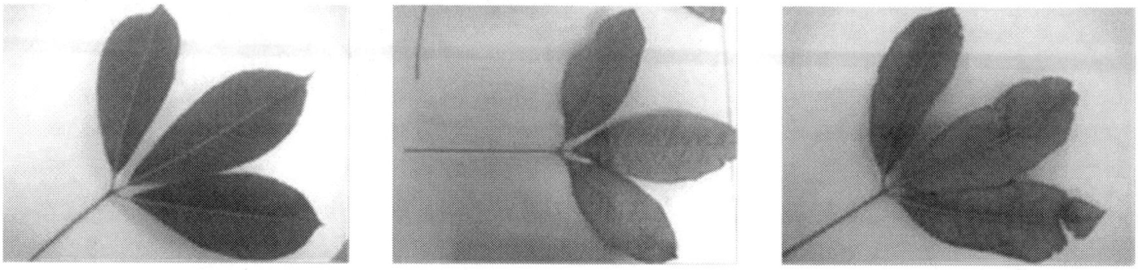

(a) Healthy (b) Medium (c)Worst

Figure 8. Leaf samples of rubber tree in (b) and (c) are infected by white root disease

2.6. Measuring Procedure

Optical measurements were taken on four different regions of interest (ROI) locations of the top side leaf sample. The four locations are the petiolule, midrib or main vein/midrib, vein and the leaf cell as shown in Figure. 9(a)-(d).

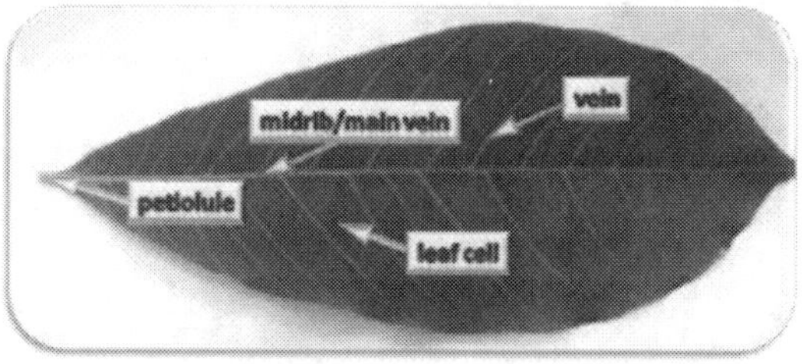

(a) Optical measurement locations on leaf

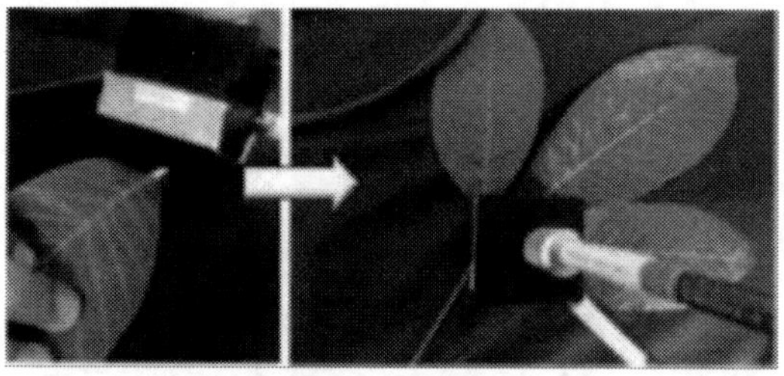

(b) OFK head positioning on petiolule

(c) OFK head positioning on main vein/midrib

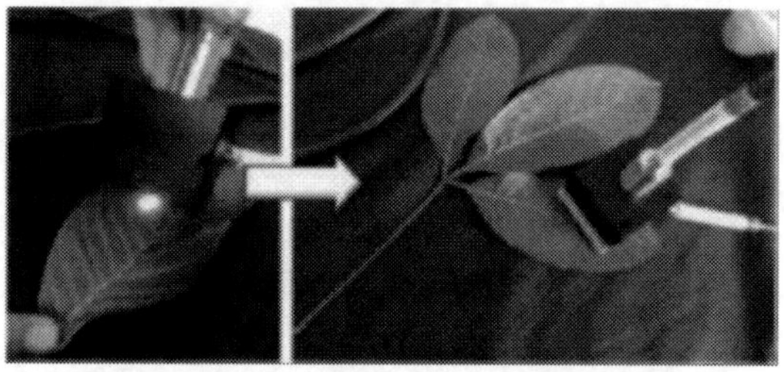

(d) OFK head positioning on vein/cell

Figure 9. MCS600 Spectrometer & OFK head measurements on selected ROI locations

The top side was preferred compared to the underside of the leaf in order to be consistent with other research works on investigating leaf color spectrum [7,9]. A total of 70 leaf samples of 2025 rubber tree clone were measured for each respective ROI's VIS spectrum reflectance reading that would represent three cases; healthy, medium and worst case. The numbers are enough for producing frequency plot that follow at least a normal distribution [12]. Healthy tree is the one not being infected by any diseases while medium case is where the tree is being infected by white root disease but categorized as early stage of infection. Finally, the worst case belongs to trees where their roots are badly infected by the white root disease and need to be treated urgently [10].

3. Results

3.1. Normality Test

The outcomes of the normality tests applied to all the data sets, are then being recorded and analysed. Measurement outcomes by the Kolmogorov-Smirnov (K-S) tests have produced significant p-values in the range from a minimum of 0.061 to a maximum of 2.00. Since these values are greater than 0.05, there is evidence that each ROI data set follow the null hypothesis of normal distribution. The summarized K-S test result for all data set is tabulated in Table 1. From the table, only healthy ROI group belongs to vein and leaf cell shows very minimal p-value of slightly equal to 0.05. Thus, indicating that these groups show weak evidence that their data are normally distributed. Such interpretations are consistent with the visual analysis made upon the graphical plots in the previous section. Overall, all of these measurements conclusively suggest that the 70 data sets can be used in the next experimental stage of inference between population either through parametric graphical methods (error bar plot) or independent t -test.

Table 1. Summarized K-S normality test results for ROI of leaf samples

ROI	*Statistic*	*Kolmogorov-Smirnov*	
		df (degree of freedom)	Sig. (2-tailed) (p-value)
petiolule_healthy	*.068*	*70*	*.200*
petiolule_medium	*.098*	*70*	*.092*
petiolule_worst	*.085*	*70*	*.200*
main vein/midrib_healthy	*.070*	*70*	*.200*
main vein/midrib_medium	*.069*	*70*	*.200*
main vein/midrib_worst	*.074*	*70*	*.200*
vein_healthy	*.128*	*70*	*.065*
vein_medium	*.069*	*70*	*.200*
vein_worst	*.050*	*70*	*.200*
leafcell_healthy	*.103*	*70*	*.061*
leafcell_medium	*.061*	*70*	*.200*
leafcell_worst	*.095*	*70*	*.197*

3.2. Error Bar Plots Analysis

The error bar or interval estimator needs to be analysed after the previous normality test has being concluded. This plot provides the possible location of the group population mean (i.e. the μ) and is used whenever graphical information about the range of the predicted population mean is required from the available known samples observation of a single group [12]. In this work, error bar of all the data sets were plotted as shown in Figure 3. There are 3 types of symbols for each ROI with respect to healthy, medium and worst condition in the y-axis, representing the range of group's population estimated of mean value. Also, each symbol has its lower confidence limit (LCL) and upper confidence limit (UCL) where the range between these limits indicates the predicted location of the

IOP Conf. Series: Materials Science and Engineering **99** (2015) 012023 doi:10.1088/1757-899X/99/1/012023

population mean, μ [13]. Measurements of the error bar plots discussed above are summarized in Table 2.The rightmost column is the range for the 95% CI between UCL and LCL while the mean values demonstrate the locations of the ROI groups as illustrated in Figure 10. From this table and with respect to the 95% Confident Interval (CI) ranges, only the main vein/midrib and leaf cell group has shown distinct for the three conditions *i.e.* there is difference in the mean values between healthy, medium and worst. Thus, their specifications printed in Table 2 can be used as reference for development of sensor instrumentation in the future.

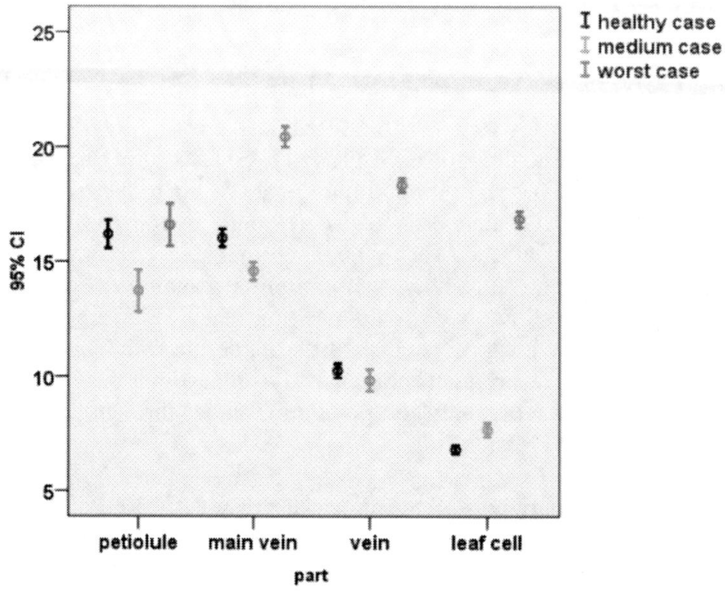

Figure 10. Respective locations of error bars for each ROI group

Table 2. Measurements of the respective range of error bar plot

ROI	Descriptive		
	Mean ± Std. Error	95% Confidence Interval	
		LCL	UCL
petiolule_healthy	16.19±0.30	15.58	16.80
petiolule_medium	13.73±0.46	12.82	14.64
petiolule_worst	16.80±0.46	15.88	17.72
mainvein_healthy	16.02±0.19	15.64	16.41
mainvein_medium	14.59±0.19	14.20	14.97
mainvein_worst	20.43±0.22	19.99	20.88
vein_healthy	10.24±0.15	9.93	10.54
vein_medium	9.82±0.24	9.35	10.30
vein_worst	18.34±0.15	18.04	18.64
leafcell_healthy	6.82±0.09	6.64	7.00
leafcell_medium	7.68±0.15	7.37	7.98
leafcell_worst	16.85±0.17	16.51	17.20

3.3. Independent *t*-Test

Independent t -test is suitable for comparing means between observed samples group (conditions set in this work) extracted from two populations. The findings are based on the *p*-value. As elaborated in [12,14], *p*-value of less than 0.5 indicates there is an overwhelming evidence to infer that the two groups are significantly or can be discriminated from each other. In this experiment, all data sets were tested for equality of means with respect to each ROI group conditions. The preferred *t*-test outcomes are summarized and tabulated in Table 3. It can be concluded from the table that petiolule and vein groups are not recommended for optical measurement in pre-detecting of white root disease towards rubber tree clone 2025 based on leaf discolouration. The tests have proven that there is overwhelming evidence that any measured or estimated measurements for at least two conditions (in which, healthy-worst for petiolule (*p*-value=0.495) and healthy-medium for vein (*p*-value=0.115) would occupy the same range or location. Alternatively, the tests applied for main vein/midrib and leaf cell group has produced *p*-values less than 0.05, implying that these groups are overwhelmingly significant different between healthy, medium and worst case. Thus, *t*-test has numerically reinforced the findings from the error-bar plot investigation.

Table 3. Independent *t*-test for various conditions with respect to each ROI group

ROI	(I) category	(J) category	Mean Difference (I-J)	Std. Error	Sig. (2-tailed) (p-value)	95% Confidence Interval	
						LCL	UCL
petiolule	healthy case	medium case	2.46*	0.59	0.000	1.31	3.62
		worst case	-0.40	0.59	0.495	-1.56	0.75
	medium case	healthy case	-2.46*	0.59	0.000	-3.62	-1.31
		worst case	-2.86*	0.59	0.000	-4.02	-1.71
	worst case	healthy case	0.40	0.59	0.495	-0.75	1.56
		medium case	2.86*	0.59	0.000	1.71	4.02
main vein/midrib	healthy case	medium case	1.44*	0.29	0.000	0.87	2.01
		worst case	-4.41*	0.29	0.000	-4.98	-3.84
	medium case	healthy case	-1.44*	0.29	0.000	-2.01	-0.87
		worst case	-5.85*	0.29	0.000	-6.42	-5.28
	worst case	healthy case	4.41*	0.29	0.000	3.84	4.98
		medium case	5.85*	0.29	0.000	5.28	6.42
vein	healthy case	medium case	0.41	0.26	0.115	-0.10	0.93
		worst case	-8.10*	0.26	0.000	-8.61	-7.59
	medium case	healthy case	-0.41	0.26	0.115	-0.93	0.10
		worst case	-8.51*	0.26	0.000	-9.03	-8.00
	worst case	healthy case	8.10*	0.26	0.000	7.59	8.61
		medium case	8.51*	0.26	0.000	8.00	9.03
leaf cell	healthy case	medium case	-.86*	0.20	0.000	-1.25	-0.46
		worst case	-10.04*	0.20	0.000	-10.43	-9.64
	medium case	healthy case	.86*	0.20	0.000	0.46	1.25
		worst case	-9.18*	0.20	0.000	-9.58	-8.78
	worst case	healthy case	10.04*	0.20	0.000	9.64	10.43
		medium case	9.18*	0.20	0.000	8.78	9.58

*. The mean difference is significant at the 0.05 level.

4. Conclusion

In this research, the instruments used were reliable and the measuring procedures were set with professional guidance from the expert domain and also from previous research work guidelines [7]. The optical measurements of the identified ROIs of rubber tree leaves were done by using the measuring head OFK 30 that was connected to MCS 600 instrument. A total of 640 measurements were collected in this work from the identified four ROIs. Each ROI has 210 reading samples with 70 each representing the healthy, medium and worst condition of the rubber tree with respect to white root disease infection. Initially, the measuring head scanned for 120 reflectance percentage samples that correspond to the visible spectrum (VIS) range from 367.88nm to 781.62nm during measurement of each ROI. However as far as this research is concerned, only reflectance reading that represent green wavelength (546nm) was considered and hence, each ROI with the respective condition has 70 recorded readings for numerical analysis.

Graphical methods such as the interval estimator or error-bar plots were then used for visual analysis of ROI groups' discrimination. The error bar plots displayed the 95% CI range of the possible location of population mean representing each ROI group with respect to types of conditions. It was observed that only main vein/midrib and leaf cells show clear discrimination amongst the healthy, medium and worst condition. Multiple comparison using independent *t*-test for equality of means from two populations, was implemented in the last stage of the statistical tests for more definitive conclusion. In the test, condition samples from each ROI were compared with other. The measured *p*-values equal to 0.000, thus implying that there is an overwhelming evidence only the main vein/midrib and leaf cell ROI group are able to give significant discrimination between healthy, medium or worst condition. The test results has reinforced the findings through observations of the error-bar plots where other than these two ROI groups, petiolule and vein group are not suitable to be referred for developing optical measurements in pre-detecting infection of white root disease towards rubber tree clone 2025 based on leaf discolouration.

5. Acknowledgement

This work would not be accomplished without the generosity of Malaysian Rubber Board (MRB) of permitting usage of facilities at the Crop Improvement and Protection Unit (CIPU), Rubber Research Institute of Malaysia (RRIM), Sg. Buloh, Selangor. Thus, the authors would like to thank the RRIM personnels particularly Ms. Azimah Izhar and Mr. Segar A/L Meenah@Raman for their contribution and support. Not to forget many thanks to Faculty of Electrical Engineering and Research Management Institute (RMI), Universiti Teknologi MARA Shah Alam in facilitating this project. Also, appreciation to Ministry of Education for funding our research work under Fundamental Research Grant Scheme (FRGS) with the code 600-RMI/FRGS 5/3 (83/2013). Last but not least, special appreciation should also be given to Industrial Based Instrumentation and Technology (I-BIT) Group members for their help in completing this project.

6. References

[1] M. R. Board. (2012, Official Portal, *Malaysian Rubber Board (Lembaga Getah Malaysia)*. Available: http://www.lgm.gov.my

[2] M. J. M. Hassan, *Economics of Rubber Forest Plantation*, Malaysian Rubber Board December 2005.

[3] B. S. Rao, *Maladies of Havea in Malaysia*, Rubber Research Institute of Malaysia, p. 62, October 1975.

[4] V. K. Rajalakshmy and K. Jayarathnam, *Root Diseases and Non-microbial Maladies*, Rubber Research Institute of India, Kottayam.2000.

[5] W. H. Chan, CP Wong and CC Wong, *Control of White Root Disease in immature rubber with 3 systemic fungicides,*. The Planter. Incorporated Society of Planters, Kuala Lumpur 1991.

[6] F. N. Osman, H. Hashim, S. A. M. Al Junid, A. Haron, N. E. Abdullah and M. F. Muhammad *A*

Statiscal Approach for Rubber Seed Clones Classification using Reflectance Index, in in the Proc. of the 4th Asia Int. Conf. on Mathematical Modelling & Computer Simulation (AMS 2010), Kota Kinabalu, Sabah, Malaysia, 2010.

[7] M. A. Haron, H. Hashim, F. N. Osman, S. A. M. Al Junid, *Classification of Rubber Tree Leaf Disease using Spectrometer,* in in the Proc. of the 4th Asia Int. Conf. on Mathematical Modelling & Computer Simulation (AMS 2010), Kota Kinabalu, Sabah, Malaysia, 2010.

[8] N. Omar, H. Hashim, F. N. Osman and A. E. Darami, *A Statistical Analysis of Led Reflectance for Various Rubber Seeds Clone*, in in Proc. of the 8th WSEAS Int. Conf. on Circuits, Systems, Electronics, Control & Signal Processing (CSECS'09), Tenerife, Canary Islands, Spain, 2009.

[9] N. E. Abdullah, A. A. Rahim, H. Hashim and M. M. Kamal, *Classification of rubber tree leaf diseases using multilayer perceptron neural network*, in in Proc. the IEEE 2007 5th Student Conference on Research and Development (SCOReD2007), Bangi, Selangor, Malaysia,2007.

[10] W. C. P. a. W. C. C. Chan W.H., *Control of White Root Disease in immature rubber with 3 systemic fungicides*, The Planter. Incorporated Society of Planters, Kuala Lumpur 1991.

[11] O. Carl Zeiss AG. *Optical Systems/Spectrometer*. Available:www.zeiss.com/lenses4industry

[12] G. Keller and B. Warrack, *Inference About the Comparisons of Two Populations*, in Statistics for Management and Economics. vol. 5th., ed California: Duxbury Thompson Learning 2000, pp. 394-470.

[13] M. H. Katz, *Multivariable Analysis: A Practical Guide for Clinicians*, United Kingdom: Cambridge University Press, 2001.

[14] D. C. Montgomery, *Design and Analysis of Experiments.* New York: John Wiley and Sons Inc., 1997.

[15] G. D. Christian and J. B. Callis, *Trace Analysis* 2nd ed.: John Wiley & Sons 2010.

IOP Conf. Series: Materials Science and Engineering **99** (2015) 012024 doi:10.1088/1757-899X/99/1/012024

Statistical Discrimination of Latex between Healthy and White Root Infected Rubber Tree based on Dry Rubber Content

Mohd Suhaimi Sulaiman[2], Hadzli Hashim[1], Ahmad Faiz Mohd Sampian[1], Nina Korlina Madzhi[1], Azrie Faris Mohd Azmi[1], Noor Aishah Khairuzzaman[1], Faridatul Aima Ismail[1]

[1]Faculty of Electrical Engineering, MARA University of Technology, 40450 Shah Alam, Selangor, Malaysia.

[2]Faculty of Electrical Engineering, MARA University of Technology, 23000 Dungun, Terengganu, Malaysia.

Emails:suhaimisulaiman2009@gmail.com,hadzli66@salam.uitm.edu.my,sayaahmadfaiz@gmail.com,

Abstract. Dry rubber content (DRC) is one of main material existing inside latex. It is usually in ranged of 25% – 45% of rubber latex. Statistical analysis are done to determine the discrimination of dry rubber content of latex between healthy and white root infected rubber tree. Based on 150 rubber trees and 10 clones tested, parametric test which include normality test, error-bar plot, and paired samples test are done. The result outcomes have shown that both data of dry rubber content of latex for healthy and white root infected rubber tree are normally distributed. Error-bar plot test is clearly indicated that there is visible discrimination between both cases. Paired samples test are done to reinforce this findings in terms of numerical *p*-value which is found to be less than 0.05. Thus, this indicate overwhelming evidence that healthy group can be discriminated from white root. Conclusively, changes in DRC content in latex can be correlated with white root disease infections of rubber tree.

1. Introduction

Heavea Brasiliensis or known as Rubber was originated from Brazil introduced to South East Asia in 1876 by Farris, Cross and Hendry Wickham [1]. Rubber product is one of the main contributors to Malaysia economy growth[2]. Latex and timber from rubber trees are used in many industries such as furniture, automotive, buildings, medical and etc. Latex is a milky white fluid consisting of proteins, alkaloids, starches, sugar, oils, tannins, resins, and gums. The gums or dry rubber content plays a very important roles in determining the quality and the price of latex. Higher dry rubber content will result in much higher price for the latex. . Malaysia supply for about 60% of the world consumption of disposable rubber gloves which mainly made from latex [3].

Usually, rubber latex is consist of 50-80% water, 25-45% dry rubber content and 2-5 % non-rubber particle[4-6]. The dry rubber content of latex differs due to several factors such as its morphology and environment [6]. The latex price is higher if the dry rubber content is high as it is the only component needed in latex for rubber product manufacturing industries [5].

White root disease (WRD) is one of the main problems faced by rubber tree [1,8]. White root disease is said to be the most threatening among other rubber tree disease. The cause of this disease is from fungus named Rigidoporus lignosus which infect the surface of tree root in three stages which are penetration, colonization and degradation [7]. Throughout the infection phase, the colonization of the pathogen will affect the nutrient absorption of the rubber tree [9]. Since the white root disease infection affect the nutrient absorption of the tree, its effect on the latex should also be investigated. This paper objective was to compare the dry rubber content of healthy and white root infected rubber tree by using statistical analysis techniques.

2. Methodology

2.1. Determination of White Root Disease and Healthy Rubber Tree

The determination of white root disease were done with help of expertise from Integrated Pest and Disease Management Programme (IPDM) of Rubber Research Institute Malaysia (RRIM). Samples classes were determined by tree morphology and presence of white root disease fungus. Basically there are 5 types of infection for white root disease. It is evaluated based on fungus infection on how many segmentation (a,b,c,d) at root has occur. As example, type 3 infection is determine if any three out of four segmentation has been infected by white root fungus. The tree is said to be type 5 infected if it is unable to produce latex and dying. Figure 1 shows the segmentation of root area for infection type determination.

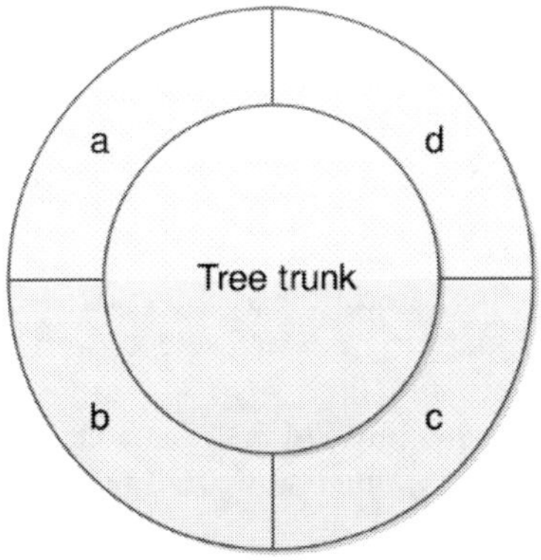

Figure 1. Tree root area (top view) segmentation for white root infection categories

2.2. Samples Collection

Experiment was conducted at Rubber Research Institute Malaysia (RRIM) Kota Tinggi on Jun 2014 and January 2015. The experiment were tested on 150 rubber trees from 10 different clones which are 2010, 2013, 2014, 2019, 2020, 2023, 2024, 2025, 2026, 3001. These clones were obtained from three main stations in RRIM Kota Tinggi named Pelepah, Permatang and Pemandi. The clones were chosen based on the availability of white root disease samples in their group. Predetermination of white root disease infected trees were done several days before sample collection by a trained RRIM officer based on the tree morphology and presence of white root disease fungus. In other words, if there was a white root disease from a specific clone, a healthy samples from the same clone was also obtained. Latex samples were collected directly from tree trunk at where the tree were usually tapped. The container used to store latex was ensured to be uncontaminated. Minimum amount of 60 ml of latex must be collected from each tree in order to do a dry rubber content test. Preservative was inserted into the samples to avoid coagulation and maintain its freshness.

2.3. Dry Rubber Content (DRC) Test

After the samples were obtained from the field, dry rubber content test were done by using RRIM Dry Rubber Content (DRC) Standard Test Method. This test were based on ISO 126: 1989 which was developed by RRIM in order to determine the DRC of the samples. This method involves process to dry the latex into dry rubber [10]. Latex with 25g in quantity is dried into dry sheet to measure DRC content. The DRC was calculated as below:

$$DRC = [(m1 - m0) \times 100]\% \quad (1)$$

Where

m0 = the mass, in grams, of the latex test portio
m1= the mass, in grams, of the dried sheet

2.3. Statistical Analysis
Statistical analysis were done in order to determine the relationship of the dry rubber content with the health condition of a tree. Comparison of DRC of latex from healthy rubber try and white root infected rubber tree was initiated using normality test in order to investigate the samples population follow normal distribution or not. The data is said to be normal if frequency of data is highest in middle and lowest towards extreme [11]. Since it follows normal distribution, then parametric test was applied using error bar plotting and followed by paired samples test for conclusive numerical findings.

3. Results
3.1 Cataloging
Tables 1- show cataloging of samples collection on June 2014 and January 2015 for this work. With respect to these figures, a total 71 suitable trees are selected on June 2014 and 79 on January 2015 to be used in conducting the DRC Test. For both session, samples taken from Pelepah has the highest quantity with 42.7 % followed by Permatang (36.7 %) and lastly, Pemandi (20.7%). Among 150 tress being sampled, there are 34 white root infected and 116 healthy rubber trees.

Table 1. Numbers of Samples Collected based on session.

Session		
	Frequency	(%)
14-Jun	71	47.3
15-Jan	79	52.7
Total	150	100

Table 2. Numbers of Samples Collected based on station

Station		
	Frequency	Percent
Pelepah	64	42.7
Pemandi	31	20.7
Permatang	55	36.7
Total	150	100.0

Table 3. Numbers of Samples Collected based on Clones.

Clone	Type		Total
	WRD	HEALTHY	
2010	2	3	5
2013	3	3	6
2014	3	26	29
2019	9	10	19

2020	4	5	9
2023	2	9	11
2024	2	9	11
2025	4	9	13
2026	2	5	7
3001	3	37	40
Total	34	116	150

3.2 Normality Test

From 150 trees that were obtained from sample collection works, 116 DRC samples for each cases were extracted. These obtained DRC values were firstly evaluated using normality test. Normality test using Kolmogorov-Smirnov algorithm was done to determine normally distribution for any sample population if the *p*-value is more than 0.05. As shown in Table 4, both healthy and white root disease samples has *p*-value of 0.2. Thus, this indicate the data follows normal distribution and with such, other parametric tests can be applied.

Table 4. Normality test of DRC value for each categories of samples.

Tests of Normality

	Kolmogorov-Smirnov		
	Statistic	df	Sig.
Healthy	.071	116	.200*
WRD	.046	116	.200*

3.3. Error Bar Plots Analysis

Figure 2 shows the result of applying error bar plot while Table 5 depicted the descriptive measurements of the test. By observation, the lower confidence limit of healthy samples (36.30) does not overlap with the upper confidence limit of white root disease (36.01). Therefore, implying both samples population means range differed from each other.

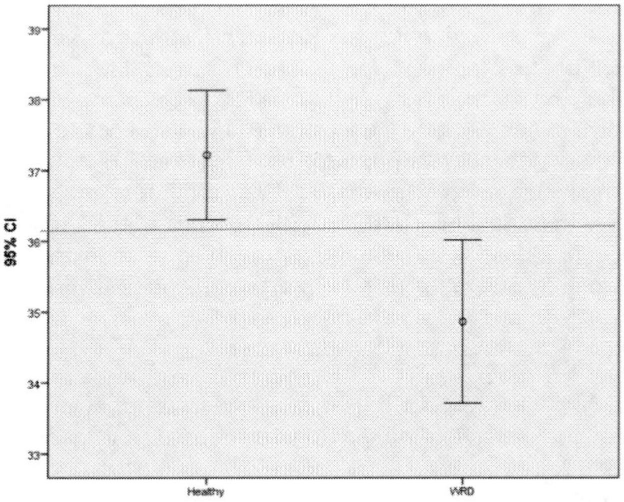

Figure 2. Error bar plot for Healthy and White Root Disease DRC Values.

Table 5. Measurements of Error Bar Plot for Healthy and White Root Infected Samples

	Mean ± Std. Error	UCL	LCL
Healthy	37.22 ± 0.46	38.13	36.30
WRD	34.87 ± 0.58	36.01	33.71

For numerical analysis, data for this work is then further tested using paired samples *t*-test and the outcomes are depicted in Table 5. With respect to table, the *p*-value for this test is less than 0.05 which indicates that there is an overwhelming evidence that both samples population can be discriminated from each other.

Table 6. Paired Samples Test for Healthy and White Root Diseases (WRD)

(I) category	(J) category	Mean Difference (I-J)	Std. Error	Sig. (2-tailed) (p-value)	95% Confidence Interval	
					LCL	UCL
Healthy	WRD	2.352	.68126	0.001	1.0033	3.7022

4. Conclusions

The investigation of discrimination between healthy and white root disease (WRD) infected rubber tree latex for 150 tress tested indicates that there are a good correlation between white root disease infection and DRC in latex. By statistical analysis, the result shown a very distinguishable value for both cases. There were no overlap in error bar between both cases, and *p*-value 0.001 of paired samples test, signal the discrimination between both cases. This indicates that white root disease infection has effect on rubber latex. In field investigated, there were many white root infected rubber trees. But, in order to get enough latex samples for DRC test, suitable trees were limited as only infected trees with mild infection would still able to produce enough latex. However, this work able to discriminate white root infected and healthy latex based on DRC.

5. Acknowledgement

The authors would like to acknowledge Rubber Research Institute of Malaysia Sungai Buloh Selangor and Kota Tinggi especially the Crop Improvement and Protection Unit staff at Kota Tinggi Station such as Mr. Amran Saari and Mr. Razak for their valuable information, suggestions, guidance and also for providing the data prior to the experiment of this research. Greatest gratitude also goes to UiTM Research Management Institute (RMI) and Faculty of Electrical Engineering, Universiti Teknologi MARA for facilitating this project work. Also, appreciation to Ministry of Education for funding our research work under Research Acculturation Grant Scheme (RAGS) with the code 600-RMI/DANA/5/3/RAGS (33/2013). Last but not least, special appreciation should also be given to Industrial Based Instrumentation and Technology (I-BIT) Group members for their help in completing this project.

6. Refferences

[1] N. W. Daud, S. J. Mokhatar, and C. F. Ishak, "Assessment of selected Hevea brasiliensis (RRIM 2000 Series) seeds for rootstocks production," *African J. Agric. Reseearch*, vol. 7, no. 21, pp. 3209–3216, Jun. 2012.

[2] Malaysian Rubber Board, "Official Portal," *Malaysia Rubber Board (Lembaga Getah Malaysia)*, 2012. .

[3] N. I. Strategy, "Rubber Products Innovation," *Malaysia, Agensi Inovasi*, 2011. [Online]. Available: http://www.innovation.my/pdf/innovating_formulation_of_nis/Rubber Products Innovation.pdf.
[4] D. C. Blackley, *Polymer Latices: Science and technology Volume 1: Fundamental principles*. Springer, 1997, p. 584.
[5] S. Julrat, M. Chongcheawchamnan, S. Member, T. Khaorapapong, O. Patarapiboolchai, M. Kririksh, and I. D. Robertson, "Single-Frequency-Based Dry Rubber Content Determination Technique for In-Field Measurement Application," vol. 12, no. 10, pp. 3019–3030, 2012.
[6] Z. Zhao, "A novel measurement system for dry rubber content in concentrated natural latex based on annular photoelectric sensor," vol. 5, no. March, pp. 251–260, 2010.
[7] V. I. Omorusi, "Effects of White Root Rot Disease on Hevea brasiliensis (Muell . Arg .) – Challenges and Control Approach," 2012.
[8] L. G. Malaysia, *Manual Teknologi Penanaman Getah*, 2nd ed. Lembaga Getah Malaysia, 2007, p. 53.
[9] K. E. Jayasuriya, "Factors affecting disease tolerance of rubber tree and research needs for developing disease-tolerant genotypes for the sustainability of rubber industry," 2004.
[10] I. S. 126-1989, "Natural Rubber Latex Concentration. Determination of Dry Rubber Content." .
[11] J. Pallant, *SPSS survival manual*. McGraw-Hill Education (UK), 2013.

PMMA microfiber coated with ZnO nanostructure for the measurement of relative humidity

H. A. Rahman, N.Irawati, T. N. R. Abdullah and S. W. Harun
Faculty of Electrical Engineering,

Universiti Teknologi MARA (UiTM),

40450 Shah Alam, Malaysia

Email: husna232@salam.uitm.edu.my

Abstract. PMMA microfibers coated with zinc oxide (ZnO) nanostructure are demonstrated for relative humidity monitoring. The fabrication of two different microfiber structures, namely the straight PMMA microfiber and the PMMA microfiber loop resonator (PMLR) were performed by using the direct drawing technique. The sensitivities of the PMMA microfibers were compared and analyzed. The results implied that the PMLR is more sensitive than the straight PMMA microfiber and the sensitivity of the sensors were further increased with the additional coating of ZnO nanostructure on the surface of the PMMA fiber.

Keywords — polymethyl methacrylate (PMMA), PMMA microfiber, relative humidity, zinc oxide nanostructure

1. Introduction

Over the last two decades, fiber optic sensors have been gaining popularity among research institutions and industries. The fiber optic sensor technology offers the ability to accomplish multiplexing capabilities, remote sensing, versatility, improved performance, high sensitivity and can easily operate under strong electromagnetic field conditions. They also offer multiple possibilities for combining a large number of different sensors (e.g., those used for detecting temperature, pH, humidity, high magnetic field, displacement, pressure and acceleration) into the same optical fiber, thus eliminating the need for multiple cables as required in traditional electronic sensing [1].

Recently, optical microfibers have been demonstrated to be useful in sensor applications including temperature, pH, concentration of molecules and humidity due to its interesting optical properties. An evanescent field is created around the decreased microfiber diameter of several micrometers, enabling 50% of the power guided through it to propagate outside the physical microfiber. This evanescent field is able to offer high light intensity due to strong confinement and is especially beneficial for sensing the surrounding that comes into contact with it. By exploiting these properties, low-cost, miniaturized and all-fiber based optical devices for various applications [2] could be developed. Furthermore, these optical microfibers have shown to be promising for the development of various kinds of micro- and nanophotonic components and devices [3-4]. Typical microfiber-based sensing structures, including the simple straight microfiber and the microfiber loop resonator have been widely used. The former can be directly fabricated by using the heat pulling method while the latter can be easily maintained by van der waals or electrostatic forces at the joint area [5]. Embedding the microfiber loop inside a polymer is another possible route to fabricate a robust micro-resonator for sensing applications [6-8].

Zinc oxide (ZnO) is an attractive, relevant and biosafe material and can be use as a sensing element. It has a wide band gap of 3.3eV at room temperature and may be integrated into a wide range of nanoscale devices such as optoelectronic devices, surface acoustic wave devices, piezoelectric devices, transparent conducting materials, solar cell and field emitters. In addition, ZnO nanostructures are one of the most promising materials for the fabrication of chemical and biosensors due to its exotic and versatile properties including biocompatibility, nontoxicity, chemical and photochemical stability, high specific surface area, optical transparency, electrochemical activities, high electron communicating features and so on [9-11]. Growing ZnO nanostructure on flat substrates has been extensively demonstrated in various platforms such as glass, silica and aluminium foil [12-14]. Recently, it has been reported that ZnO nanostructure enhances the interaction of fiber guided light for humidity sensor [15].

Here, a straight PMMA microfiber and a PMMA microfiber loop resonator (PMLR) were fabricated by direct drawing the solvated polymer. The smooth morphology of PMMA microfibers make them suitable for low loss optical wave guiding and sensing. Furthermore, PMMA optical fibers are low cost polymer fibers that are generally more physically robust and flexible than silica fibers [16]. In this paper, the straight PMMA microfiber and the PMLR were coated with ZnO nanostructure for detecting changes in relative humidity. The ZnO nanostructure coating on the microfibers induces changes to the optical properties in response to an external stimulus.

2. Methodology

Firstly, a silica tapered fiber was fabricated through a tapering process at high temperature known as flame brushing technique. This technique was able to fabricate the silica microfiber with good accuracy. Figure 1 shows a schematic illustration of the silica microfiber fabrication setup using oxy butane burner as a torch. Through this technique, the diameter of the heated glass can be reduced when the single mode fiber is stretched by two fiber holders. An amplified spontaneous emission (ASE) source is injected into one end of the single mode fiber while the other end is connected to an optical spectrum analyzer (OSA) to monitor the transmission spectrum of the silica microfiber during the fabrication process.

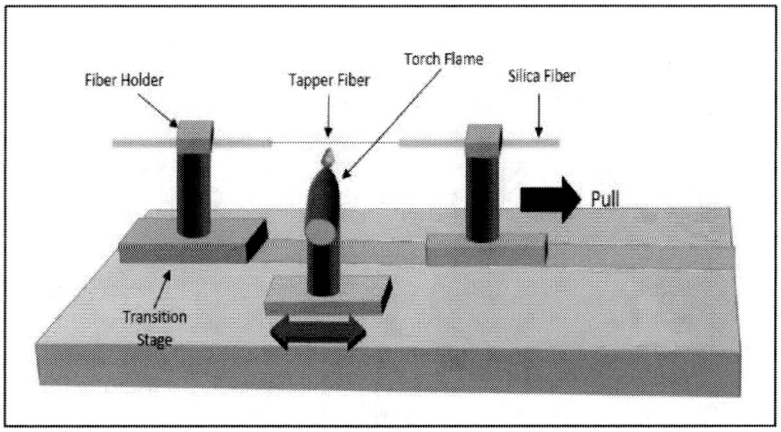

Figure 1. Schematic illustration of the silica microfiber fabrication setup.

Next, A PMMA microfiber was fabricated using direct drawing technique. The PMMA microfiber with a diameter of 6 microns was fabricated by using a one-step process as illustrated in Figure 2. PMMA was selected as the polymer wave-guiding material due to its high mechanical strength, good dimensional stability, good weather resistance, and its natural transparency above the deep ultraviolet region. A heating plate was used to melt PMMA and keep the temperature constant during the fiber drawing process. First, the tip of fabricated tapered silica fiber with a diameter of about 125 μm was immersed into the molten PMMA. Then the fiber tip was retracted from the molten polymer with a speed of 0.1−1 m/s, leaving a PMMA microfiber extending between the molten PMMA and the tip. The extended PMMA microfiber was quickly quenched in air and subsequently forming a bare PMMA microfiber. The microfiber diameter can be controlled by the pulling speed and viscosity of the polymer (which depends on the hot plate temperature). Microfibers produced by this technique are uniform in diameter over a long length with good flexibility.

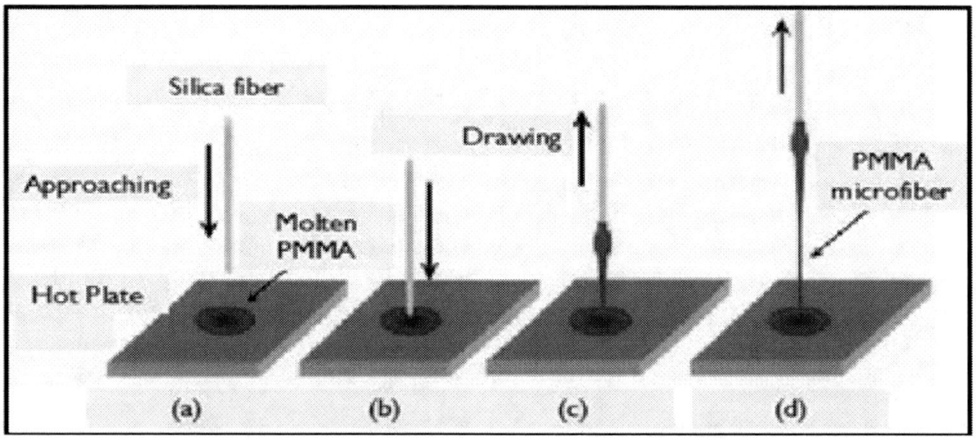

Figure 2. Schematic diagram of the PMMA microfiber fabrication setup.

As for the ZnO nanostructure, the seed layer was deposited onto the fiber to grow the ZnO nanostructures using a simple manual dip coating technique. The seeded solution was prepared by dissolving zinc acetate dehydrate (Zn(CH3COO)2•2H2O) as a precursor in isopropanol with a molarity of 0.025 M. The solution was stirred at 60° C for 2 hours at ambient temperature to yield a clear and homogenous solution. Then, the solution was cooled down to room temperature for the coating process. The fiber was manually dipped into the seeding solution and dried at 30°C to evaporate the solvent repeatedly for 5 times to increase the thickness of the fiber during the coating and drying process. Subsequently, the growing solution was prepared by dissolving 0.01M zinc nitrate hexahydrate (Zn(NO3)2.6H2O and 0.01M hexamethylenetetramine (HMTA) in 100ml deionized water. The deposition process of ZnO nanostructure on the fibers was performed using sol-gel immersion method by suspending the ZnO fibers in the growing solution at 30°C for 2 hours. The fiber was then characterized using Field Emission Scanning Electron Microscope (FESEM) to investigate the morphology of the ZnO nanostructure coated onto the PMMA microfiber (see Figure 3).

The setup used to measure relative humidity is shown in Figure 4. The amplified spontaneous emission (ASE) light from the Erbium-doped fiber amplifier (EDFA) was launched into the microfiber probe placed in a sealed chamber with a dish filled with saturated salt solution. The transmitted light was measured by the optical spectrum analyzer (OSA). In the experiment, the performance of the proposed sensor was investigated for various level of relative humidity. The output power was measured against the relative humidity ranging from 20% to 80% in steps of 5% using the omega RH-21 C temperature-relative humidity meter.

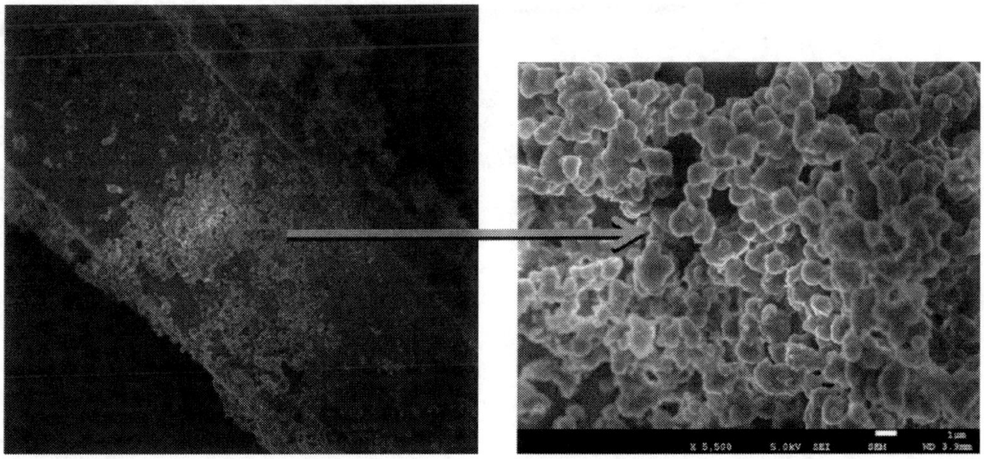

Figure 3. FESEM image of the ZnO nanostructure coating grown on the PMMA microfiber.

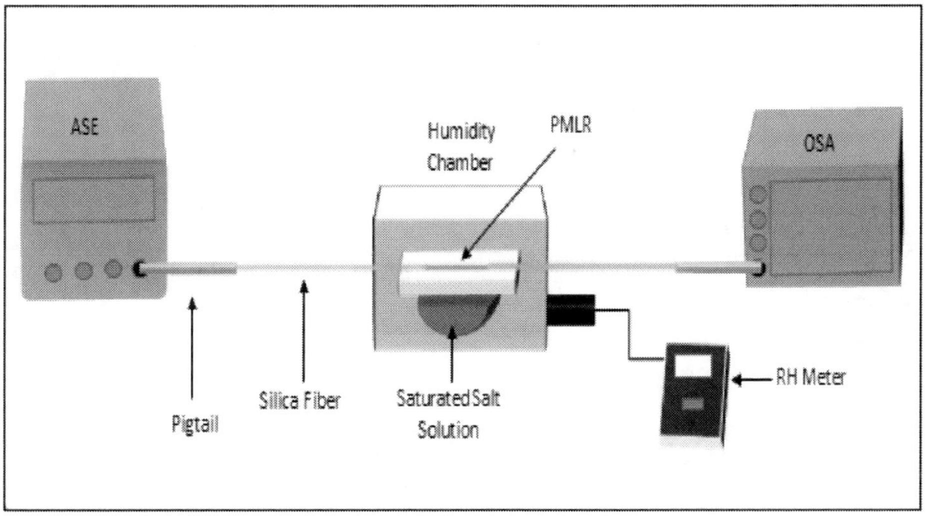

Figure 4. Experimental setup for the relative humidity sensor using the PMMA microfiber.

3. Results and discussion

Figure 5 shows the microscope image of the straight PMMA microfiber and the PMLR with a PMMA microfiber diameter and loop diameter of 6 μm and 56 μm, respectively. As shown in Figure 5, the input and output ends of the PMLR are aligned to each other. The alignment was made possible due to the surface attraction force of electrostatic and Van der Waals, which kept the ends of the PMMA to stick to each other. The van der Waals attraction force between the two adjacent PMMA microfiber is strong enough to withstand the elastic force from the bended microfiber and maintain the microfiber loop structure. Figure 6 shows the comb transmission spectrum of the PMLR, which was obtained by using an ASE source in conjunction with an OSA. The extinction ratio is about 1.0 dBm.

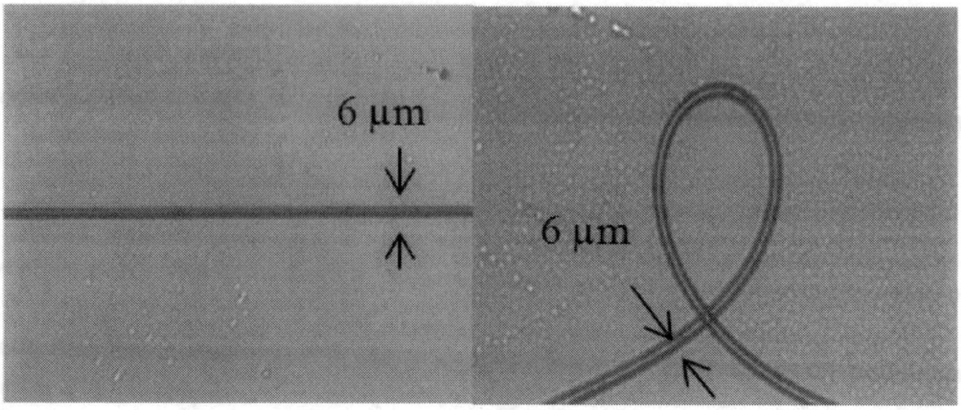

Figure 5. Optical microscope image of the straight PMMA microfiber (left) and the PMLR (right) from optical microscope image

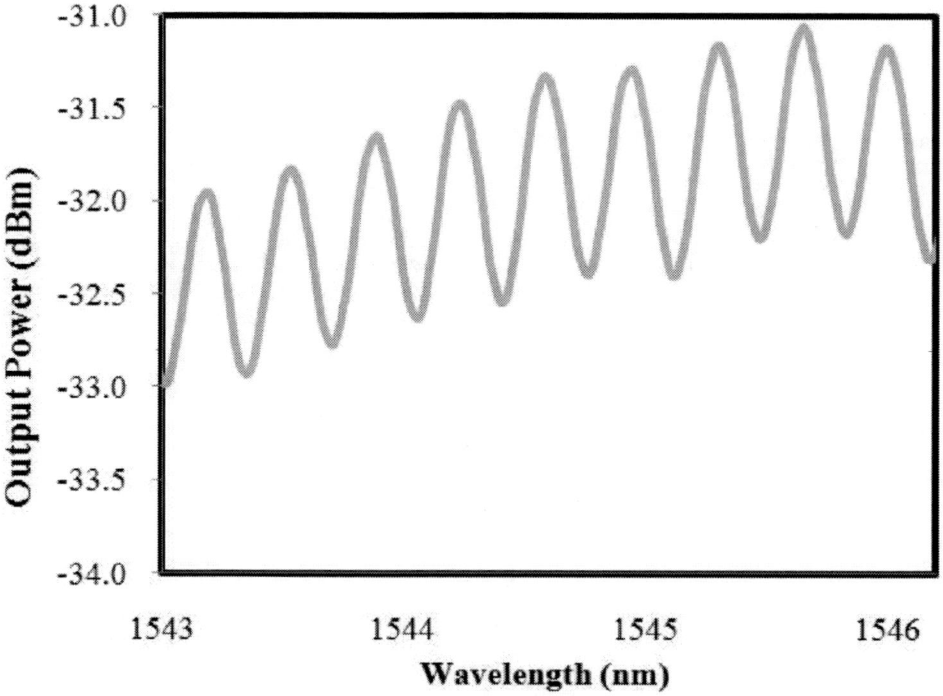

Figure 6. The ASE spectrum of the fabricated PMLR

During the experiment, the measurements were recorded throughout the relative humidity range of 20% to 80%. Figure 7 shows the output of the uncoated straight PMMA microfiber against relative humidity. The linearity, sensitivity and resolution of the sensor was obtained at 97%, 0.117 dBm/% and 8.41 %, respectively. The sensitivity refers to slope of the linear curve whereas the resolution was obtained from the ratio of the standard deviation to the sensitivity of the sensor (refer Table 1).

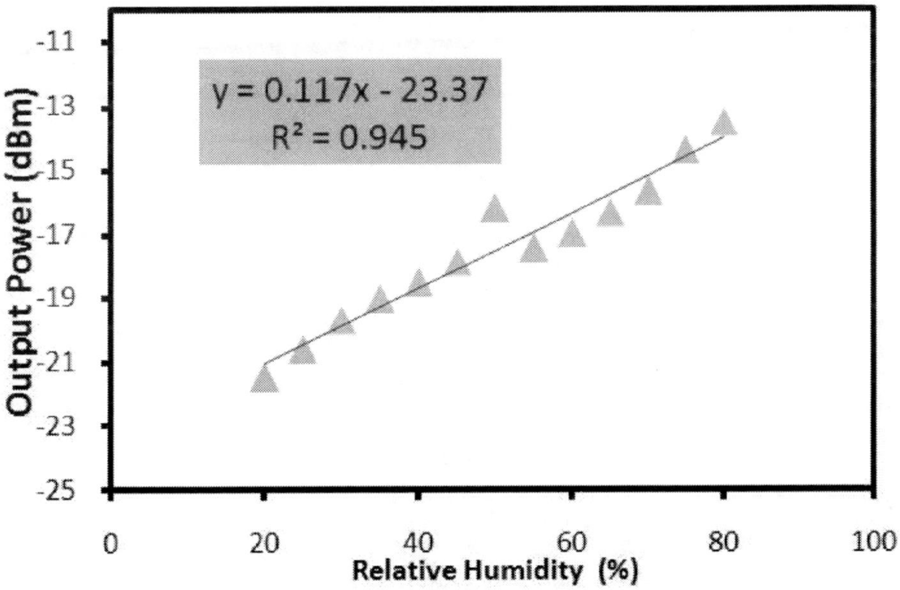

Figure 7. The output power of the ASE against relative humidity for the uncoated straight PMMA microfiber.

Table 1. The performance of the humidity sensor using the uncoated straight PMMA microfiber.

Parameter	**Value**
Sensitivity (dBm/%)	0.117
Linear Range (%)	20-80
Linearity (%)	More than 97%
Standard Deviation (dBm)	0.985
Resolution (%)	8.41

In Figure 8, the output power of the straight PMMA microfiber coated with the ZnO nanostructure is plotted against the relative humidity. It is observed that the output power increases with the increase in the relative humidity. The sensitivity of the slope is 0.1439 dBm/% for the output power versus relative humidity within the -14.55 dBm/% to -5.13 dBm/% range and a standard deviation of 0.945 dBm obtained from repeated measurements. The ZnO nanostructure coating has increased the sensitivity of the sensor 23%, hence improving its resolution by 22%. The ZnO composite that are exposed to humid environment cause a rapid surface absorption of water molecules. The increase in the water molecules cause an increase in both the effective refractive index of the surrounding medium and the absorbtion coefficient of the ZnO composite, leading to larger leakage of light [17].

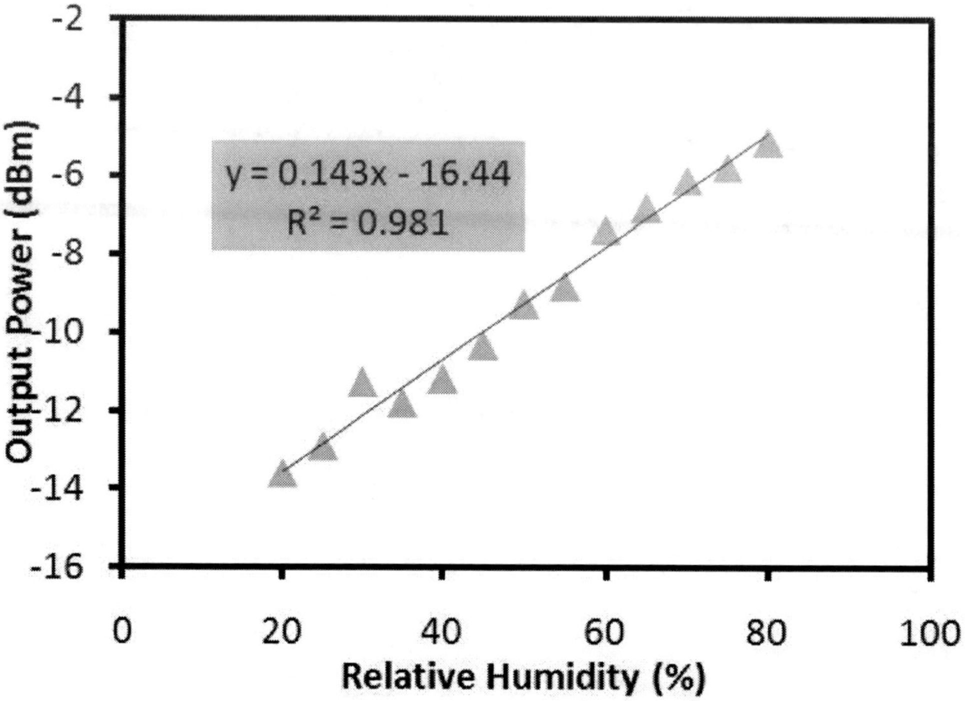

Figure 8. The output power of the ASE against relative humidity for the straight PMMA microfiber coated with ZnO nanostructure.

Table 2. The performance of the humidity sensor using the straight PMMA microfiber coated with ZnO nanostructure.

Parameter	**Value**
Sensitivity (dBm/%)	0.1439
Linear Range (%)	20-80
Linearity (%)	More than 99%
Standard Deviation (dBm)	0.945
Resolution (%)	6.57

Figure 9 depicts the output power obtained from the uncoated PMLR for various relative humidity levels. The performance of the relative humidity sensor is summarized in Table 3. Similar to the previous results, it is observed that an increase in the relative humidity level can be detected by an increase in the output power. The sensitivity and the resolution are 0.137 dBm/% and 7.88%, respectively.

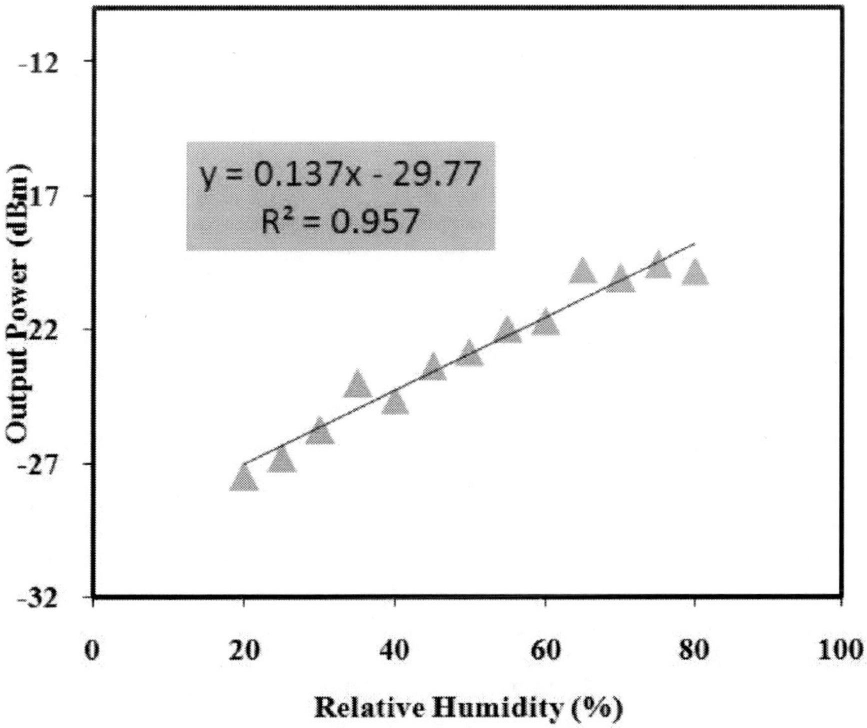

Figure 9. The output power of the ASE against relative humidity for the uncoated PMLR.

Table 3. The performance of the humidity sensor using the uncoated PMLR.

Parameter	Value
Sensitivity (dBm/%)	0.137
Linear Range (%)	20-80
Linearity (%)	More than 97%
Standard Deviation (dBm)	1.08
Resolution (%)	7.88

Figure 10 shows the output power obtained from the ZnO coated PMLR against the relative humidity. The performance of the relative humidity sensor is summarized in Table 4. The sensitivity and the resolution are 0.178 dBm/% and 6.045 %, respectively. It is observed that the coating of the ZnO nanostructure is able to increase the sensitivity of the sensor by 30%, hence improving the sensor's resolution by 23%. In general, the PMLR outperforms the straight PMMA microfiber in terms of its sensitivity and resolution. This is because the PMLR induces more portion of the evanescent wave outside of the microfiber, enabling more interaction with the relative humidity.

IOP Conf. Series: Materials Science and Engineering **99** (2015) 012025 doi:10.1088/1757-899X/99/1/012025

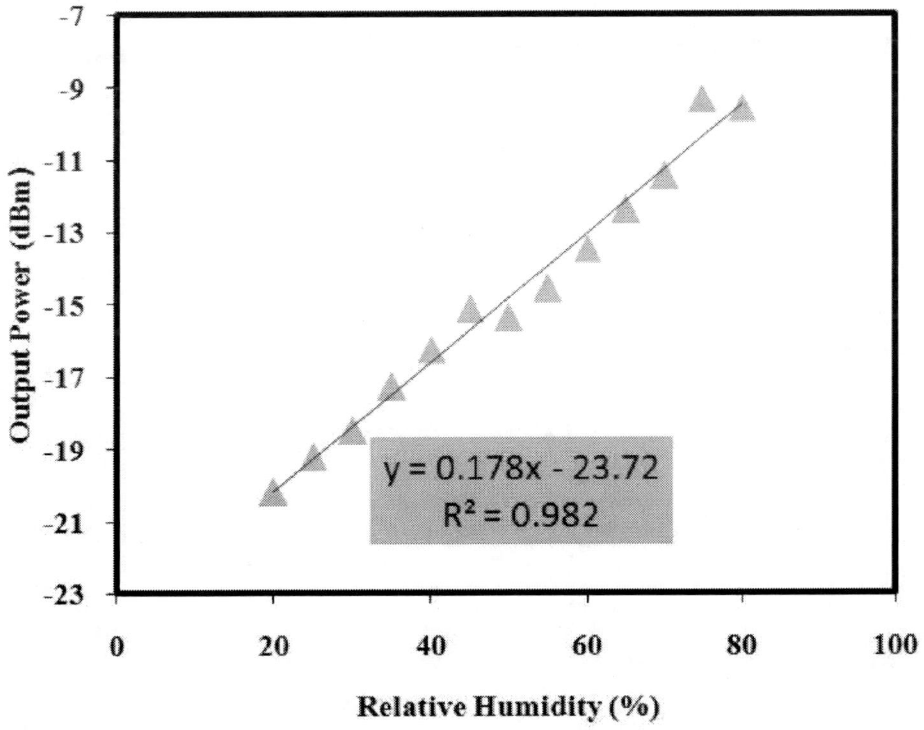

Figure 10. The output power of the ASE against relative humidity for the PMLR with ZnO nanostructure coating.

Table 4 The performance of the humidity sensor using the PMLR with ZnO nanostructure coating.

Parameter	Value
Sensitivity (dBm/%)	0.178
Linear Range (%)	20-80
Linearity (%)	More than 99%
Standard Deviation (dBm)	1.076
Resolution (%)	6.045

4. Conclusion

A simple humidity sensor based on the straight PMMA microfiber and the PMLR have been successfully demonstrated within the range of 20% to 80%. In comparison, when the PMLR is used instead of the straight PMMA microfiber, an improvement in the sensitivity and resolution of the sensor was observed. It is observed that coating of the ZnO nanostructure is able to further increase the sensitivity of the sensor hence improving the sensor's resolution. The coating of sensitive material onto the tapered fiber has successfully enhanced the performance of the microfiber as a humidity sensor.

References

[1] Yeo L, Sun T and Grattan K T V, 2008. Fibre-optic sensor technologies for humidity and moisture measurement. Sens. Actuators A: Phys. 144: 280–295.

[2] Bilodeau, F., K. O. Hill, S. Faucher, and D. C. Johnson, (1988), Low-loss highly overcoupled fused couplers: Fabrication and sensitivity to external pressure, Lightwave Technology, Journal of 6, 1476-1482.

[3] L. Tong, F. Zi, X. Guo, and J. Lou, "Optical microfibers and nanofibers: a tutorial," Optics Communications, vol. 285, pp.4641–4647, 2012.

[4] G. Brambilla, F. Xu, P.Horak et al., "Optical fiber nanowires and microwires: fabrication and applications," Advances in Optics and Photonics, vol. 1, no. 1, pp. 107–161, 2009.

[5] Lou, J., Wang, Y., & Tong, L. (2014). Microfiber optical sensors: A review.Sensors, 14(4), 5823-5844.

[6] Vienne, G., Li, Y., & Tong, L. (2007). Effect of host polymer on microfiber resonator. Photonics Technology Letters, IEEE, 19(18), 1386-1388.

[7] Xu, F., & Brambilla, G. (2007). Embedding optical microfiber coil resonators in Teflon. Optics letters, 32(15), 2164-2166.

[8] Xiao, L., Grogan, M. D., Leon-Saval, S. G., Williams, R., England, R., Wadsworth, W. J., & Birks, T. A. (2009). Tapered fibers embedded in silica aerogel. Optics letters, 34(18), 2724-2726.

[9] Q. Wan, Q. H. Li, Y. J. Chen, T. H. Wang, X. L. He, J. P. Li and C. L. Lin, Appl. Phys. Lett., 2004, 84, 3654.

[10] B. S. Kang, F. Y. Ren, W. Heo, L. C. Tien, D. P. Norton and S. J. Pearton, Appl. Phys. Lett., 2005, 86, 112105.

[11] F. Zhang, X. Wang, S. Ai, Z. Sun, Q. Wan, Z. Zhu, Y. Xian, L. Jin and K. Yamamoto, Anal. Chim. Acta, 2004, 519, 155.

[12] Wei, M., Zhi, D., & MacManus-Driscoll, J. L. (2005). Self-catalysed growth of zinc oxide nanowires. Nanotechnology, 16(8), 1364.

[13] Lee, J. H., Leu, C., Chung, Y. W., & Hon, M. H. (2006). Fabrication of ordered ZnO hierarchical structures controlled via surface charge in the electrophoretic deposition process. Nanotechnology, 17(17), 4445.

[14] Umar, A., Kim, B. K., Kim, J. J., & Hahn, Y. B. (2007). Optical and electrical properties of ZnO nanowires grown on aluminium foil by non-catalytic thermal evaporation. Nanotechnology, 18(17), 175606.

[15] Batumalay, M., Harun, S. W., Irawati, N., Ahmad, H., & Arof, H. (2014). A Study of Relative Humidity Fiber-Optic Sensors. IEEE sensor journal.

[16] Lye, P. G., Boerkamp, M., Ernest, A., & Lamb, D. W. (2005, January). Investigating the sensitivity of PMMA optical fibres for use as an evanescent field absorption sensor in aqueous solutions. In Journal of Physics: Conference Series (Vol. 15, No. 1, p. 262). IOP Publishing.

[17] Liu, Y., Zhang, Y., Lei, H., Song, J., Chen, H., & Li, B. (2012). Growth of well-arrayed mZnO nanorods on thinned silica fiber and application for humidity sensing. Optics express, 20(17), 19404-19411.

IOP Conf. Series: Materials Science and Engineering **99** (2015) 012026 doi:10.1088/1757-899X/99/1/012026

Fabrication of Poly (methyl methacrylate) and Poly(vinyl alcohol) Thin Film Capacitors on Flexible Substrates

Bindu Salim[1], Jaisree Meenaa Pria KNJ[2], M. Alagappan[2], A Kandaswamy[2]

[1] *Nanotech research facility, PSG Institute of Advanced Studies, Coimbatore.*
[2] *Dept of Biomedical Engineering, PSG College of Techology, Coimbatore.*

Abstract— Flexible electronics is becoming more popular with introduction of more and more organic conducting materials and processes for making thin films. The use of polymers as gate dielectric has over ruled the usage of conventional inorganic oxides in Organic Thin Film Transistors (OTFTs) on account of its solution process ability and ease of making highly insulating thin film. In this work Capacitance is fabricated with polymeric dielectrics namely poly (methyl methacrylate) - PMMA and poly (vinyl alcohol) – PVA. The electrodes used for these capacitors are Indium Tin Oxide (ITO) and Aluminium. Capacitance value of 9.5nF/cm^2 and 33.12nF/cm^2 is achieved for thickness of 510 nm of PMMA and 80 nm of PVA respectively. This study on capacitance can be used for assessing the suitability of these polymers as gate insulators in OTFTs.

1. Introduction

The choice of dielectric in the development of a thin film transistor is very crucial. The use of polymer as dielectric has the advantage that they can be made into a solution and spin coated or drop casted. Spin coating of polymers has been considered to be a simple and reliable method of forming thin films by several researchers. Although several polymers are used as gate dielectric in OTFTs, poly(methyl methacrylate) and poly(vinyl alcohol) have drawn great attention by the researchers. S. Mandel *et al* [1] have made a comparative study on the capacitance and dielectric constant of PVA, PMMA and poly(vinyl phenol)-PVP thin films and found that PMMA exhibited a higher capacitance of 13.3nF/cm^2 with dielectric constant of 4.5 at 100 KHz. PMMA has been extensively used as gate insulator in pentacene Thin Film Transistors(TFTs) as growth of pentacene over PMMA was highly favoured than on SiO_2 and that the grains of pentacene grew bigger, thereby increasing the field effect mobility [2]. Capacitances of PMMA layer of different thicknesses have been measured using an LCR meter on ITO/PMMA/Au parallel plate capacitor structure in the development of pentacene photo detectors [3]. Impedance Spectrometer studies have been carried out on varying concentrations of PMMA thin films sandwiched between ITO and aluminium [4] for a frequency range of 0 to 10 KHz. Dielectric composite made of PMMA with 40% $CaCu_3Ti_4O_{12}$ has been proved to have very low dielectric loss at 100 KHz, making it a promising choice for capacitor application [5]

In this paper, we present the morphological study of PMMA and PVA coated on glass and the fabrication of thin film capacitors with PMMA and PVA as dielectric over a flexible substrate. Capacitance values were measured using an LCRQ bridge. This study is useful for choosing dielectric material for development of flexible OTFT.

2. Material and Methods

The capacitor fabrication used spin coating for dielectric and physical vapour deposition for electrode deposition. In the process of fabricating PMMA and PVA capacitors on ITO coated PET, thickness measurements were done apriori on spin coated polymeric films on chemically cleaned glass substrates by RCA treatment. PMMA and PVA are dissolved in toluene and deionized water with 300mg PMMA in 10ml toluene and 0.5g PVA in 10ml deionized water respectively by magnetic stirring at room temperature. These solutions were spin coated on RCA cleaned, flat glass slide using the SPECTROSPIN spin coater-SC8000 at different spin rates for 60 sec. PMMA and PVA spin coated films were then vacuum dried at 100°C and 80°C respectively. The dried polymeric thin films on glass substrates were investigated using the Atomic Force Microscopy (AFM) for observing the morphology and Spectroscopic Ellipsometer for measuring the thickness, which have been detailed in section 4. ITO coated PET is chosen as the flexible substrate where ITO acts as the bottom electrode and aluminium is deposited on the dielectric to form the top electrode. The schematic of the capacitor structure is shown in Fig 1.

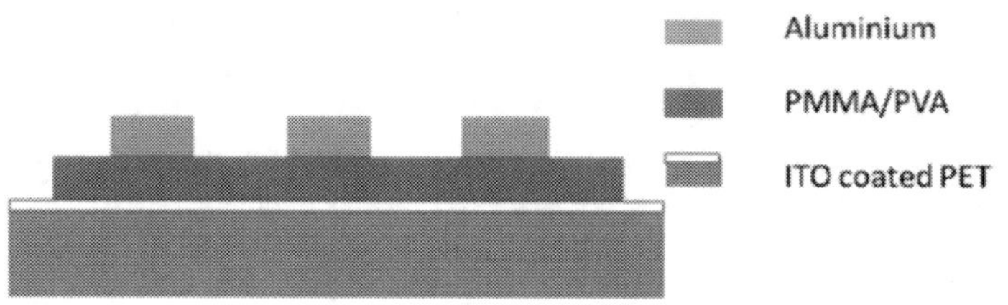

Figure 1. Structure of capacitor

The various masks used for fabrication of the capacitor is shown in Fig 2. ITO coated PET sheets (Sigma Aldrich) were cut to $4.5cm^2$ area. They were cleaned by ultra-sonication with acetone, ethanol, and deionized water and then dried. The flexible substrates with ITO on one side were masked on all the four sides with a width of 5mm to form the bottom electrode contact for the capacitor. Then, the polymeric dielectric – PMMA or PVA was spin coated in the middle unmasked area as shown in Fig 2b, subsequently vacuum annealed at 100°C or 80°C for 1 hour.

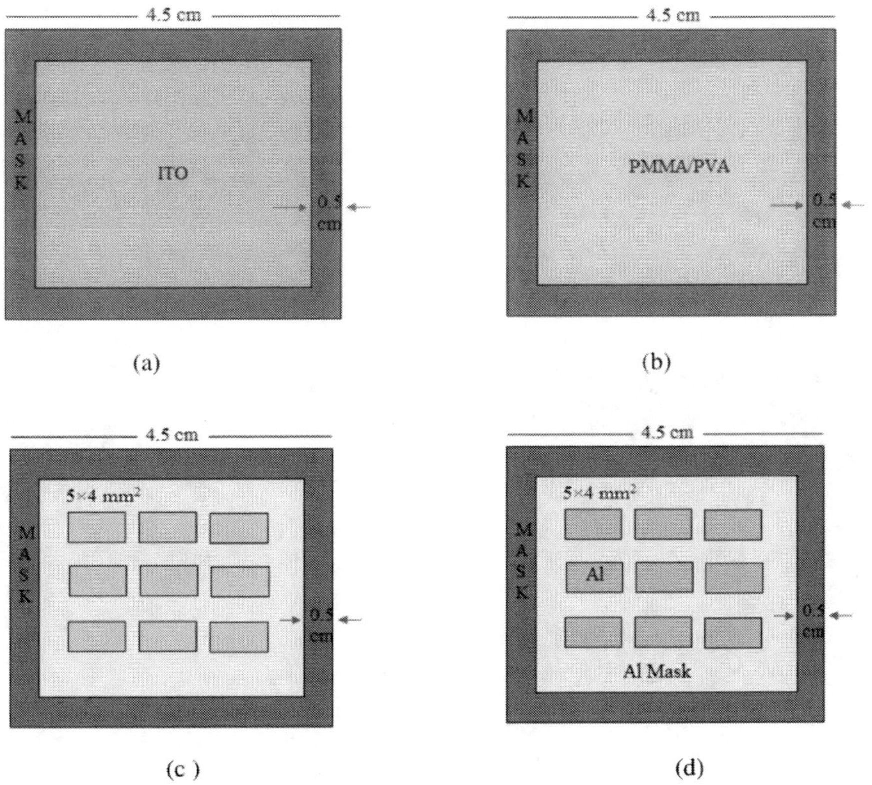

Figure. 2. Steps in Fabrication of PMMA/PVA Flexible Capacitor – Top View (a) Masking ITO of flexible ITO coated PET substrate; (b) Spin Coating PMMA/PVA; (c) Shadow mask with open windows of $20mm^2$ each; (d) Aluminium deposition

The deposition of aluminium was carried out with a mask of window dimensions 5×4 mm^2. The metal film is annealed at 100°C for half an hour to enable the proper adhesion of the metal atoms to the polymeric film.

3. Theory/calculation

As spin coating is the method adopted to make the polymeric thin film, a general equation that relates the thickness of spin coated films, concentration of solution and spin speed has been proposed by C.B.Walsh [6] is given as follows.

$$h = Ac^n \omega^{-\alpha} \tag{1}$$

where h is the thickness of the spin coated film, c is the solution concentration, ω is the spin speed, A and α are constants due to other parameters that influence thickness and n is concentration dependent system parameter. Further, an exponential relation governing the thickness of spin coated thin films with the solution concentration has been given by T.Tippo *et al* [7] and is given by Eq. 2

$$h = Ae^{nc} \omega^{-\alpha} \tag{2}$$

With decreasing film thickness, an increasing trend in capacitance was observed which is in correlation with capacitance equation given in equation 3.

$$C = \frac{\varepsilon_0 \varepsilon_r A}{d} \tag{3}$$

where, ε_0 is the permittivity of free space, εr dielectric constant, A is area of the capacitor, d thickness of the dielectric.

4. Results and Discussion

(a) AFM analysis:

Surface morphology observations of the two chosen dielectrics – PMMA and PVA spin coated on glass substrates using AFM over a 7μm² scan area are described here. Fig. 3 and Fig. 4 shows the 2D and 3D topographical profiles of spin coated PMMA at two different spin speeds: 3500 and 4000 rpm respectively.

(a)

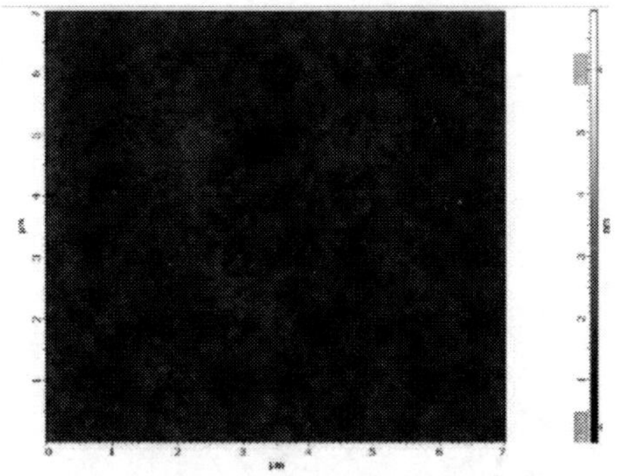

(b)

Fig. 3. a) 2D AFM Image of Spin Coated PMMA at 3500rpm; b) 3D AFM Image of Spin Coated PMMA at 3500rpm.

Fig. 3a shows the 2D micrograph of PMMA spin coated at 3000rpm with very few bright spots that denote the peaks in the 3D image of Fig. 3b. It is highly desirable that in case of an OTFT, the organic semiconductor coating to be done above the polymeric dielectric should be pore free. The presence of pores in the dielectric surface would make poor insulation between the gate electrode and the organic semiconductor. In addition appropriate surface roughness must be extended by the dielectric for the proper adhesion of the above lying semiconducting film.

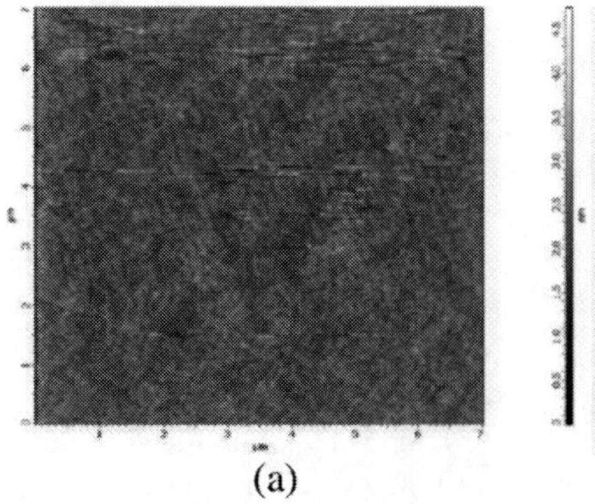

(a)

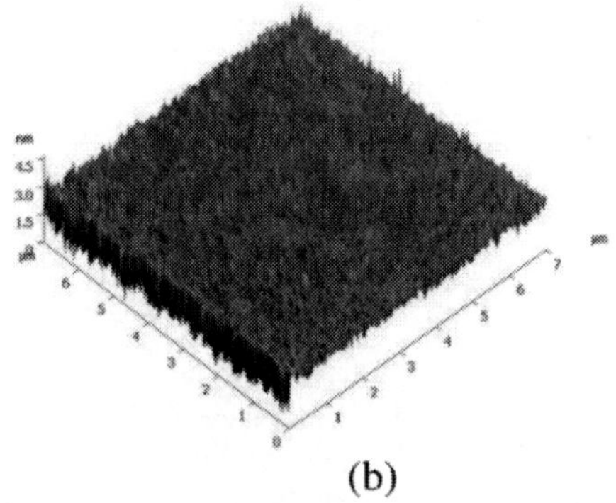

(b)

Fig. 4. a) 2D AFM Image of Spin Coated PMMA at 4000rpm; b) 3D AFM Image of Spin Coated PMMA at 4000rpm.

IOP Conf. Series: Materials Science and Engineering **99** (2015) 012026 doi:10.1088/1757-899X/99/1/012026

The 2D AFM image of Fig. 4a with PMMA spin coated at 4000rpm appears to have smooth morphology and there occurs less number of peaks in Fig. 4b when compared to Fig. 3b.
The surface topography in terms of flatness, asymmetry, presence of grooves and pits can be well explained by roughness, skewness and kurtosis tabulated in Table 1.

Table 1.Morphological Parameters of PMMA Thin Films at Selected Spin Rates.

Spin Speed (*rpm*)	Surface Roughness (*nm*)	Skewness	Kurtosis
3500	0.19	0.375848	7.77016
4000	0.24	-0.517825	2.58208

The average roughness value which is an estimate of the average deviation in height above and below an imaginary central plane is found to be more for the spin coated PMMA at a higher spin rate. The degree of symmetry of a surface about a mean line, denoted by skewness is positive for 3500rpm spin coated film, representing a surface that has more peaks than valleys or pits in the film and can be confirmed from the countable number of protruding peaks in 3D image of Fig 3b. Skewness being negative for the film spin coated at 4000rpm denotes a surface having more valleys than peaks as seen in Fig 4b. As defined in [8], kurtosis indicates the sharpness of a surface and supports the skewness results. Kurtosis above a value of 3 means the sample has more peaks than pits and when goes below 3, as in the case of 4000rpm spin coated thin film, the number of pits are more than the number of peaks.

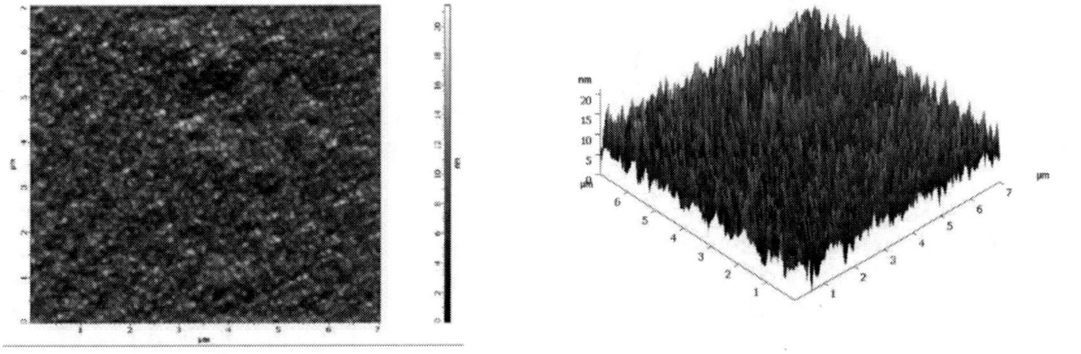

(a) (b)

Fig. 5. a) 2D AFM Image of Spin Coated PVA at 1000rpm; b) 3D AFM Image of Spin Coated PVA at 1000rpm.

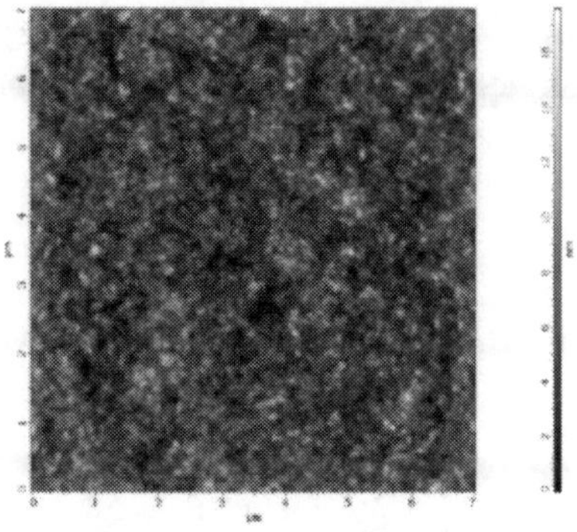

(a)

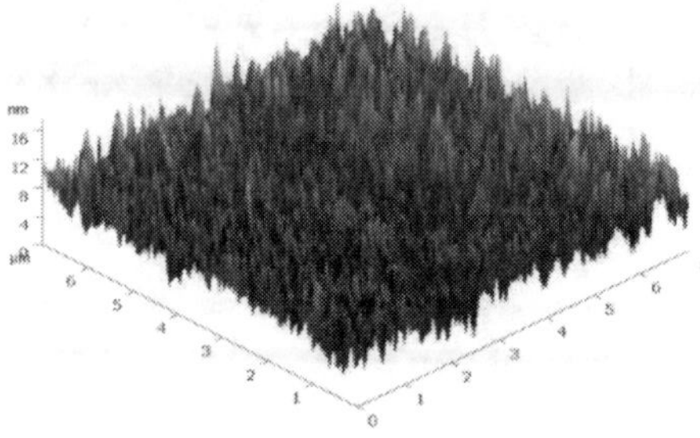

(b)

Fig. 6. a) 2D AFM Image of Spin Coated PVA at 1500rpm; 6.b) 3D AFM Image of Spin Coated PVA at 1500rpm.

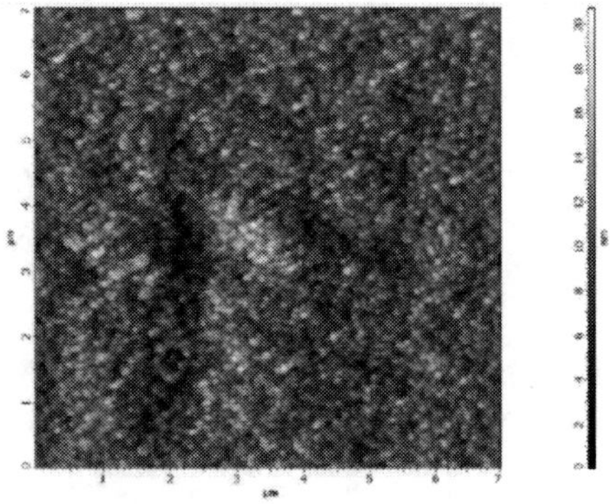

(a)

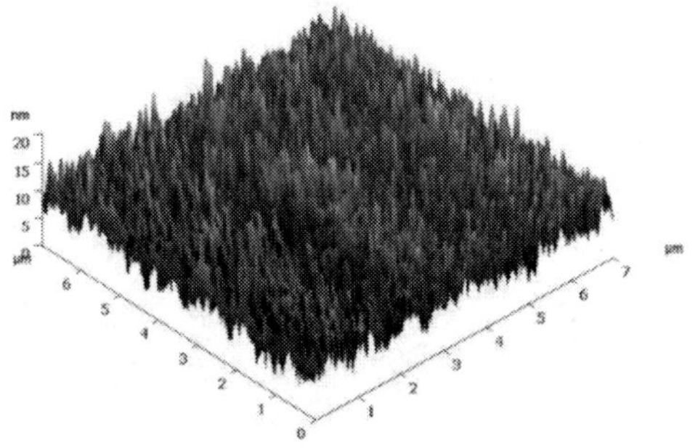

(b)

Fig. 7. a) 2D AFM Image of Spin Coated PVA at 2000rpm; 7 b) 3D AF.M Image of Spin Coated PVA at 2000rpm.

The spin coated PVA films more or less show similar morphological nature at the three different spin rates: 1000, 1500 and 2000rpm. These films are found to have spherical structures with very small variations in average surface roughness, skewness and kurtosis tabulated in Table 2.

Although no significant inference can be drawn from the grainy patterns of the 2D images of PVA thin films, there are sharper peaks in all the three cases of spin coated PVA films confirmed by the 3D images of Fig. 5b, Fig.6b and Fig.7b. At the three selected spin rates: 1000,1500 and 2000rpm, skewness and kurtosis values listed in Table 2, are close to zero which indicate the number of peaks and valleys in the film are almost equal. This indicates a planar thin film of PVA formed over glass.

Table 2. Morphological Parameters of PVA Thin Films at Selected Spin Rates.

Spin Speed (*rpm*)	Surface Roughness (*nm*)	Skewness	Kurtosis
1000	2.11242	0.241732	0.0688455
1500	1.67592	0.13162	0.0458512
2000	2.0561	0.0915515	0.0275083

(b)Thickness and Capacitance measurement:

The thickness measurements of PMMA and PVA thin films were done using Spectroscopic Ellipsometer and listed in Table 3 and Table 4 respectively. Table 3 and 4 also shows the measured capacitance for various film thickness. A decreasing trend in thickness of the film with increasing spin speed was observed. As discussed in section 3, the thickness values of the spin coated thin films hold an inverse relation with spin speed as defined by Eq. 1 and 2.

Table 3.Thickness of PMMA Thin Films at Various Spin Rates

Spin Speed (*rpm*)	Thickness (*nm*)	Capacitance (nF/cm^2)
3500	1556.02	5.35
4000	1274.45	6.35
6000	713.25	6.94
6500	610.25	8.74
7000	560.76	9.015
7500	510.32	9.5

Table 4. Thickness of PVA Thin Films at Various Spin Rates

Spin Speed (*rpm*)	Thickness (*nm*)	Capacitance (nF/cm^2)
1000	150.00	18.33
1500	105.00	30.745
2000	80.00	33.12

When thickness of PMMA films are exponentially fitted as shown in Fig.8, the error bars are reduced at higher spin speeds. So, the same Eq.2 represents the thickness of spin coated PMMA thin films and the same holds good for the PVA thin films.

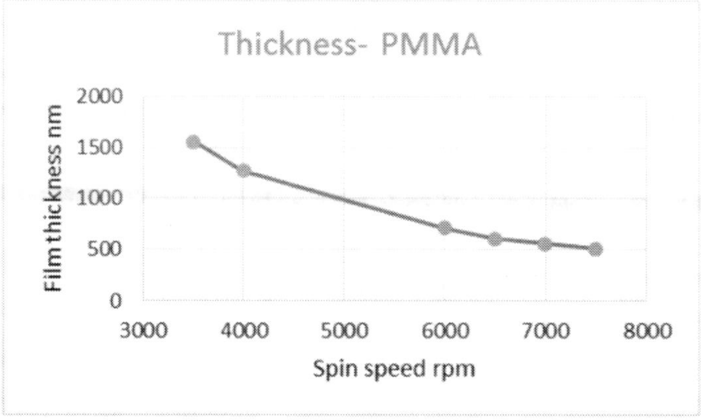

Fig. 8. Thickness vs. Spin Speed of PMMA Thin Films.

The three layer capacitor structures fabricated as described in section 1, with nine capacitors of same dimensions formed on a single substrate at a particular spin rate is shown in Fig. 9.The average capacitance of all the capacitors were measured using an LCR-Q Bridge 6018 at a fixed input frequency of 1 KHz in parallel mode.

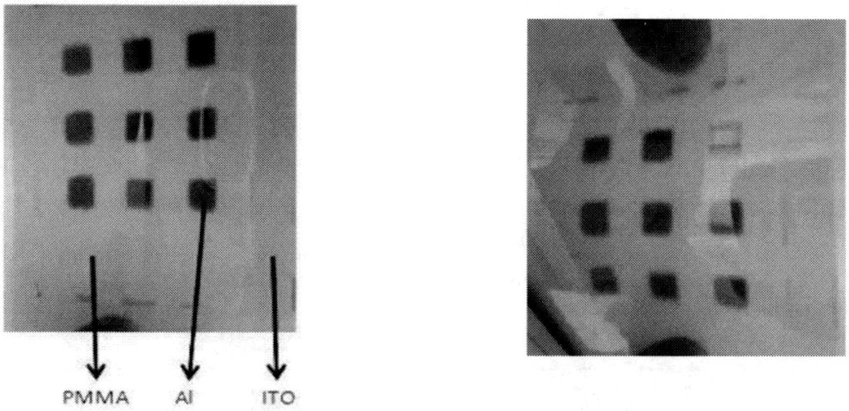

Fig. 9. Flexible PMMA Thin Film Capacitor.

The capacitance values measured for nine capacitors averaged is plotted in Fig. 10 for PVA as dielectric and in Fig. 11 for PMMA as dielectric. The capacitance values obtained obeys equation 3.

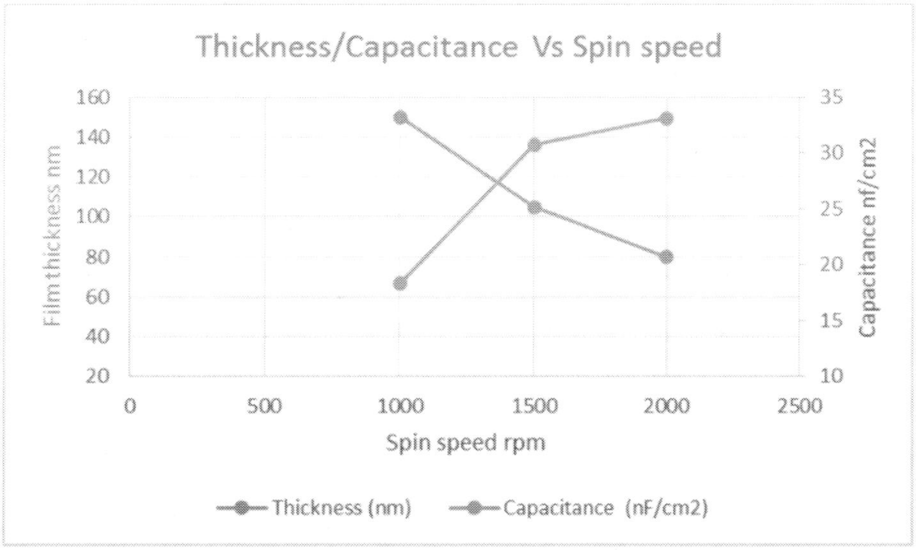

Fig. 10. Plot of Thickness/ Capacitance vs. Thickness of PVA Thin Films.

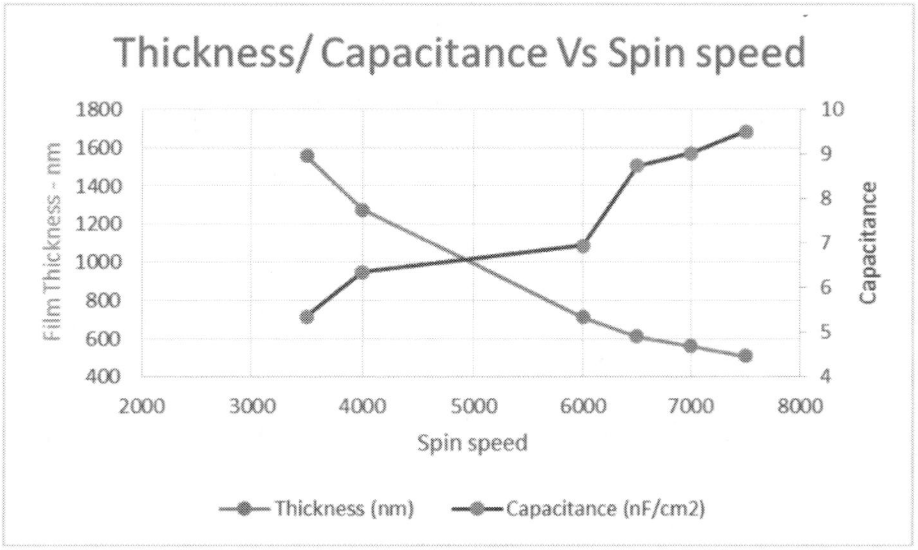

Fig. 11. Plot of Capacitance vs. Thickness of PMMA Thin Films.

As capacitance decreases with increase in frequency, it is necessary to study the capacitive behaviour of PMMA for large frequency range and varying applied gate bias by C-V measurements. This will give an insight into the accumulation, depletion and inversion regimes of the final OTFT device. Further, the decreasing capacitance with increasing frequency at a fixed bias voltage may be attributed to trapping and detrapping of charge carriers, increasing inability of the dipoles to orient

themselves in a rapidly varying electric field and slow release of charge carriers from relatively deep traps in the amorphous PMMA film [9].
However at higher frequencies, there is some amount of energy being absorbed by the dielectric from the applied voltage which is referred to as the dielectric loss. Hence in addition to the capacitance measurements at varying voltages, leakage current studies on the dielectric for varying electric field will help in understanding the breakdown field strength of the material. This will decide if the polymer will make a good dielectric in the transistor. The capacitance offered by the dielectric greatly influences the operation of Organic Thin Film Transistors (OTFTs).

5. Conclusions

We have presented step by step procedure for fabricating flexible capacitors using organic dielectrics namely PMMA and PVA. Both are coated by the spin coating process. A very simple method to measure the capacitance using an LCRQ bridge was adopted which yielded 9.5nF/cm^2 and 33.12nF/cm^2 for lowest thickness of PMMA and PVA respectively. Also the surface analysis of these thin films on glass substrates is investigated under AFM confirms their dielectricity at a very low thickness.

References

[1] Samuel Mandal, Monica Katiyar, *"Processing and performance of organic insulators as a gate layer in organic thin film transistors fabricated on polyethylene terephthalate substrate", Bull. Mater. Sci.,* Vol.36, No.4, pp. 653-660, August 2013.

[2] J.Puigdollers, C.Voz, A.Opella, R.Quidant, I.Martin, M.Vetter, R.Alcubilla, *"Pentacene thin-film transistors with polymeric gate dielectric", Organic Electronics*, 5, pp.67–71, 2004.

[3] D.Yang, L.Zhang, S.Y.Yang, B.S.Zou, *"Influence of the Dielectric PMMA Layer on the Detectivity of Pentacene-Based Photodetector With Field-Effect Transistor Configuration in Visible Region" IEEE Photonics Journal*, vol. 5, No. 6, December 2013.

[4] Lyly Nyl Ismail, Habibah Zulkelfe, Sukreen Hana Herman and Mohamad Rusop Mahmood, *"Influence of Doping Concentration on Dielectric, Optical and Morphological Properties of PMMA Thin Films", Advances in Material Science and Engineering*, 2012.

[5] P.Thomas, R.S. Ernest Ravindran and K.B.R.Varma, Editors. *Dielectric properties ofPoly(methyl methacrylate) (PMMA)/CaCu3Ti4O12 Composites. Proceedings of the 10th International Conference on the Properties and Applications of Dielectric Materials,* (2012) July 24-28; Bangalore, India

[6] Christopher B. Walsh, Elias I. Franses, *"Ultrathin PMMA films spin-coated from toluene solutions", Thin Solid Films*, 429, pp. 71–76, 2003.

[7] T. Tippo, C. Thanachayanont, P. Muthitamongkol, C. Junin, M. Hietschold, A. Thanachayanont, *"The effects of solvents on the properties of ultra-thin poly (methyl methacrylate) films prepared by spin coating", Thin Solid Films* ,546, pp.180–184, 2013.

[8] M. Raposo, Q. Ferreira and P.A. Ribeiro, *"A Guide for Atomic Force Microscopy Analysis of Soft- Condensed Matter"*, Modern Research and Educational Topics in Microscopy, 2007.

[9] B. Chandar Shekar, S. Sathish, R. Sengoden, *"Spin coated nano scale PMMA films for organic thin film transistors", Physics Procedia*, 49, pp.145 – 157,2013.

Acknowledgment

The authors acknowledge Dr. P Radhakrishnan, Director, PSG Institute of Advanced Studies and staff of nanoPSG Institute of Advanced Studies for their support for fabrication and characterization of thin film capacitors. We specially thank Dr. B Geetha Priyadarshini, Mr. Ramprashanth, and Mr. Karthikeyan for their support on coating and AFM characterization.

Effect of Human Movement on Galvanic Intra-Body Communication during Single Gait Cycle

I W Ibrahim, A H A Razak, A Ahmad and M K M Salleh

Center of Electronic Engineering Studies (CEES), Faculty of Electrical Engineering (FKE), Universiti Teknologi MARA, 40450 Shah Alam, Selangor

Email: izaidi_wanibrahim@yahoo.com

Abstract. Intra-body communication (IBC) is a communication system that uses human body as a signal transmission medium. From previous research, two coupling methods of IBC were concluded which are capacitive coupling and galvanic coupling. This paper investigates the effect of human movement on IBC using the galvanic coupling method. Because the human movement is control by the limb joint, the knee flexion angle during gait cycle was used to examine the influence of human movement on galvanic coupling IBC. The gait cycle is a cycle of people walking that start from one foot touch the ground till that foot touch the ground again. Frequency range from 300 kHz to 200MHz was swept in order to investigate the signal transmission loss and the result was focused on operating frequency 70MHz to 90MHz. Results show that the transmission loss varies when the knee flexion angle increased. The highest loss of signal at frequency range between 70MHz to 90 MHz was 69dB when the knee flexion angle is 50° and the minimum loss was 51dB during the flexion angle is 5°.

1. Introduction

With the busy lifestyle of people today, the important to always continuous monitoring the patients are become a must. A system where a patient vital sign can always be monitored without visiting the hospital can solve this problem. The system that no need for manual self-administered health system consists of body area network (BAN) and infrastructure area network [1]. A new wireless body area network (WBAN) standard, IEEE 802.15.6 was ratified in 2012. This WBAN consist of RF and Non-RF WBAN which is intra-body communication (IBC). Nowadays, the miniature sensor that capable of monitoring vital sign such as heart beat, glucose level, and blood pressure already applies in bio-medical system monitoring using the RF propagation. Figure 1 show the modern hospital monitoring network consists of implantable sensor and using the IBC system [2].

Intra-body communication (IBC) was first introduced by Zimmerman in 1995 which is called as personal area network (PAN) [3]. IBC is a wireless communication system where human body acts as a medium for the signal transmission between transmitter and receiver. There are two coupling methods for IBC which are capacitive and galvanic coupling. For the capacitive coupling, the signal transceiver electrode is attached at human body and the ground electrode is floating. In galvanic coupling method, both transceiver electrodes are attached at human body.

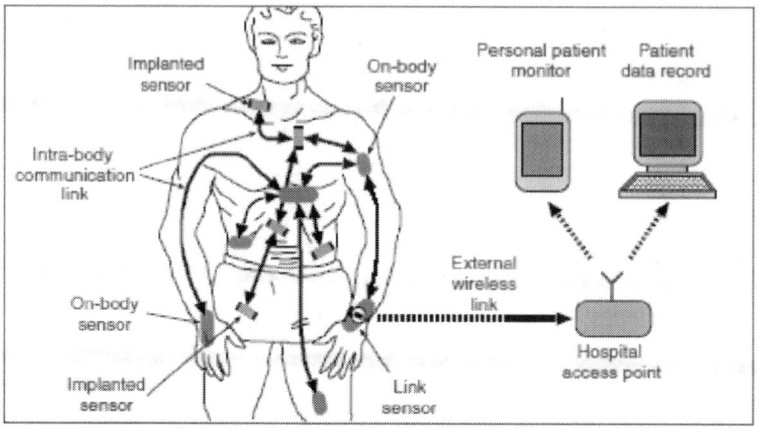

Figure 1. Modern hospital monitoring network.

Most research done in IBC is focused on electrical properties of human body, modeling method of body tissues and also hardware design of the IBC system. In order to design the IBC transceiver, the understanding of communication channel is important. Therefore, the effect of human body movement as communication channel was investigated in this paper. The galvanic coupling IBC approach was used to measure the signal loss when there is movement. Since the subject will bends their knee in order to walk, the measurement was related with gait cycle analysis that also called walking cycle.

There also some research that relate with body movement was study previously. The impedance change when the body is in motion was measured by Hachisuka et al. in 2003 [4].Then Schenk et al. also reported that there were small channel attenuation variation when the body moving [5]. Recently, the influence of sitting and standing position to the capacitive coupling method also were investigated by Lucev et al. [6]. The main objective of this paper is to investigate the effect of knee joint existence and flexion angle on galvanic coupling IBC signal attenuation using single gait cycle analysis. A portable vector network analyzer (miniVNA pro) was used to transmit the signal from transmitter and measure the signal at the receiver.

The frequency range in this study is from 300 kHz to 200 MHz and mainly focused on 70 MHz to 90 MHz. This paper is organized into 4 sections. Section 2 briefly explains the method of how experiment was done. Section 3 presents the experimental result and discussion during no movement and during movement and the section 4 give the final conclusion of the empirical measurement.

2. Method

2.1. Galvanic Coupling Approach

In the galvanic coupling method, an electrical signal is applied differentially between two transmitter electrodes [7]. Majority of current will flow between the two transmitters electrodes while only small portions of current arriving at the receivers electrodes. The signal attenuation that arrives at the receivers' electrode was measured for different knee flexion angle. Figure 2 below shows the setup for galvanic coupling IBC. The galvanic coupling method was choose based on previous study that concluded galvanic coupling proved to have superior data transmitting compare with capacitive [8].

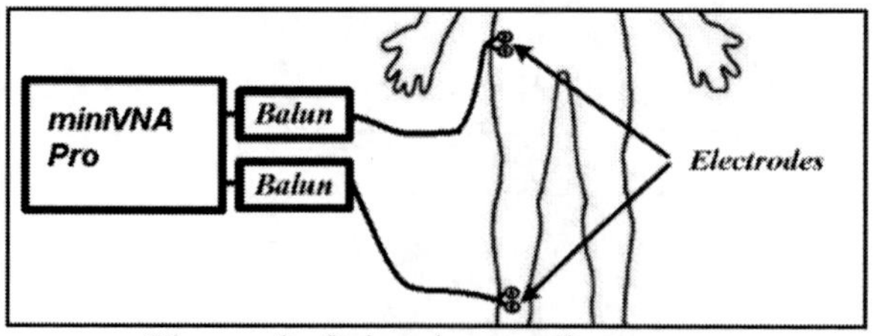

Figure 2. Galvanic coupling setup.

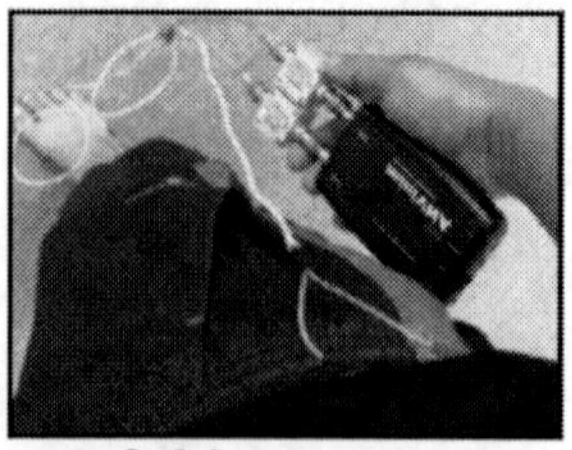

Figure 3. Measurement setup.

2.2. Experiment Setup

The experiment for effect of human movement on galvanic IBC was setup as in Figure 3. To generate signal at transmitter and measure signal at receiver module, a portable vector network analyser (miniVNA Pro) was used. A pair of baluns also placed on the transceiver to provide realistic signal transmission path.

For the frequency, the generate signal from miniVNA Pro was sweep from 300 kHz to 200MHz. A pair of receiver (RX) electrodes was placed on the subject's right waist and transmitter (TX) electrodes placed above subject's right ankle. The distance between transmitter and receiver electrodes was set to 90 cm calculated from the centres of electrodes.

In order to measure the effect of human movement, the experiment was related with gait analysis: specifically, knee flexion during a gait cycle. Figure 4 shows the sagittal plane joint angles (degree) during a single gait cycle of knee flexion. Referring to Figure 4 it shows that at 70% of single gait cycle the knee flexion is at its maximum nearly 50° of the joint angle. These maximum degrees of joint angle happen in feet adjacent (FA) position.

The reading was repeated based on the positions of the legs during a single gait cycle by the right leg [8] shown in Figure 5. The positions are initial contact (IO), opposite toe off (OT), heel rise (HR), opposite initial contact (OI), toe off (TO), feet adjacent (FA), and tibia vertical (TV). The signal transmission losses were measured at each leg position and were repeated three times in order to get the average signal value. The average was compared and reported.

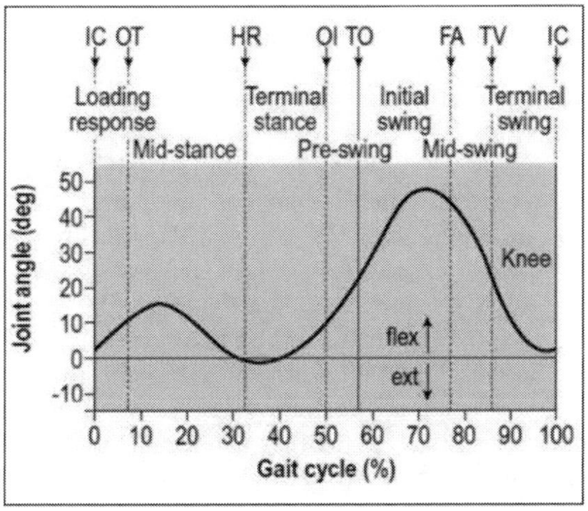

Figure 4. Sagittal plane joint angles (degree) during single gait cycle of knee (flexion) [9].

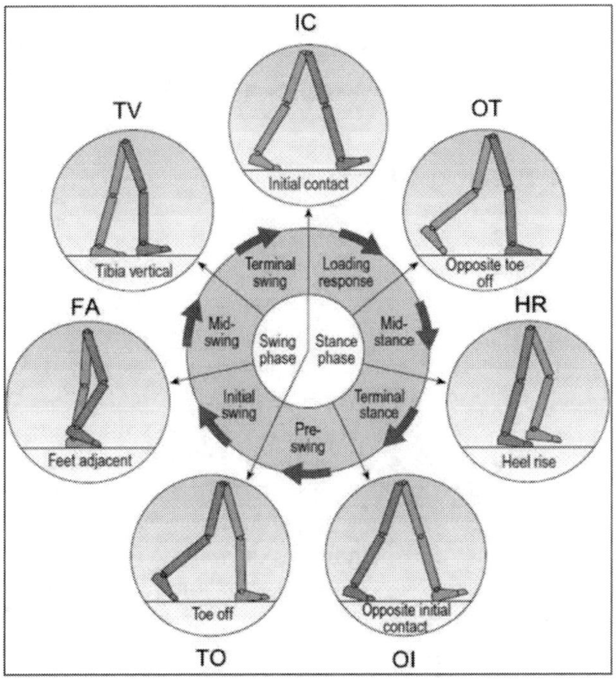

Figure 5. Positions of the legs during a single gait cycle by the right leg [9].

2.3. Protocol
The attenuation signal was measured for the frequency sweep of 300 KHz to 200 MHz. The result was focused between our decided operating frequency 70 MHz to 90 MHz for each of the measured transmission losses. This range frequency was choose based on previous study that concluded within this frequency range, the capacitive and galvanic method have minimum attenuation as shown in Figure 6 [7].

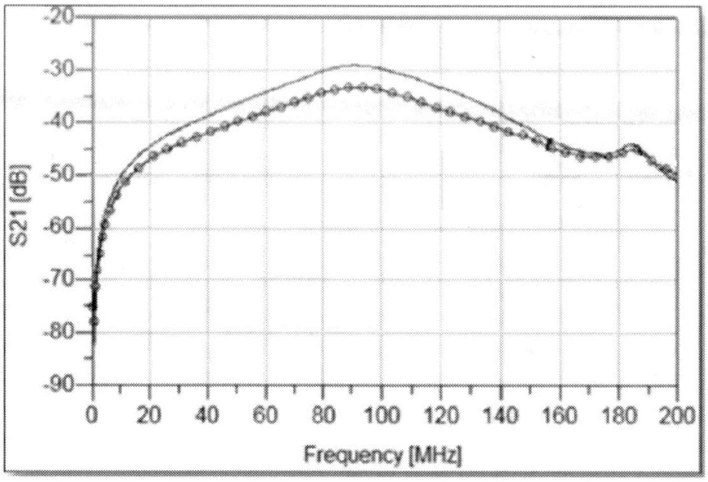

Figure 6. Signal attenuation for galvanic method from 0 to 200MHz [7].

3. Result and Discussion

3.1. During no Movement.
Human body consist of limb joint that allow the movement of body from one place to another and also a motion like stretching and bending. When people move or walking, the limb joint segment also will bend and the angle between limbs joint also keep changing. Hence, an experiment was done to investigate the effect of knee joint existence and flexion angle on galvanic coupling IBC signal attenuation with a fixed distance between transceiver and receiver, which is between ankle and waist that are 90 cm. First, we test the loss that came from the utilized cable. We connect the transmitter cable to the receiver cable and run the attenuation sweep from 300 kHz to 200 MHz.

Figure 7(a) show signal loss from the cable. From figure 7(b), it can be seen that the loss at 80 MHz was 7 dB. Referring to Figure 8 during no human movement (no knee flexion) the maximum loss was 40 dB at our operating frequency. We also monitor that between 20 MHz and 40 MHz, the loss is higher at 65 dB. Another major thing can be notice that at each stage of the gait cycle, below 40 MHz the results' variation in attenuation between – 50 dB and – 70 dB, which further support why in this design, frequency around 80 MHz was focused. Thus, in our experiment, we look at the attenuation between 70 and 90 MHz.

IOP Conf. Series: Materials Science and Engineering **99** (2015) 012027 doi:10.1088/1757-899X/99/1/012027

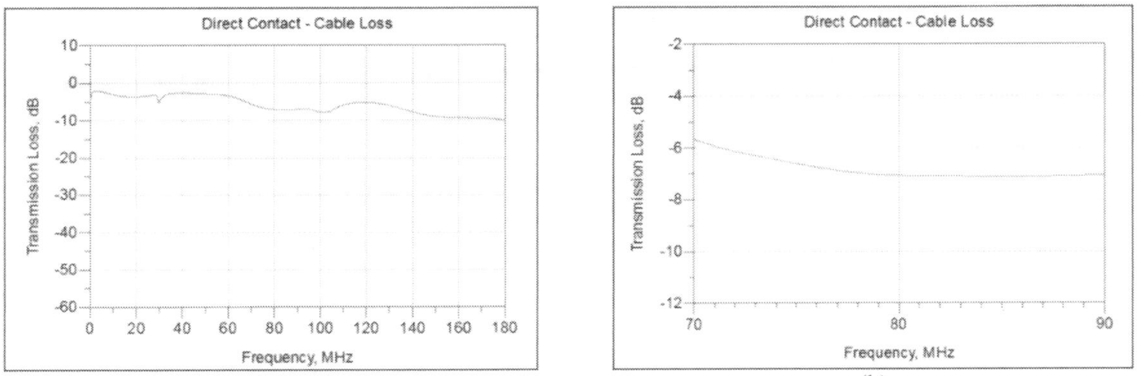

(a) (b)

Figure 7. Cable loss on transmitter: (a) frequency 0 MHz to 180 MHz, (b) frequency 70 MHz to 90 MHz.

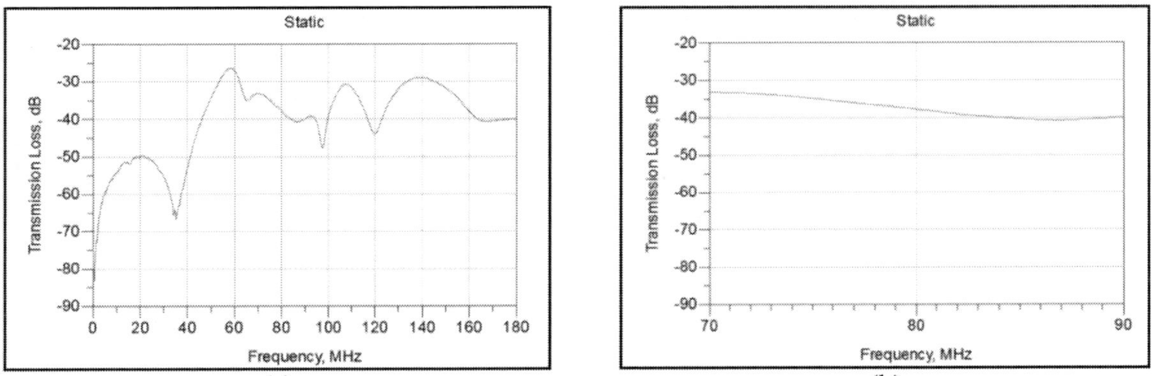

(a) (b)

Figure 8. IBC movement effect – no movement (static): (a) transmission loss between 0 MHz to 180 MHz,(b) transmission loss between 70 MHz to 90 MHz

3.2. During Gait Cycle.

Figure 9 shows attenuation sweep from 300 kHz to 200 MHz for initial contact, the maximum transmission loss between 70 and 90 MHz was 58 dB. Figure 10 reveals that the maximum transmission loss between 70 and 90 MHz was 54 dB for opposite toe off position of the gait cycle. For heel rise the loss was 61 dB as shown in Figure 11. The attenuation for opposite initial contact was 61 dB as revealed in Figure 12. Figure 13 shows attenuation sweep from 300 kHz to 200 MHz for toe off, the maximum transmission loss between 70 and 90 MHz was 56 dB. Figure 14 reveals that the maximum transmission loss between 70 and 90 MHz was 69 dB for feet adjacent position of gait cycle and in Figure 15, the maximum transmission loss between 70 and 90 MHz was 51 dB for tibia vertical position.

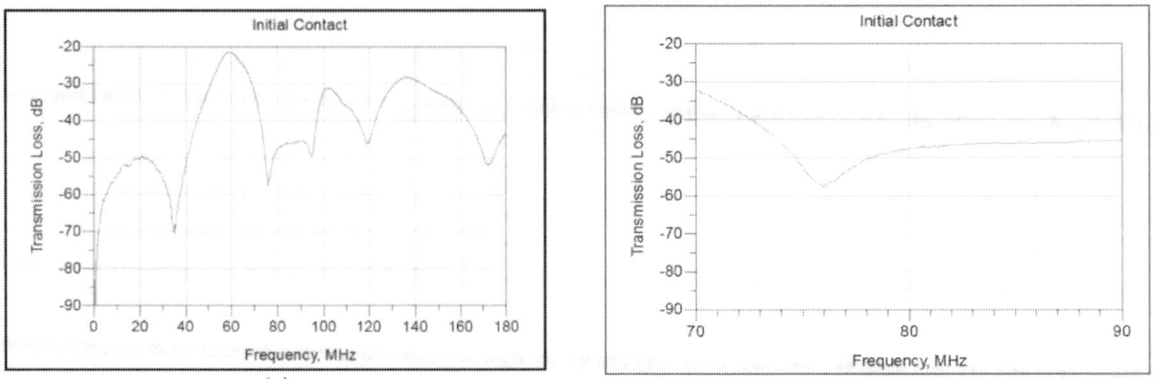

(a) (b)

Figure 9. IBC movement effect – initial contact position: (a) transmission loss between 0 MHz to 180 MHz, (b) transmission loss between 70 MHz to 90 MHz.

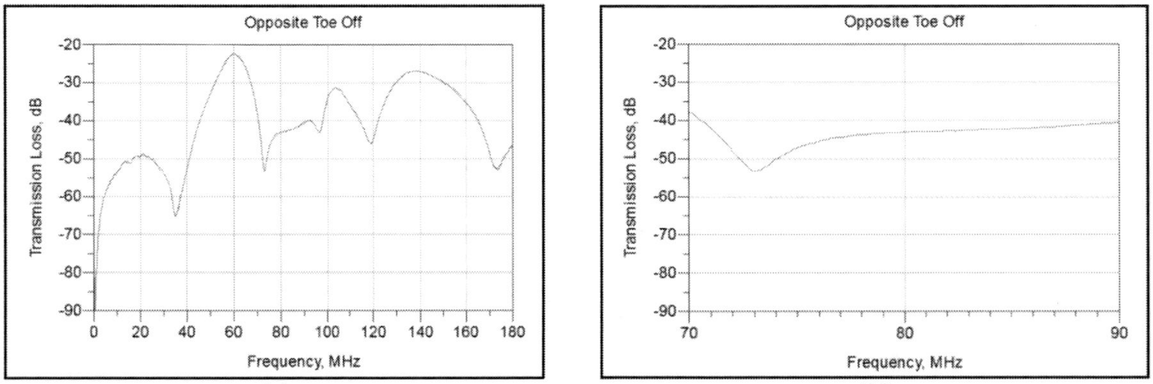

(a) (b)

Figure 10. IBC movement effect – opposite toe off position: (a) transmission loss between 0 MHz to 180 MHz, (b) transmission loss between 70 MHz to 90 MHz.

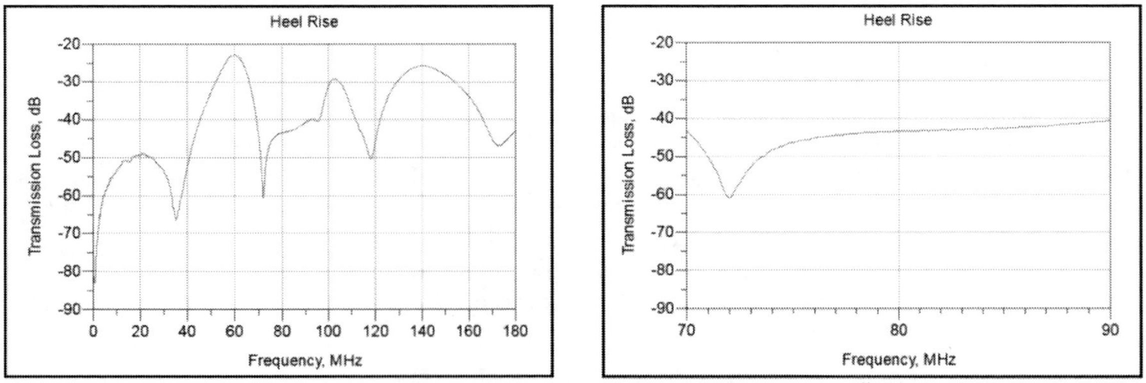

(a) (b)

Figure 11. IBC movement effect – heel rise position: (a) transmission loss between 0 MHz to 180 MHz, (b) transmission loss between 70 MHz to 90 MHz.

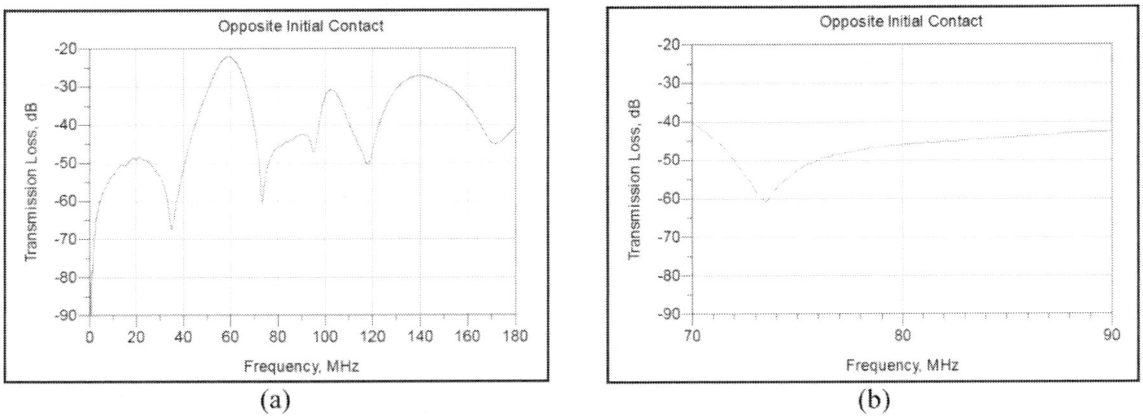

Figure 12. IBC movement effect – opposite initial contact position: (a) transmission loss between 0 MHz to 180 MHz , (b) transmission loss between 70 MHz to 90 MHz.

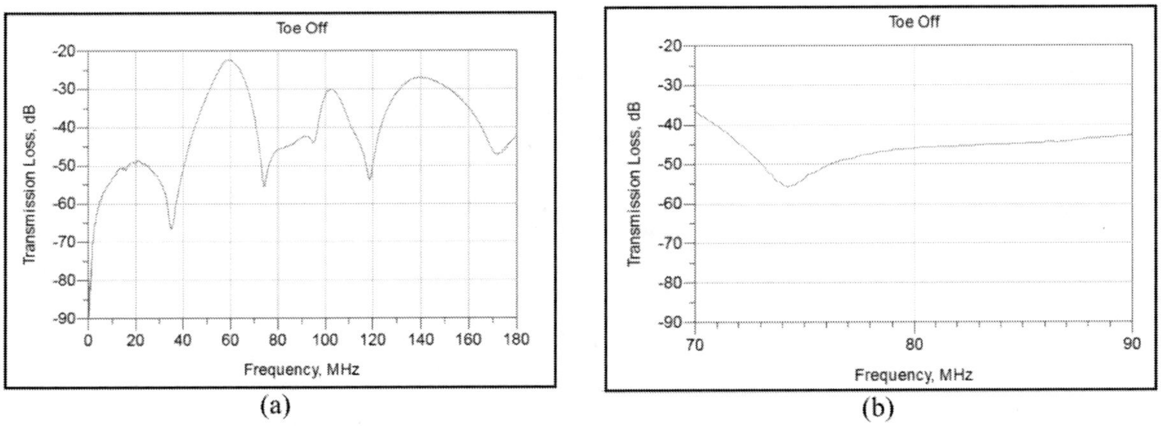

Figure 13. IBC movement effect – toe off position: (a) transmission loss between 0 MHz to 180 MHz, (b) transmission loss between 70 MHz to 90 MHz.

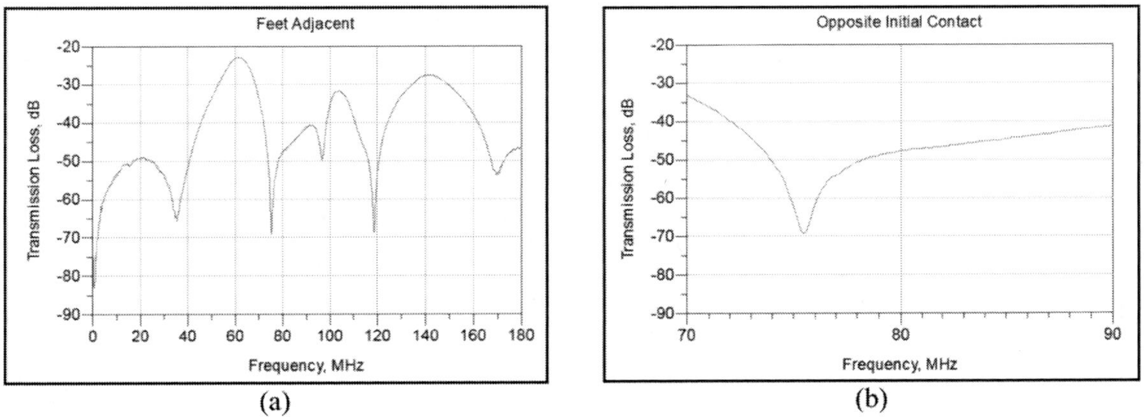

Figure 14. IBC movement effect – opposite initial contact position: (a) transmission loss between 0 MHz to 180 MHz, (b) transmission loss between 70 MHz to 90 MHz.

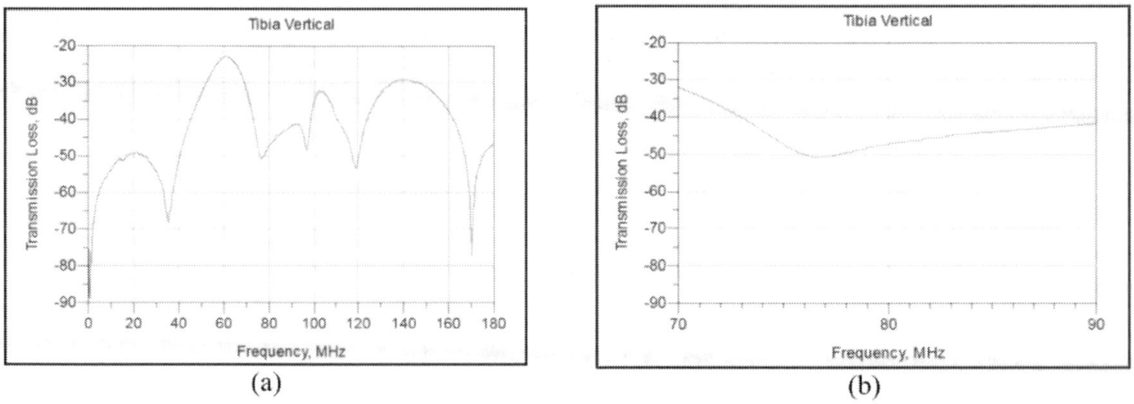

Figure 15. IBC movement effect – tibia vertical position: (a) transmission loss between 0 MHz to 180 MHz, (b) transmission loss between 70 MHz to 90 MHz.

4. Conclusion

From the result gathered we can conclude that first the frequency range between 70 to 90 MHz is suitable for galvanic approach using our system design. Second, based on the gait cycle study, we conclude that the transmission loss varies when the position of the gait cycle changes. The transmission loss increase as the knee flexion angle increased. We monitored that the highest loss was 69 dB at feet adjacent position of the gait cycle. The knee flexion angle at feet adjacent position is 50°. The lowest loss was 51 dB at the tibia vertical position. Here the knee flexion angle is 5°. These high losses affected the quality of the transmission of galvanic coupling IBC.

References

[1] Lai D T H and Begg R 2011 *Healtcare sensor networks: Challenges Toward Practical Implementation.* (oca Raton, FL, USA: CRC Press)
[2] Zurich E T H 2007 *Intra-Body Communication for Biomedical Sensor Networks* (Switzerland)
[3] Zimmerman T G 1995 *Personal Area Networks (PAN): Near-Field Intra-Body Communication.*
[4] Hachisuka K, Nakata A, Takeda T, Shiba K, K. Sasaki, Hosaka H, and Itao K 2003 *Development of wearable intra-body communication devices* **105** 109–115
[5] Schenk T C W, Mazloum N S, Tan L and Rutten P 2008 *Experimental characterization of the body-coupled communications channel* p 234–239
[6] Lucev Ž, Krois I, and Cifrek M 2012 *Effect of Body Position and Movement in A capacitive intrabody communication channel from 100 kHz to 100 MHz* **61(12)** 3280-3289
[7] Seyedi M, Kibret B, Lai D T H, and Faulkner M 2013 *An empirical comparison of limb joint effects on capacitive and galvanic coupled intra-body communications* **1** 213–218
[8] Razak A H A, Zayergh A and Begg R K, Seyedi M, Lai D T H 2013 *BFSK Modulation to Compare Intra-Body Communication Methods for Foot Plantar Pressure Measurement* 172-176
[9] Whittle M W 2007 *Gait analysis: an introduction, 4th ed.* (Butterworth-Heinemann)

IOP Conf. Series: Materials Science and Engineering **99** (2015) 012028 doi:10.1088/1757-899X/99/1/012028

Health Monitoring System Based on Intra-Body Communication

A H A Razak, I W Ibrahim, A H Ayub, M F Amri, M H Hamzi, A K Halim, A Ahmad and S A M Al Junid

Center of Electronic Engineering Studies (CEES), Faculty of Electrical Engineering (FKE), Universiti Teknologi MARA, 40450 Shah Alam, Selangor

Email: hadi@iee.org, izaidi_wanibrahim@yahoo.com

Abstract. This paper presents a model of a Body Area Network (BAN) health monitoring system based on Intra-Body Communication. Intra-body Communication (IBC) is a communication technique that uses the human body as a medium for electrical signal communication. One of the visions in the health care industry is to provide autonomous and continuous self and the remote health monitoring system. This can be achieved via BAN, LAN and WAN integration. The BAN technology itself consists of short range data communication modules, sensors, controller and actuators. The information can be transmitted to the LAN and WAN via the RF technology such as Bluetooth, ZigBee and ANT. Although the implementations of RF communication have been successful, there are still limitations in term of power consumption, battery lifetime, interferences and signal attenuations. One of the solutions for Medical Body Area Network (MBANs) to overcome these issues is by using an IBC technique because it can operate at lower frequencies and power consumption compared to the existing techniques. The first objective is to design the IBC's transmitter and receiver modules using the off the shelf components. The specifications of the modules such as frequency, data rate, modulation and demodulation coding system were defined. The individual module were designed and tested separately. The modules was integrated as an IBC system and tested for functionality then was implemented on PCB. Next objective is to model and implement the digital parts of the transmitter and receiver modules on the Altera's FPGA board. The digital blocks were interfaced with the FPGA's on board modules and the discrete components. The signals that have been received from the transmitter were converted into a proper waveform and it can be viewed via external devices such as oscilloscope and Labview. The signals such as heartbeats or pulses can also be displayed on LCD. In conclusion, the IBC project presents medical health monitoring model that operates at the range of 21 MHz frequency and reduce the power consumption for a longer battery lifetime.

1. Introduction

This project presents a model of a Body Area Network (BAN) health monitoring system based on Intra-Body Communication (IBC). A health monitoring system is a system that monitors the health signal of the human such as heartbeats, heart rates, and respiratory rates. It is a comprehensive monitoring system that beneficial to the human healthcare. IBC was originally proposed by T.G. Zimmerman[1]. The concept of IBC is to utilize the human body as a medium for transmission. The signals will pass through the human body using induced electric near field which is exists at around the human body. One of the main advantages of using IBC is that electromagnetic noise and interference have a little influence on the transmissions of the signal because the signals do not leak through the skin of the human body [2].

The main objective of the project is to design IBC modules that can operate at the 21 MHz. The first objective is to design the IBC's transmitter and receiver modules using the of – the – shelf components. Next objectives are to model and implement the digital parts of the transmitter and receiver modules on the Altera's FPGA board.

IBC is suitable solution for MBANs in a mobile health care system because of the low power requirement of the IBC itself [3]. IBC has the characteristics of high transmission quality with high data rates, easy network access and no communication bandwidth problem which has made the IBC are better compared to the near field electromagnetic waves [4]. It is also suitable for the mobile terminals and the other terminals embedded in environments.

The main problem of IBC is to use a lower frequency such as 21 MHz in order to transmit the signal through the human body and display it in a proper waveform. It is because the frequency of 21 MHz is the centre of the operating frequency band of IBC with scalable data rates of 164 – 1312.5 kB/s according to the IEEE 802.15.6 standard of Wireless Body Area Network (WBAN)[5]. The challenge of this project is to design a hardware setup, which are the transmitter and the receiver of IBC that can satisfy the IEEE 802.15.6 standard of WBAN.

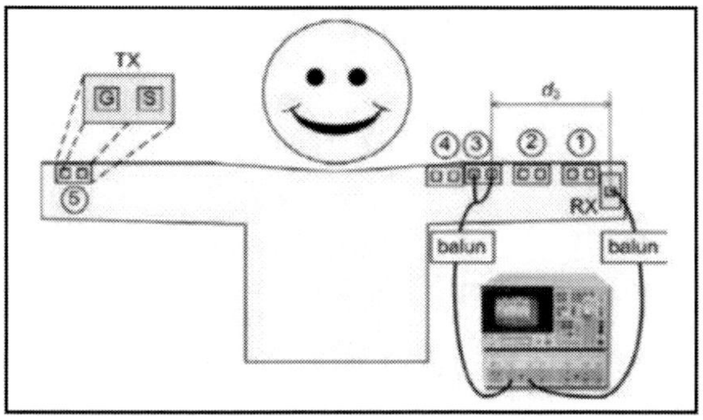

Figure 1. Example of Measurement setup for IBC [6].

IBC is categorized in a wireless communication field. However, IBC has several benefits over traditional wireless schemes. Some of the advantages are as follows: First, the use of cables is no longer needed. Second, the transmission of the signal area easily to conduct at will since the transmission of the signal can be formed only when the human body is touching the terminal. Third, the electromagnetic noise and interference have a little influence on the transmissions of the signal because the signal does not leak through the skin of the human body. Fourth, the human body can use the communication channel exclusively, which make the large-throughput data communication is highly possible [7].

The health monitoring system consists of two main modules which are the transmitter module and the receiver module. The sensor will detect the health signal such heart beats, respiratory rates or heart sound. After that, the signal will be transmitted via the transmitter and the balun. The human body will become the transmission medium of the signal as the signal will be travelled from the balun of transmitter to the balun of receiver. The external device such as oscilloscope will be displaying the signal that has been received from the receiver.

It is anticipated that the success of this project will help in the advancement of wireless biomedical and wireless sensor network market which encompasses every layer of the society.

The benefit will include in wearable system, health-coming monitoring devices and even sport performance analysis just to name a few. Therefore, it is reasonable to pursue designing such device due to its potential contribution to the community.

IOP Conf. Series: Materials Science and Engineering **99** (2015) 012028 doi:10.1088/1757-899X/99/1/012028

2. Design and Methodology

There are 3 different designs sections that need to be focused concurrently in this IBC project. The first section is to design the whole system using off the shelf components. The second and third sections will focus on designing the digital block in the transmitter and receiver modules using the FPGA Altera board. It also includes interfacing with the FPGA on-board modules and the discrete components. The design task for each section has been equally divided and simultaneously executed among the team.

2.1. Off-the-shelf component transmitter and receiver module

The main idea of this project is design a transmitter and receiver module for intra-body communication. In each module consists of some sub-modules such as analog-digital converter (ADC), parallel load 8-bit shift register, memory, BFSK modulator, and BFSK demodulator. All sub-modules are combined and implemented in this project to perform intra-body communication.

2.1.1. Analog-digital converter (ADC). The input signal read from the human body is in analog waveform. The ADC function is to convert the analog signal into digital signal. The input signal read at human body by the sensor is in analog signal. In order to make it digital signal ADC is used convert the signal. The output ADC reading can be calculated using this formula [8]:

$$\frac{Resolution\ of\ the\ ADC}{System\ Voltage} = \frac{ADC\ reading}{Analog\ Voltage\ Measured} \quad (1)$$

2.1.2. Parallel load 8-bit shift register. The output ADC produced 8 bit simultaneously which mean all 8 bit are data transmitted. 74LS165 IC used to shifted out all 8 bit parallel data into series which is compatible to the BFSK modulator.

2.1.3. Memory. The data shifted from the shift register stored inside memory so that the data not lose for every clock cycle.

2.1.4. BFSK modulator. The data from memory is sent to BFSK modulator. The data is manipulated base on its setting which is high input it produce 23.9MHz and low input it produce 21.3MHz.

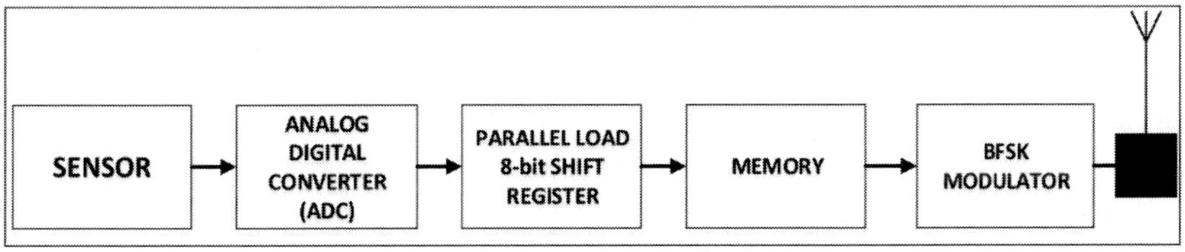

Figure 2. Block diagram transmitter module using off-the-shelf component.

2.1.5. BFSK demodulator. The data produced by modulator been demodulated using BFSK demodulator. Before that the signal been amplified because the output signal from modulator is small and in order BFSK demodulator to recognize the data transmitted.

IOP Conf. Series: Materials Science and Engineering **99** (2015) 012028 doi:10.1088/1757-899X/99/1/012028

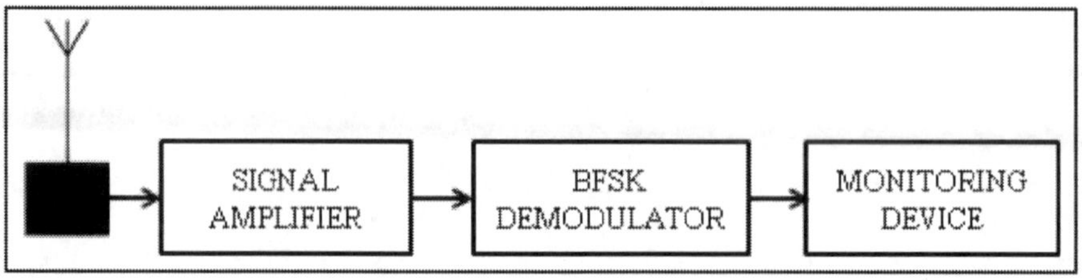

Figure 3. Block diagram receiver module using off-the-shelf component.

2.2. Digital Block of the Transmitter Module

The FPGA Altera board has been used for development of digital block in the transmitter modules of the IBC. The Quartus II software has been used to design the digital block for the transmitter modules. The Verilog HDL language has been implemented in order to develop the digital block in the transmitter modules of the IBC. The main block modules have a several separated component modules that operates a specific task for the transmission purpose.

The component modules that had been used to develop the transmitter modules are sensor, Phase Lock Loop (PLL), analog to digital converter (ADC), modulator, amplifier and patches. The sensor is used to sense or detect the pulse of the human body. The function of the PLL module is to lock the clock frequency to the specific value. The function of the ADC module is to convert the analog signal from the output sensor to a digital signal that can be read by the modulator. The function of modulator module is to modulate the frequency of the signal from the ADC. The modulation method that has been implemented in order to modulate the signal is BFSK modulation method. The function of the amplifier inside the transmitter modules is to amplify the power of the signal. The patches for the transmitter will be acted as the antenna to transmit the signal. The result of the transmission can be observed using the Oscilloscope.

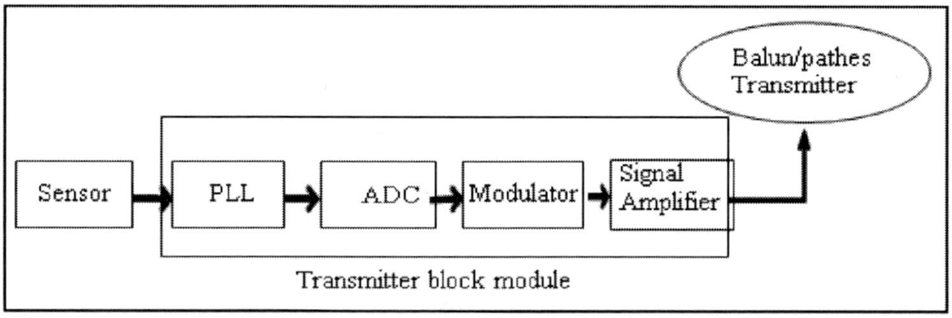

Figure 4. Block diagram of transmitter modules of a health monitoring system based on IBC.

2.3. Digital Block of the Receiver Module

There are several components modules that will be used to design the receiver modules like the binary frequency shift keying (BFSK) demodulator, amplifier and patches or balun. The function of BFSK demodulator is to demodulate the receiving signal from the patches into the data result such as the data of the human heart beats. The receiving signal is a signal that was transmitted through the human body via the transmitter modules of the IBC. The function of the amplifier inside the receiver modules is to amplify the power of the signal. The patches or balun for receiver will be acted as the antenna to receive the signal that was transmitted via transmitter module.

The demodulation method that has been implemented in order to demodulate the signal is BFSK demodulation method. Figure 5 shows the block diagram of receiver modules of health monitoring system based on IBC. There are several combination modules that have been used to develop the

IOP Conf. Series: Materials Science and Engineering **99** (2015) 012028 doi:10.1088/1757-899X/99/1/012028

receiver modules such as clock divider, Selector, Multiplexer and Series Input Parallel Output (SIPO) as shown in Figure 6.

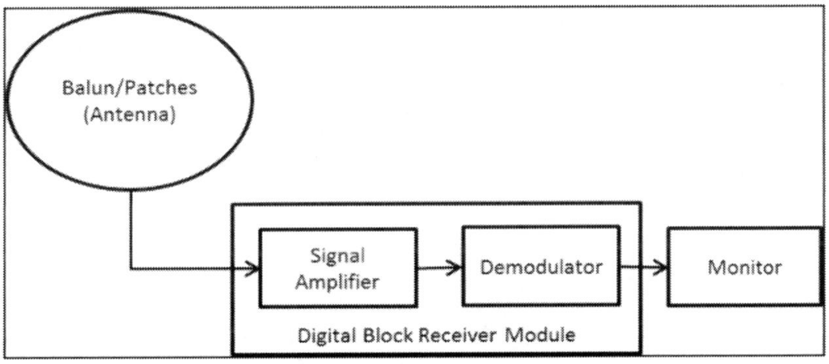

Figure 5. Block diagram of receiver modules of a health monitoring system based on IBC.

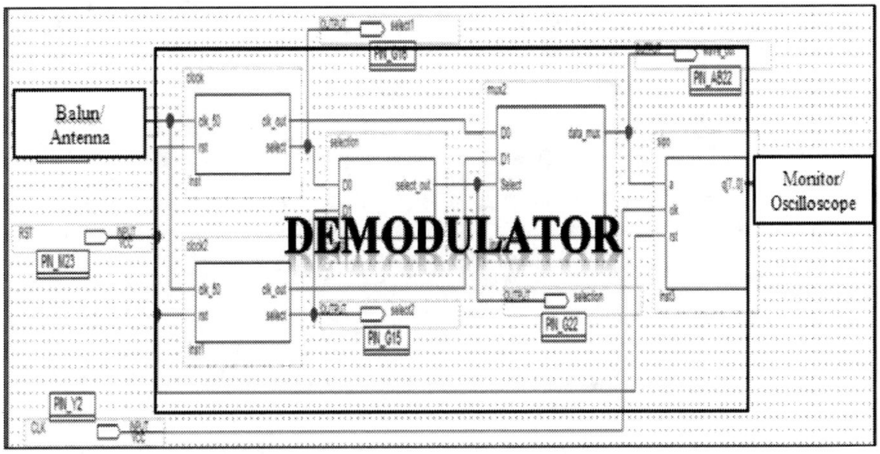

Figure 6. Block module of BFSK demodulation for receiver module inside Quartus II software.

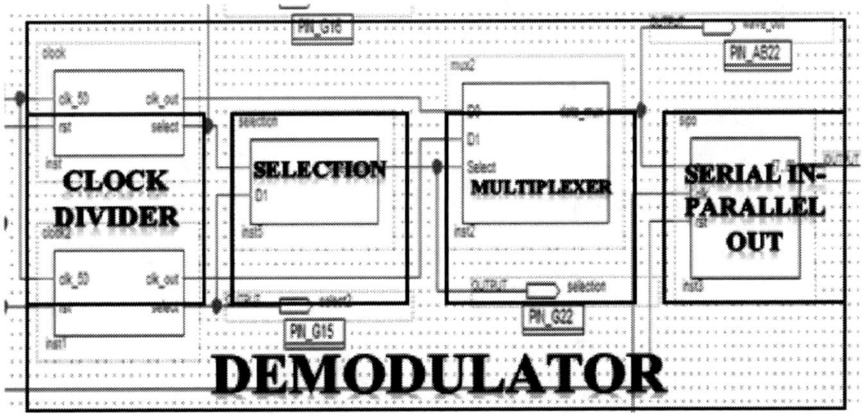

Figure 7. Combination blocks inside the BFSK demodulation for receiver module inside Quartus II software.

Figure 7 shows the combination block inside the BFSK demodulation for receiver module. Clock divider module used to lock the clock frequency to the specific value which is 12 MHz and 8 MHz for this design. Then, the selector module function as a selector input for multiplexer module. Multiplexer module will create a combination of two value of frequency from the clock divider such as high bit is 12 MHz and low bit is 8 MHz in-term of digital waveform. SIPO was used to demodulate the combination of two signals from multiplexer and change it into the 8 bits binary number. The 8 bits binary number will determine that the receiver module either successful read the input data at the transmitter module or not successful read the input data that was transmitted through human body. The result of the receiver can be observed using the Oscilloscope.

3. Result and Discussion

3.1. Off-the-shelf component transmitter module

The design had been tested and simulated in the Proteus software to perform the design behaviour of each module before proceed to design using off-the-shelf component.

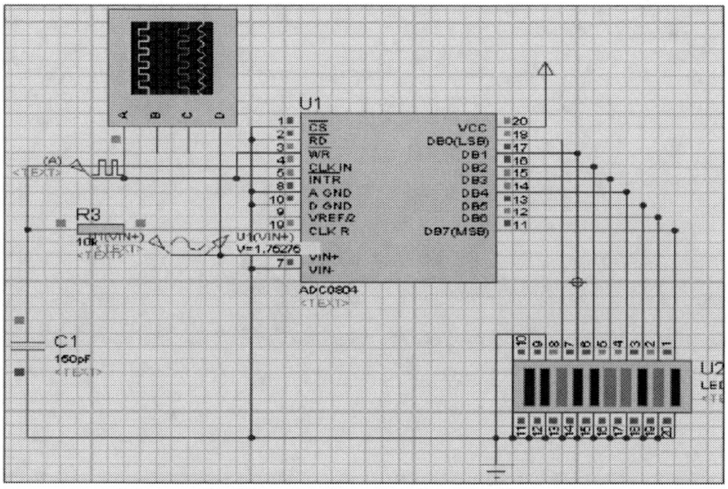

Figure 8. Circuit diagram for ADC module.

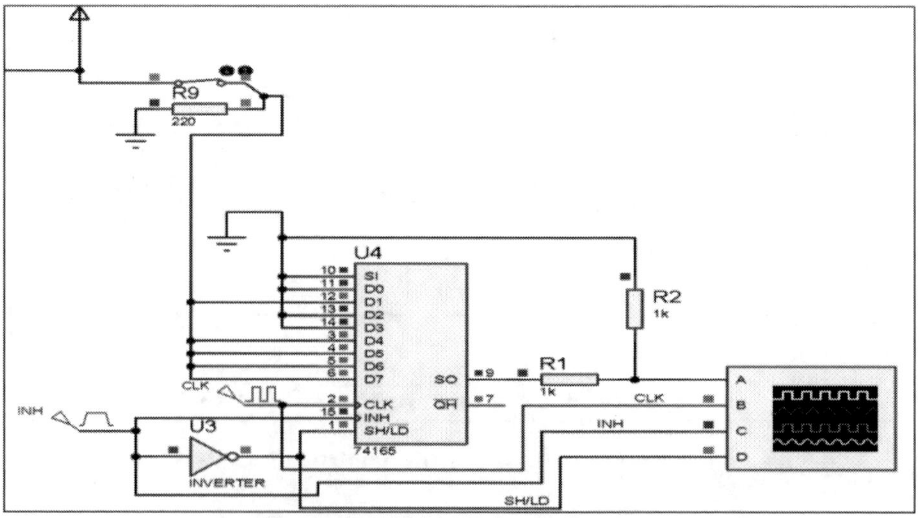

Figure 9. 74LS165 circuit diagram.

3.1.1. Analog-digital converter (ADC) module. ADC module used to convert the analog signal from input sensor into digital signal.

3.1.2. Parallel load 8-bit shift register module. 74LS165 IC is used in to convert the 8-bit parallel data into series output data. This performed provide a proper data reading for the modulator input. Figure 9 shows 74LS165 circuit diagram. Figure 10 shows the result simulation of the shift register module. The first, second, and third waveform is produced by the function generator to provide clock for the shift register to perform the load and shift data. The fourth waveform is the output data shifted from parallel input into series output.

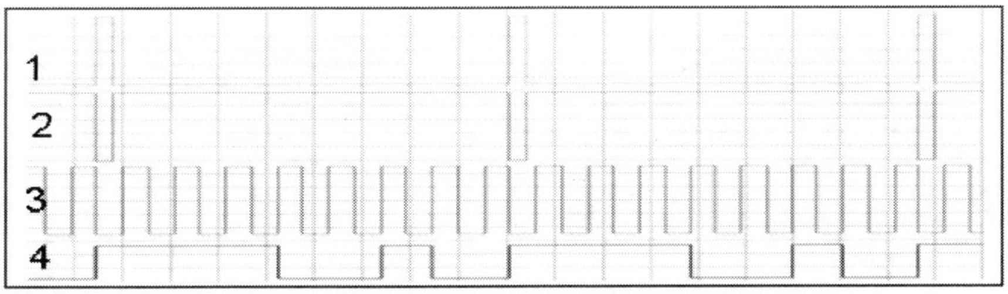

Figure 10. Result simulation of the shift register module.

3.1.3. Binary frequency shift keying modulator (BFSK) module. BFSK modulator is used to manipulate the input signal either high or low into two different frequencies. From the simulation the signal produced by the BFSK is 23.9MHz for high input signal and 21.3MHz for low input signal.

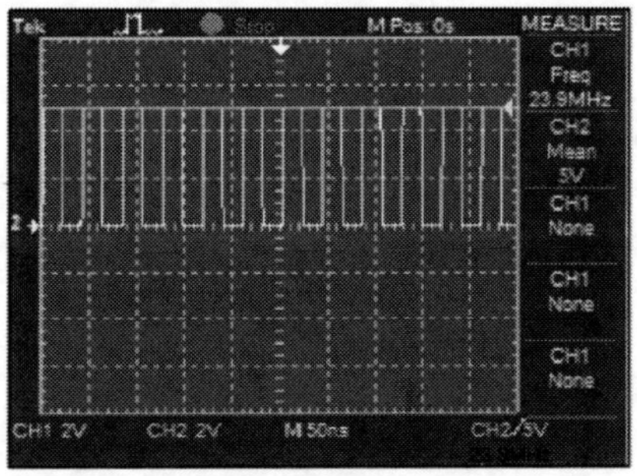

Figure 11. 23.9MHz frequency produced at high input signal.

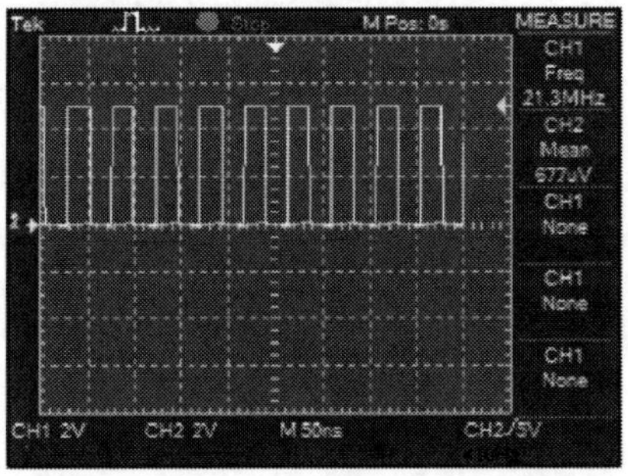

Figure 12. 21.3MHz frequency produced at low input signal.

3.2. Digital Block of the Transmitter Module

A simple BFSK modulation module block has been design inside the Quartus II software. The design had been tested at the FGPA Altera DE2 board. The waveform can be observed using the oscilloscope. The BFSK modulation method is using two different values of frequencies in order to modulate the signals. In this project, the frequencies that were applied to the modulation method are 21 MHz and 24 MHz. The PISO module has been simulated using the VWF simulation in order to test the behavior of the PISO. The frequencies that had been produced from the module block are in the range of 21 MHz until 24 MHz.

Both of the frequency had been successfully locked into the specific value, which mean that the PLL module had successfully developed. The waveform that has been produced from the BFSK module block is not in the fully square wave waveform. Due to the limitation of the components of the DE2 board itself that is not harmonic enough.

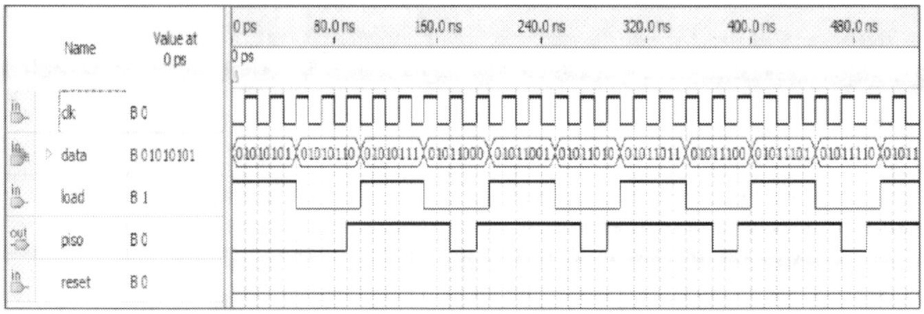

Figure 16: Simulation result for PISO block using VWF simulation method.

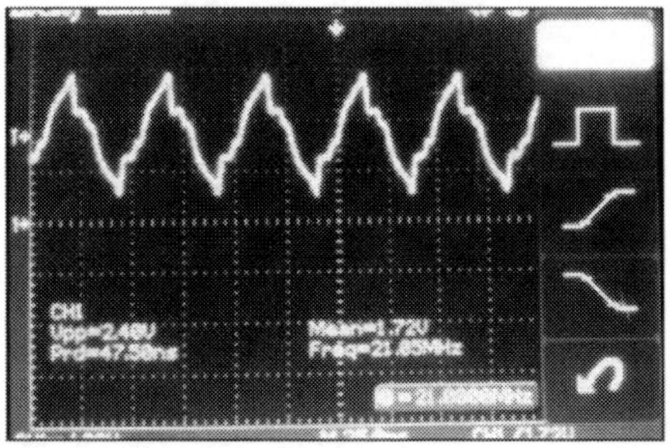

Figure 16. First frequency that have been locked at 21 MHz.

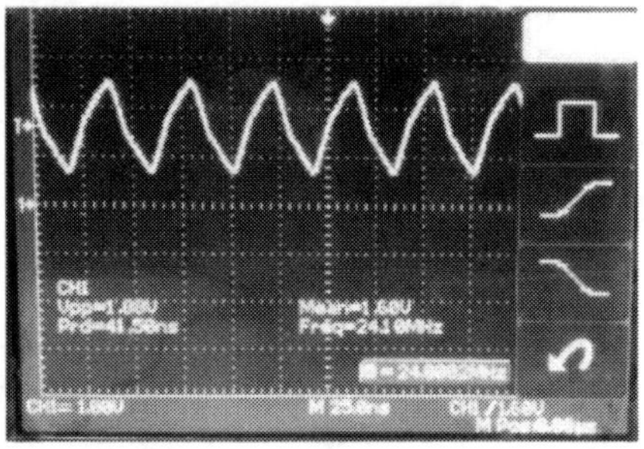

Figure 17. Second frequency that have been locked at 24 MHz.

3.3. Digital Block of the Receiver Module

The design had been tested on the FGPA Altera DE2 board. The waveform can be observed using the oscilloscope. The BFSK demodulation method is using two different values of frequencies in order to demodulate the received signals. In this project, the frequencies that were applied to the demodulation method are 12 MHz and 8 MHz. The selector module and SIPO module has been simulated using the University Program VWF in Quartus II software in order to test the behaviour of both modules. The selector module and SIPO module also had been simulated in ModelSim Altera software to identify timing violation using testbench for both modules. The frequencies that had been produced from the clock divider block are in the range of 12 MHz and 8 MHz. Both of the frequency had been successfully locked into the specific value, which mean that the clock divider had successfully developed. The waveforms that have been produced from the BFSK demodulation block are not in the fully square wave waveform due to the limitation of the components of DE2 board itself that is not harmonics to generate a proper square wave for output waveform.

Figure 18 shows the simulation result for selection or selector block using University Program VWF simulation method inside Quartus II software. The result shows that the output of selector is toggle between 1 and 0 based on which first input either D0 or D1 are high. If D0 is high, the output of selector will high but if D1 is high, the output selector will low. This scenario will toggle the output of the selector. Figure 19 shows the simulation for selection or selector block from testbench using ModelSim software to identify the time violation of the selector block. The similarity result from both simulation shows that the selector block has been designed successfully.

IOP Conf. Series: Materials Science and Engineering **99** (2015) 012028 doi:10.1088/1757-899X/99/1/012028

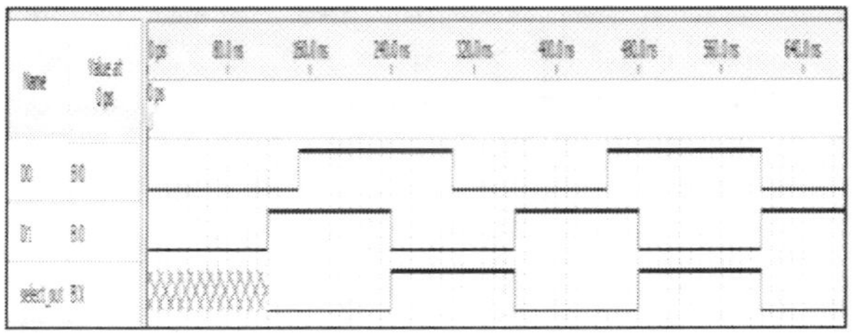

Figure 18. Simulation result for selector block using University Program VWF simulation method.

Figure 19. Simulation result for testbench of selector block using ModelSim software.

Figure 20 shows the simulation result for Serial In – Parallel Out (SIPO) block using University Program VWF simulation method inside Quartus II software. Result shows that the output of SIPO block has generated 4-bits output from the 1-bit input. The 4-bits output are based on current input when the clock of the SIPO is high. The 4-bits output can be declare as successful generated when the output get the same input after the input has entered 4 times from one input port. Figure 19 shows the simulation result for testbench of SIPO block using ModelSim software to identify the time violation of SIPO block. The result of the simulation is not same from University Program VWF because the time hold in testbench is not same with the University Program VWF and this effected to generate the 4-bits output properly.

The output signal from receiver module are shown in figure 22 for high frequency is 12 MHz and figure 23 for low frequency is 8MHz has successful generated.

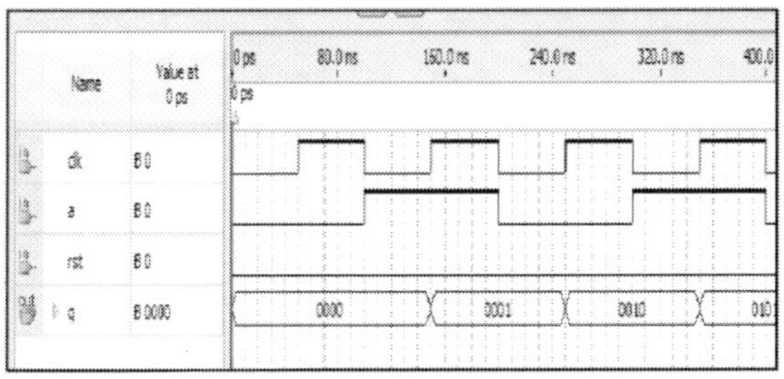

Figure 20. Simulation result for SIPO block using University Program VWF simulation method.

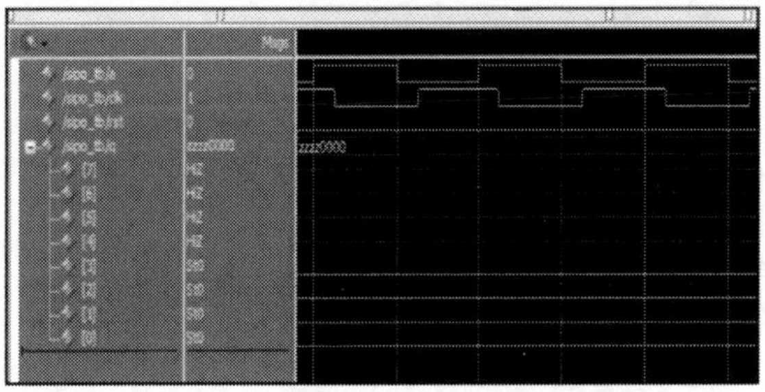

Figure 21. Simulation result for SIPO block using ModelSim software.

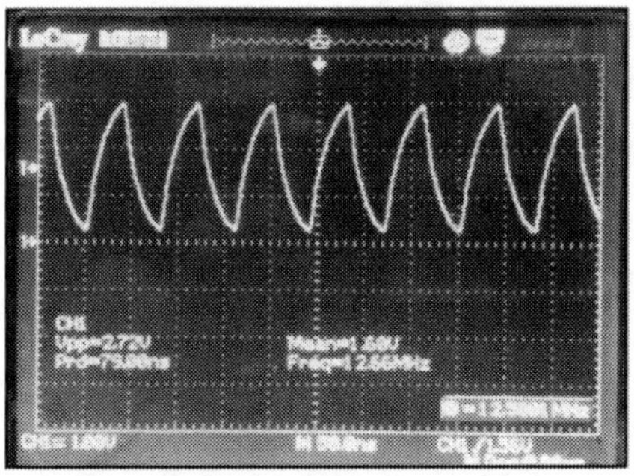

Figure 22. first frequency for high bits that have been locked at 12 MHz.

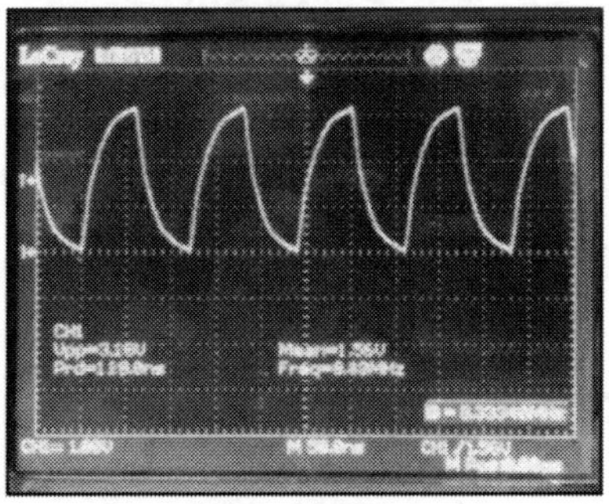

Figure 23. second frequency for low bits that have been locked at 8 MHz.

IOP Conf. Series: Materials Science and Engineering **99** (2015) 012028 doi:10.1088/1757-899X/99/1/012028

4. Conclusion

In conclusion, the low power of health monitoring system had been designed, developed and implemented based on IBC method. The IBC has a characteristic of low power consumption and low communication frequency without enlarging antenna size compare to RF technology. The signals do not leak through the skin of the human body thus the electromagnetic noise and interference has little influence on the transmissions of the signal. For the future recommendation work, the health monitoring system can be improved by enabling it to monitor the human health more efficiently and display the result to the external device such as oscilloscope or mobile phones.

References

[1] Zimmerman T G 1995 *Personal Area Networks (PAN): Near-Field Intra-Body Communication.*

[2] Alshehab A, Wu C T, Kobayashi N, Sok S and Shimamoto S 2012 *Intra-bodyhybrid communication scheme for healthcare systems* **2(1)** 1–12

[3] Callej´on M A, Roa L M, Reina-Tosina J and Naranjo-Hern´andez D 2012 *Study of Attenuation and Dispersion Through the Skin in Intrabody Communications Systems* **16(1)**

[4] Seyedi M, Kibret B, Lai D T H and Faulkner M 2013 *A Survey on Intra-body Communications for Body Area Network Application* **60(8)** 2067-2079

[5] Committee L S 2012 *IEEE Standard for Local and metropolitan area networks - Part 15.6: Wireless Body Area Networks* p 271

[6] Lucev Ž, Krois I, and Cifrek M 2012 *A capacitive intrabody communication channel from 100 kHz to 100 MHz* **61(12)** 3280–3289

[7] Iyer A, Tilak D, Rahate K and Sawant P 2012 *Comparison of Approaches to Intra-body Communication* **3(5)** 731-736

[8] Gray N 2006 *ABCs of ADCs Agenda - ABCs of ADCs* pp. 1–64

Wireless Sensors System for Broken Rotor bar Fault Monitoring using Wavelet Analysis

Mohammad Rezazadeh Mehrjou [1,2], Norman Mariun [1,2], Mahdi Karami [1,2], Norhisam Misron [1], Mohd Amran Mohd Radzi [1,2]

[1]Centre for Advanced Power and Energy Research (CAPER), Faculty of Engineering, Universiti Putra Malaysia, Serdang, Selangor 43400, Malaysia
[2] Department of Electrical& Electronic Engineering, Faculty of Engineering, Universiti Putra Malaysia, Serdang, Selangor 43400, Malaysia

E-mail: mehrjou.mo@email

Abstract. Accurate condition monitoring prevents unexpected failures in electrical systems including induction machines, and hence improves their performance significantly. To enhance the reliability of condition monitoring systems, wireless sensor systems are developed. In the recent years, researchers have placed considerable emphasis on developing cost-effective scheme using wireless sensor systems for fault diagnosis of equipments in industry. As broken rotor bar is one of the main causes of malfunction in electrical motors, this paper proposes a method for early detection of this failure in induction machines using wireless sensor system. In this respect, a test bed is developed where a sensor measures the motor current and then a microcontroller connected to this current sensor read and send the data to wireless sensor for remote real time data analysis. In the receiver unit, a Lab VIEW based program is developed to store data in a database and MATLAB is used for signal processing and fault.

1. Introduction

Induction machines (IMs), usually referred as the heart of industrial plants, are ubiquitous in industries. They facilitate production processes, operations and related services in various plants. However, during their operation, they are under inevitable stresses (electrical, mechanical, environmental, and thermal stresses) in the workplaces. These stresses create several Faults in different parts of IMs. These faults are one of the major reasons for poor performance of the industrial plants as they perturb the secure operation of the motor, imperil the standard manufacturing, and thus bring about the substantial expenditure [2].An effective maintenance technique can avoid generation of sever failures that leads to malfunction of these devices, and therefore decrease unwanted expenses by preventing unscheduled downtimes. Fault diagnosis of IM is a common maintenance problem and industries are seeking for reliable maintenance systems that offer improved performance of the electrical motor through minimizing the risk of catastrophic failures and thus decrease both operational and maintenance costs. Such maintenance systems include precise assessment of machine performance through suitable condition monitoring and fault prediction techniques. Accurate monitoring of IMs parameters like current, vibration and torque, is thus a key for reliable maintenance schemes.

There are two common strategies applied for managing the equipment maintenance in industries, which are preventive and reactive maintenance [3]. In the reactive maintenance strategy, device is let to run until it fails, and then it is repaired if it is repairable. Preventive maintenance is a scheduledbased strategy that assesses the lifetime of any device using failure statistics obtained from the similar system. Both these traditional maintenance strategies may result undesired outcomes including unnecessary downtime, offensive risk, high maintenance costs and inefficient usage of resource. The conventional maintenance methods for electrical motors are typically based on using portable or handheld data collection equipments [3]. Performance of electrical motor is checked manually by sending labor to the area where these devices are installed. This method is then time consuming and cumbersome and more importantly cannot provide precise condition monitoring of the equipment, which in turn results in inaccurate assessment of the site. This approach also impedes frequent monitoring of the device, where unexpected failures can threat safe performance of the motor. On-line condition monitoring of electrical motors based on predictive maintenance can offer accurate fault detection in the device and thus avoid unforeseen failures in them. Due to availability of data achieved through on-line monitoring, the reliability of the maintenance program is also improved [4]. Online predictive maintenance for IMs can be integrated, through wireless data collection technologies [3,5].Wireless sensing can enhance the availability of system by connecting it to the selected equipment. This capability make them appropriate for different applications, particularly where there is difficulty in access to the device or cabling to it [6,7]. Another advantage of wireless sensing is a number of different locations can be monitored through one station. Accordingly, using wireless sensor will minimize both fixed investments and operational costs of the plants by reducing the amount of cables installed and the number of labors needed to manually check the devices and accessories in the (working place) sites. Wireless sensors are reliable, smart, cost-effective, energy-saving and easy-to-install sensors with capability of fast data acquisition [5].

To design a good maintenance scheme, the first step includes identifying the type of equipment, its operational characteristics and type of failures may occur. Based on this information, one or more suitable condition monitoring techniques using wireless sensors are then selected. The final step involves applying precise and reliable signal processing for the data achieved through sensors. With the aim of developing a cost-effective system for maintenance, this article presents a research using wireless sensor for detection of rotor bar breakage in squirrel cage IM. Current signals from IMs are acquired and then, wavelet analysis of signal is employed for fault diagnosis. In this research, a wireless sensor will play a critical role in monitoring and supervising electrical motor in remote areas by reporting the measured data to find the fault in IMs. This system has also feasibility to apply for intelligent fault diagnosis.

1.1. Rotor fault

Different parts of IM operating in industrial plants are subjected to inevitable stresses that damage the component and may cause further malfunctions. Rotor faults have attracted a great consideration since these failures result in serious subsequent failures in other parts of motor. Bonnett and Soukup, in their comprehensive study, addressed various stresses that create failures in rotor and identified the causes [1]. There are three common faults in the rotor including breakage or cracking of bars or end-rings, rotor eccentricity and rotor bow [8].

Among these failures, it is fair to note that breakage of rotor cage bars is the main one. Any rotor bar, during operation of motor, can be partially or completely broken because of improper geometry design and/or stresses of rotor. Presence of broken rotor bar, like any other failures, causes a decrease in efficiency of motor and that increase operational cost. Breakage of rotor bar also deteriorates the state of neighboring bars progressively and brings other secondary faults that developed serious malfunctions in the motor [8].Once a bar breaks, the current flow in the neighboring bars boarding the broken bar rises up to 50% of rated current [9] and thus these bars are overheated [10]. Such extra heating in the bars causes the rotor bow and the eccentricity [11], and hence unbalanced currents and torque pulsation that consequently decrease average torque of motor [12]. Accordingly, early detection of breakage in rotor bar is crucial not only for protection of rotor but also for impede secondary faults in the motor.

IOP Conf. Series: Materials Science and Engineering **99** (2015) 012029 doi:10.1088/1757-899X/99/1/012029

1.2. Motor Current Analysis

The key for successful failure diagnosis of electrical motor relies on the availability of information that allows diagnosis of a motor condition. Condition monitoring techniques have been continuously developed over the years, which resulting a range of available methods for diagnosis of failures in IMs. Various techniques have been explored by using acoustic analysis, electromagnetic field monitoring, motor current signature analysis, induced voltage, instantaneous power, vibration and etc. The most common technologies used in the condition monitoring of rotor fault are presented in [8,13]. Among all condition monitoring techniques addressed, stator current analysis is the most popular one due to its simplicity in measurement, high reliability and accuracy. Besides that, this method is cost-effective as current sensors are easy to be implemented and do not require costly additional transducers.

When a failure is generated in the electrical motor, depends on the severity of this fault, some parameters of the machine change. For instance, any irregularity in the rotor causes an uneven distributed current in it. An ideal IM contains a single component corresponding to the supply frequency in current spectrum. Any asymmetry in IM causes other components to be appeared in spectrum of stator current. Current does not flow through broken rotor bar, and hence no magnetic flux creates around the breakage bar. Consequently, a non-zero backward rotating field that rotates at the slip frequency speed with respect to the rotor is generated. Any variation in the magnetic field of rotor induces a specific harmonic in the currents spectrum, which is added on the stator current. These harmonics have frequencies as $(1\pm2ks)f_s$, where $k=1,2,3\ldots$ is an integer number, and s and f_s are slip and fundamental frequency, respectively [14].

1.3. Signal Processing Technique

A raw signal is normally presented as time domain graphs that specify how parameters vary over time. For extracting the information from raw signal, suitable signal processing technique needs to be applied. Signal processing provides necessary information extracted from a raw signal by applying a mathematical correlation. Signal processing techniques can provide principal information about the specific failure in the motor that makes the detection of the failure feasible. Basically, signal processing can be accomplished either in time analysis, frequency analysis or time-frequency analysis.

Frequency analysis based on fast Fourier transform is the most usual signal analysis technique utilized for fault detection in IMs [8]. However, there are some inconsistencies in ability of using Fourier analysis for fault detection [15,16] and thus advanced signal processing techniques are proposed for more accurate and reliable fault detection. These advanced techniques, like Short-time Fourier transform and wavelet transform are dependent upon time-frequency domain analysis of signal that provides simultaneous time and frequency analysis.

In overcoming the problem of resolution in Short-time Fourier transform, wavelet transform was introduced, this technique has been widely selected in different areas of signal processing. Wavelet analysis provides both the frequency and time information of a signal by expressing original signal in an oscillatory functions series. These functions are defined with respect to both frequency and time. In wavelet transform, the signal is divided into time-scale space where the dimension of window in both time and frequency (scale) are flexible [17]. The Wavelet transform localizes the information of signal in the time–scale plane which makes it suitable for non-stationary analysis signals. A practical version of discrete wavelet transform, called wavelet multi-resolution analysis, was introduced by Mallat [18]. In this algorithm, original signal is disintegrated into a series of minor waves which is a member of wavelet family. The energies associated to the disintegrated signal for each detail and approximation coefficient can be calculated and refer in [19].

The multi-resolution analysis makes use of discrete dyadic wavelet and excerpts the approximations of the raw signal at various levels of resolution. The low-resolution and high resolution part of the raw signal represents approximation and detail respectively. The approximations and details can be determined by applying the low and high pass filters. In the multi-resolution analysis, decomposition procedure is repeated as the approximations are divided consecutively, while the analyses in details never proceed further. Accordingly, in wavelet analysis, a signal is decomposed into many lower resolution components through successive decomposition of approximations. This procedure is called "wavelet decomposition tree" and the corresponding algorithm is shown in Figure 1 [20].

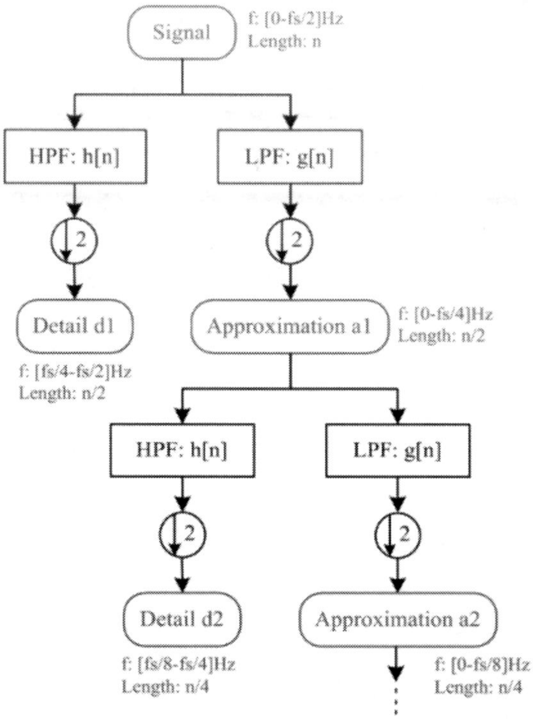

Figure 1. Dyadic Wavelet Decomposition Algorithm

Wavelet analysis is a developed and enhanced technique for fault diagnosis in many devices including IM. However, its performance strongly depends on some parameters that need to be considered. For instance, type of wavelet function applied determines the result of signal decomposition. Different wavelet functions (such as Daubechies 8 [21,22], Daubechies10 [23], Daubechies12 [24], Daubechies 23 [25], Coiflet [26,27], Coiflet4 [28,29] and Meyer [16]) have been used for decomposition of current signal with purpose of rotor bar breakage diagnosis. Besides the effect of mother wavelet on the results of signal decomposition, various features, like energy of decomposed signal can be extracted from decomposed signal and used as fault characteristic feature [30,31,32]. Despite the multiple existing references, the scattering and the diversity of wavelet functions and features used in signal processing procedures makes a reliable comparison of the results difficult. Therefore, a comparative study that concentrates on the outcomes of different wavelet functions for early fault detection is essential. This study then intends to investigate the effects of the wavelet function for early broken rotor bar detection using energy feature of stator current signal in different load.

2. Experimental Section

2.1. Setup Configuration

In order to examine the capability of wireless system for early broken rotor bar detection, IMs under healthy and faulty situation were observed under different levels of load. The IM was joined to a generator, which works as a load simulation. Experimental data including stator current was acquired through a current sensor. The architecture of experimental setup used in this study is shown in Figure 2.

IOP Conf. Series: Materials Science and Engineering **99** (2015) 012029 doi:10.1088/1757-899X/99/1/012029

Figure 2. Architecture of experimental setup

The characteristics of IM used in this study are: rated power 1hp, star connection, rated voltage 415 V, rated current 2.2 A, 6 poles, speed 1000 rpm. The stator slots and rotor bars number are 36 and 28 respectively. The bar breakage was done artificially in the workshop by drilling the bar. The stator current signal of one phase was measured under a fixed condition using a sensor (ACS756) that provides precise and economical solutions for AC or DC current measurement in industrial systems. The instruments for checking the condition of data collection in each test are torque and speed sensor. An 8 bit processor of ATmega328 was employed to control 16 bit analog digital conversion and wireless system. A wireless sensor checked the status of system in real-time. Base-station includes server and fault diagnosis system. The server was used for receiving data and feeding it into the data analysis part. In this program, the incoming data was considered and read as string. Therefore, the data was converted to decimal value before it was sent to a database. In this database portion, the connectivity between LabVIEW and database was prior to be established. Once the connection was ensured, the incoming data would be fed in to the database table. The specification of wireless sensor system is exhibited in Table 1. Figure3 demonstrates the experimental setup used in this study.

Table 1. Wireless sensor specification

Processor (ATmega328)	Wireless Bridge
Communication: UART	Name: XBEE-PRO-S1
Embedded: IST, JTAG	Company: MaxStream
Memory: 32 K	Networking: 802.15.4
Clock: 16 MHz	Standard: IEEE 802.15.4

2.2. Fault detection algorithm

In order to examine the capability of wireless sensor for early broken rotor bar detection, IMs under healthy and faulty situation were observed at various levels of load. An IM with one broken rotor bar is considered as faulty one. Before data analysis, the raw current signal was re-sampled by synchronizing at phase 0. This preprocessing on the raw current signal is needed since the unsynchronized signal influences the results of diagnosis [16,30]. The total re-sampled cycles of the signals were five cycles. In the following step, all wavelet functions provided by wavelet tool box in MATLAB were examined to find the suitable Wavelet function. In multi-resolution analysis, a signal is disintegrate into various levels and each level enclosed by definite range of frequency. To determine the appropriate decomposition level, the original signal was first decomposed into 8 levels using wavelet decomposition technique. Each decomposition level has its own selected frequencies and detailed coefficients. Table 2 presents the frequency ranges corresponding to each detail. It is clear that the fundamental frequency as well as sidebands frequency is existed in the frequency band matching to level 8 of detail. These sidebands are commonly used to detect rotor bar breakage in IM. Three different levels of load, the nearest left and right sidebands, when k equals to one, were obtained and summarized in Table 3. The results indicate that when load increases, the sidebands frequency also changes. This change is attributed to the broken bar in the motor and then 8th level of decomposition can be chosen for wavelet analysis of current signal with purpose of broken rotor bar detection.

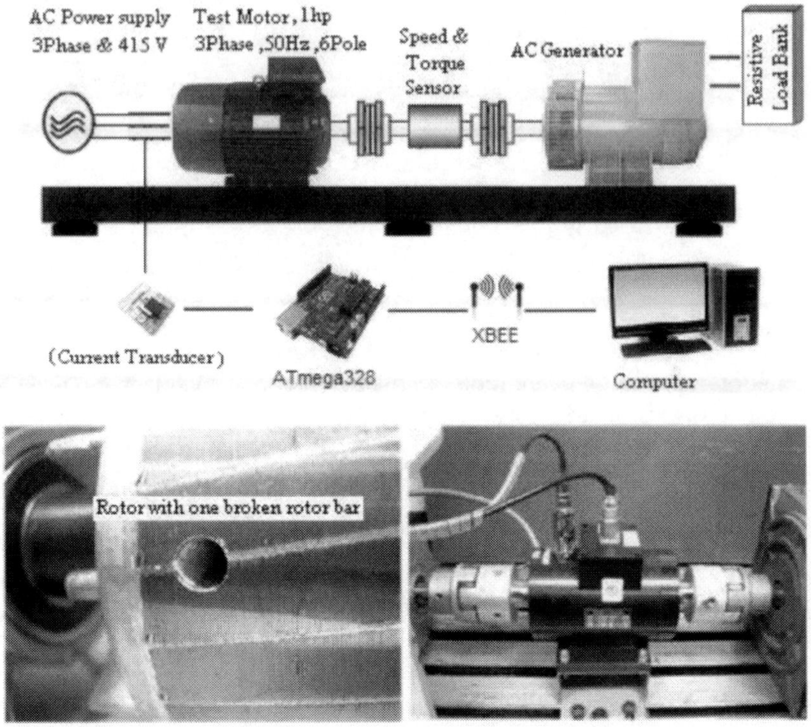

Figure 3. Experimental setup

Table 2. Frequency bands for signal decomposition using wavelet.

Level of Decomposition	Frequency Bands (Hz)
Detail 1	10000-5000
Detail 2	5000-2500
Detail 3	2500-1250
Detail 4	1250-625
Detail 5	625-312.5
Detail 6	312.5-156.25
Detail 7	156.25-78.12
Detail 8	**78.12-39.06**
Approximation 8	39.06-0

Table 3. Broken rotor bar sidebands frequency at different load levels.

Torque (Nm)	**Speed (rpm)**	**Slip**	**(1-2s)f_s (Hz)**	**(1+2s)f_s (Hz)**
2.70	984	0.016	48.4	51.6
4.00	976	0.024	47.6	52.4
6.25	960	0.040	46.0	54.0

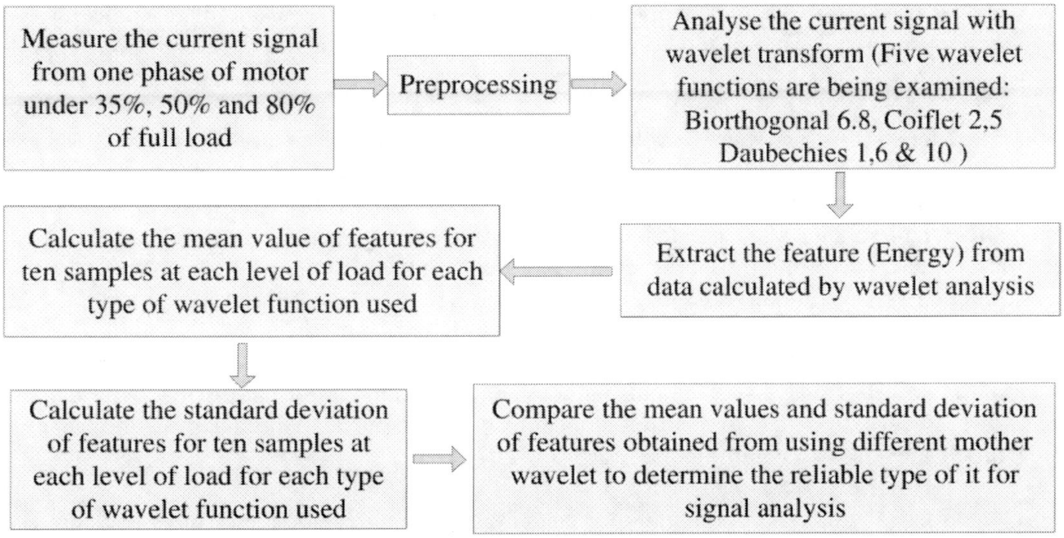

Figure 4. The flow chart diagram for broken rotor bar detection

Those wavelet functions that generate a decomposed signal with higher energy value in determined level of decomposition were selected for broken rotor bar detection. Energy coefficient, as characteristic feature, was extracted from wavelet decomposition of current signal. Each test was repeated 10 times and the average value of the features were calculated and used for further investigation. Figure 4 depicts the procedure used for wavelet analysis of current signal.

3. Results and Discussion

The energy value of current signal decomposed at 8th detail was determined and investigated as a fault characteristic feature for diagnosis of broken rotor bar. Tables 4 and 5 present the mean values for the characteristic features of the 8th detail decomposition for healthy and faulty machines under different types of mother wavelet when various levels of load were applied.

Table 4. Characteristic features of the 8th detail decomposition for healthy motor under different levels of load when various types of wavelet function were used

Feature	Load (%)	Bior 6.8	Coiflet 2	Coiflet 5	Daub 1	Daub 10
Energy detail 8	35	4658	**4803**	7029	2504	7898
	50	4763	**4911**	**7196**	2570	8096
	80	5242	5411	7929	2849	**8939**

Table 5. Characteristic features of the 8th detail decomposition for faulty motor under different levels of load when various types of wavelet function were used

Feature	Load (%)	Bior 6.8	Coiflet 2	Coiflet 5	Daub 1	Daub 10
Energy detail 8	35	4713	**4767**	7057	2513	8006
	50	4782	**4900**	**7194**	2585	8152
	80	5729	5921	8637	3095	**7854**

It has been firmly confirmed that if any failure occurs in the machine, the faulty situation has higher characteristic features values than the healthy ones [24]. However, the data presented in Tables 4 and

5 indicates some inconsistencies. For instance, when wavelet function of Coiflet 2 and Daubechies 10 were used for signal analysis, the energy values of faulty conditions were smaller than healthy.
No such inconsistencies were observed when Daubechies 1 and/or Biorthogonal 6.8 were applied for wavelet decomposition of the signal. These observations express that not only the type of wavelet function used for signal analysis influence the result but also the characteristic feature needs to be taken into account. As just two wavelet functions, Daubechies 1 and/or Orthogonal 6.8, showed reliable information, these two were further investigated. Therefore, the standard deviations of all features for these two wavelet functions were computed as presented in Table 6. Comparatively, Daubechies1 had smaller standard deviation that indicates the sampled data had smaller dispersion from the average. Therefore, the characteristic features obtained from decomposition of current signal using Daubechies 1 were more reliable to be used for early detection of fault in squirrel-cage IM than Biorthogonal 6.8. This study proved that there is a significant difference in ability of wavelet functions for accurate signal analysis and interpretation. Therefore, the arbitrary selection of wavelet function for wavelet analysis of signal cannot supply reliable information for fault diagnosis in machine. It is recommended that different characteristic features to be selected and observed for increasing the accuracy of condition monitoring.

Table 6. Statistical parameters computed for characteristic features obtained from wavelet analysis of the current signal using Biorthogonal 6.8 and Daubechies 1

			Healthy				Faulty (one BRB)			
			Biorthogonal 6.8		Daubechies 1		Biorthogonal 6.8		Daubechies 1	
Feature	Torque (Nm)	Load (%)	Mean	STD	Mean	STD	Mean	STD	Mean	STD
Energy detail 8	2.70	35	4658	6.85	2505	**4.93**	4713	26.54	2513	**19.28**
	4.00	50	4763	9.55	2570	**7.85**	4782	59.89	2585	**29.95**
	6.25	80	5242	8.21	2850	**6.45**	5729	48.94	3095	**24.66**

4. Conclusions

This article develops a method for diagnosis of broken rotor bar in squirrel-cage IM using a wireless sensors system. Different types of wavelet functions were applied to decompose the current signal. These functions were then compared in screening the feature in related to the present fault. The characteristic feature observed was the energy value of the decomposed signal. Wireless sensor used indicates a good performance in fault detection problem of IM. This study also proved that the types of mother wavelet used for signal processing considerably influence the reliability of the diagnostic method. Among different wavelet functions examined in this research for current signal decomposition, Daubechies 1 provided more reliable information for early diagnosis of rotor bar breakage in IM. An increase in the value of characteristic feature was observed for faulty situation as compared to the healthy situation that worked under the same level of the load.

Acknowledgments

The authors would like to express their gratitude to Ministry of Education Malaysia for financial support through grant number FRGS-5524356 and Universiti Putra Malaysia for the facilities provided during this research work.

References

[1] Bonnett A H and Soukup G C 1988 Analysis of rotor failures in squirrel cage induction motors *IEEE Trans. Ind. Applicat* **24** 1124.

[2] Akhondi M R, Talevski A,Carlsen S and Petersen S 2010 Applications of wireless sensor networks in the oil, gas and resources industries *Proc. of the 24th IEEE International Conference on Advanced Information Networking and Applications (AINA)* pp 941-948.

IOP Conf. Series: Materials Science and Engineering **99** (2015) 012029 doi:10.1088/1757-899X/99/1/012029

[3] Hashemian H M 2011 Wireless sensors for predictive maintenance of rotating equipment in research reactors *Annals of Nuclear Energy* **38** 665.

[4] Hashemian H M 2006 Aging Management through On-Line Condition Monitoring," *Proc.of the PLIM + PLEX Conference* (Paris, France).

[5] Son J D, Niu G, Yang B S, Hwang D H and Kang D S 2009 Development of smart sensors system for machine fault diagnosis *Expert Syst Appl*, **36** 11981.

[6] McLean C and Wolfe D 2002 Intelligent Wireless Condition-Based Maintenance *Sensors Magazine* **19** 142.

[7] Spencer B F, Ruiz-Sandoval M and Kurata N 2004 Smart sensing technology: opportunities and challenges *Struct. Control. Health Monit.* **11** 349.

[8] Mehrjou M R, Mariun N, Hamiruce Marhaban M and Misron N 2011 Rotor fault condition monitoring techniques for squirrel-cage induction machine-A review Mech Syst Signal Pr **25** 2827.

[9] Bellini A, Filippetti F, Tassoni C and Capolino G A 2008 Advances in diagnostic techniques for induction machines *IEEE Trans. Ind. Electron.* **55** 4109.

[10] Fiser R and Ferkolj S 2001 Application of finite element method to predict damaged induction motor performance *IEEE Trans. Magn.* **37** 3635.

[11] Finley W R, Hodowanec M M and Holter W G 2000 Analytical approach to solving motor vibration problems *IEEE Trans. Ind. Applicat.* **36** 1467.

[12] Ilonen J, KamarainenJ K, Lindh T, Ahola J, Kälviäinen H and Partanen J 2005 Diagnosis tool for motor condition monitoring *IEEE Trans. Ind. Applicat.* **41** 963.

[13] Nandi S, Toliyat H A and Li X 2005 Condition monitoring and fault diagnosis of electrical motors-A Review *IEEE Trans. Energy Conversion* **20** 719.

[14] Kliman G B, Koegl R A, Stein J, Endicott R D and Madden M W 1988 Noninvasive detection of broken rotor bars in operating induction motors *IEEE Trans. Energy Conversion* **3** 873.

[15] Antonino-Daviu J A, Riera-Guasp M, Folch J R and Palomares M P M 2006 Validation of a new method for the diagnosis of rotor bar failures via wavelet transform in industrial induction machines *IEEE Trans. Ind. Applicat* **42** 990.

[16] Lee S H, Cheon S P, Kim Y and Kim S 2006 Fourier and Wavelet Transformations for the Fault Detection of Induction Motor with Stator Current, *Proc. International Conference on Intelligent Computing* (ICIC 2006) pp 557-569 (Kunming, China).

[17] Daubechies I 1990 The wavelet transform, time-frequency localization and signal analysis *IEEE Trans. Inform. Theory.* **36** 961.

[18] Mallat S G 1989 A theory for multiresolution signal decomposition: the wavelet representation *IEEE Trans. Pattern Anal. Machine Intell.* **11** 674.

[19] Hedayati Kia S, Henao H and Capolino G A 2009 Diagnosis of Broken-Bar Fault in Induction Machines Using Discrete Wavelet Transform Without Slip Estimation," *IEEE Trans. Ind. Applicat* **45** 1395.

[20] Polikar R, Udpa L, Udpa S S and Taylor T Frequency invariant classification of weld inspection signals *IEEE Trans. Ultrason., Ferroelect., Freq. Contr.* **45** 614.

[21] Faiz J, Ebrahimi B M, Akin B and Asaie B 2009 Criterion Function for Broken-Bar Fault Diagnosis in Induction Motor under Load Variation Using Wavelet Transform *Electromagnetics* **29** 220.

[22] Arashloo R S and Jalilian A 2010 Design, Implementation and Comparison of Two Wavelet Based Methods for the Detection of Broken Rotor Bars in Three Phase Induction Motors *Proc.of the 1st Power Electronic & Drive Systems & Technologies Conference* (PEDSTC), pp 345-350.

[23] Cusidó J, Romeral L, Ortega J A, Rosero J A and Espinosa A G 2008 Fault Detection in Induction Machines Using Power Spectral Density in Wavelet Decomposition *IEEE Trans. Ind. Electron.* **55** 633.

[24] Abbaszadeh K, Milimonfared J, Haji M and Toliyat H A 2001 Broken bar detection in induction motor via wavelet transformation *Proc. of the 27th annual conference of the IEEE industrial electronics society* pp 95-99.

[25] Cusido J, Romeral L, Espinosa A G, Rosero J A and Ortega J A 2007 Fault detection in Induction Machines by using Continuous and Discrete Wavelet Decomposition *Proc. of the European conference on power electronics and applications* pp 1-8.

[26] Bae H, Kim Y T, Lee S H, Kim S and Lee M H 2005 Fault diagnostic of induction motors for equipment reliability and health maintenance based upon Fourier and wavelet analysis *Artif. Life and Robot* **9** 112.

[27] Mariun N, Mehrjou M R, Hamiruce Marhaban M and Misron N 2011 An experimental study of induction motor current signature analysis techniques for incipient broken rotor bar detection. *Proc. of International Conference on Power Engineering, Energy and Electrical Drives* pp 1-5 (Malaga,spain).

[28] Sadeghian A, Ye Z and Wu B 2009 Online Detection of Broken Rotor Bars in Induction Motors by Wavelet Packet Decomposition and Artificial Neural Networks *IEEE Trans. Instrum. Meas.* **58** 2253.

[29] Ye Z, Sadeghian A and Wu B 2006 Mechanical fault diagnostics for induction motor with variable speed drives using Adaptive Neuro-fuzzy Inference System *Electr. Pow. Syst. Res.* **76** 742.

[30] Bae H, Kim Y T, Lee S H, Kim S and Lee M H 2005 Fault diagnostic of induction motors for equipment reliability and health maintenance based upon Fourier and wavelet analysis *Artif. Life and Robot* **9** 1125.

[31] Mehrjou M R, Mariun N, Hamiruce Marhaban M and Misron N 2011 Evaluation of fourier and wavelet analysis for efficient recognition of broken rotor bar in squirrel-cage induction machine *Proc. of IEEE International Conference on Power and Energy* pp 740-743 (Malaysia).

[32] Lee S H, Wang Y Q and Song J 2010 Fourier and wavelet transformations application to fault detection of induction motor with stator current *J Cent South Univ T* **17** 93.

IOP Conf. Series: Materials Science and Engineering **99** (2015) 012030 doi:10.1088/1757-899X/99/1/012030

Effect of annealing in hydrogen atmosphere on ZnO films for field emission display

Zurita Zulkifli[1,2*], Subash Sharma[1], Sachin Shinde[1], Golap Kalita[1], and M. Tanemura[1]

[1]Department of Frontier Materials, Graduate School of Engineering, Nagoya Institute of Technology, Gokiso-cho, Showa-ku, Nagoya, 466-8555, Japan,
[2]Faculty of Electrical Engineering, Universiti Teknologi Mara, 40450 Shah Alam, Selangor, Malaysia.
E-mail: z_zurita@yahoo.com

Abstract. Surface morphology, crystallinity, conductivity and optical transmittance of ZnO films can be modified by annealing process. Hydrogen is one of the popular annealing gases as well as nitrogen, argon, oxygen and air which are commonly used for thin film cleaning or the removal of native oxide. In general, annealing is done at high temperatures (> 600degC) to improve the film properties. From a view point of environment, however, lower annealing temperature is preferable. In this work, low annealing process was challenged to understand the effect of annealing temperature on properties of ZnO thin films and nanostructured film grown on glass substrates for transparent field emission device applications. The annealing temperature employed was 100, 200 and 450°C at 100 sccm hydrogen flow rate. ZnO thin films were deposited by RF magnetron sputtering. The ZnO thin films were characterized by X-ray diffraction analysis (XRD), Atomic Force Microscopy (AFM), UV-VIS and Raman spectroscopy. The sheet resistances reduced about 15 kohm/sq at low annealing temperature. By contrast, the optical transmittance did not show any significant changes after annealing. The FE current density increased after the ZnO nanostructures film was annealed in 100°C. The results obtained could motivate a surface treatment for flexible ZnO thin film since the substrate is always suffered by heat.

1. Introduction

ZnO is a semiconductor material which has been widely prepared as a thin film in many electronic applications. However, semiconductor material has a low electrical property. Thus, due to the limitation of a semiconductor, several techniques have been employed to improve the semiconductor properties to be more conductive and can enhance the device performance in myriad applications. One of the techniques to improve the quality of the thin film is annealing. Annealing is known to be an effective method to improve the crystalline qualities, conductivity and surface properties of the thin film with an appropriate annealing parameter [1-5]. Annealing process also could induce the thin film to become more conductive or can turn to be more insulating than before annealing. This could be happened due to the reaction on the surface of the thin films such as moisture removal, oxygen defect creation, molecules dissociations and changes in surface roughness. However, some of the reactions such as molecules dissociations, coalescence process will only take place in higher annealing process as reported elsewhere in various type of atmosphere [6-7]. From the literature, the annealing process at higher temperature tends to be more effective in improving surface and crystalline of thin film compared to low annealing temperature [8-9]. However, there is also a few reports on longer annealing time that affects the properties of ZnO thin films [10]. Due to some limitation in higher annealing temperature on some substrate materials such as plastic substrate and polymer substrate, the effect on low temperature will be an attractive issue to be tackled. Thus, by annealing at relatively low temperature, the morphology structures and properties of ZnO thin films may be altered. Another factor to be considered in the annealing process is the ambient condition. Annealing can be done in various type of atmosphere such as in air, vacuum, argon, nitrogen, oxygen and hydrogen. Some of the atmosphere will be only effective at higher temperature process but hydrogen has an advantage in which it can easily diffuse in solid film even at low temperature [11]. Thus, in this research, the annealing process was done by the influence of high hydrogen flow rate (100 sccm) at 100°C, 200°C and 450°C for a significant period of time at 60 minutes. We presume that higher flow rate of hydrogen may change the properties of the ZnO thin films at low annealing temperature. Therefore, the effect of hydrogen on the surface morphology, sheet resistance, crystalline quality, transmittance and field emission (FE) of the ZnO thin film was investigated.

2. Experimental Details

Three samples of ZnO thin films were deposited on glass substrate by RF sputtering. The substrate holder was heated up to 200°C prior to the deposition process. ZnO thin films were annealed in the quartz tube with hydrogen atmosphere for 60 minutes. The annealing temperature was varied at 100°C (Sample A), 200°C (Sample B) and 450°C (Sample C) at a fixed hydrogen flow rate of 100 sccm. Surface morphology of the ZnO thin films were characterized by AFM measurement and the crystalline quality was analyzed from the diffraction pattern obtained by x-ray diffraction spectroscopy. The sheet resistance of the thin films was measured by four point probe. For the optical properties of the thin films, the transmittance spectra were obtained from Ultraviolet-Visible (UV-VIS) spectroscopy from 300-800 nm. The thickness of films was measured by surface profiler. Nanostructured ZnO sample prepared by ion beam irradiation was annealed for 60 minutes and the field emission measurement was done in a vacuum condition of $5x10^{-4}$ Pa to obtain the field emission properties after annealing. Characterizations properties were taken before and after annealing for each sample.

3. Results and Discussion

3.1 Surface morphology

Figure 1(a) and (b) shows the surface roughness of ZnO thin films for sample A. The surface roughness average was decreased after the annealing process at 100°C from 43.5 nm for as-grown thin film to 39.2 nm after 60 minutes annealing process. The decreasing of the surface roughness at low temperature indicates that hydrogen had removed unwanted particles on the thin film surface and improves the surface morphology.

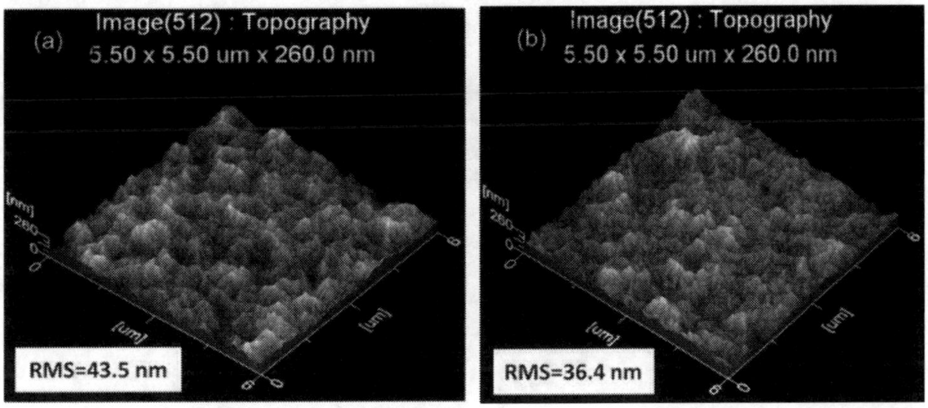

Figure 1: Surface morphology of ZnO thin film (a) before and (b) after annealed in hydrogen at 100°C

Figure 2 (a) and (b) exhibit the surface roughness of sample B before and after annealing, respectively. The result does not show any significance changes after the 60 minutes annealing at 200°C. Even though moisture can be evaporated, we assume some of the OH species on the surface also has been dissociated and thus maintain the properties of the thin film as before.

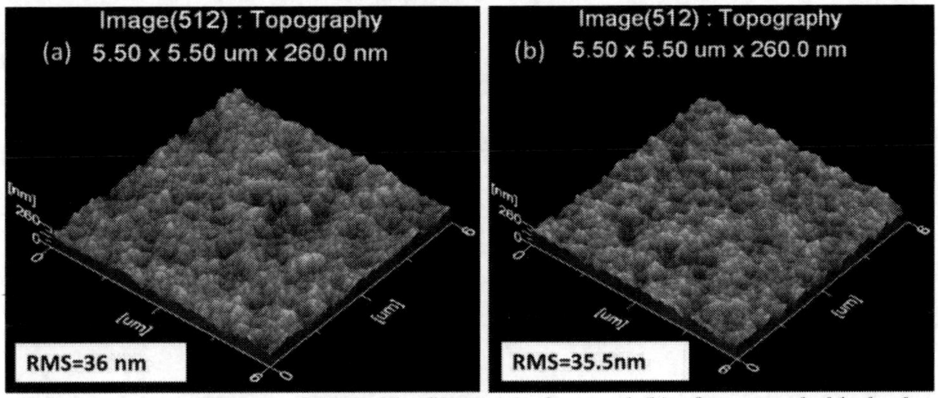

Figure 2: Surface morphology of ZnO thin film (a) before and (b) after annealed in hydrogen at 200°C

Figure 3 (a) shows the surface roughness of sample C before annealing and figure 3 (b) shows the higher surface roughness of sample C after annealing in high temperature, 450°C. Increasing of surface roughness was attributed to the increasing of grain size by coalescence process at high temperature. Some voids are arises due to the movement of grain boundaries. As the surface roughness increases, it is more likely that negatively charged oxygen-related species are adsorbed on the surface of the crystallites which in turn act as electron traps.

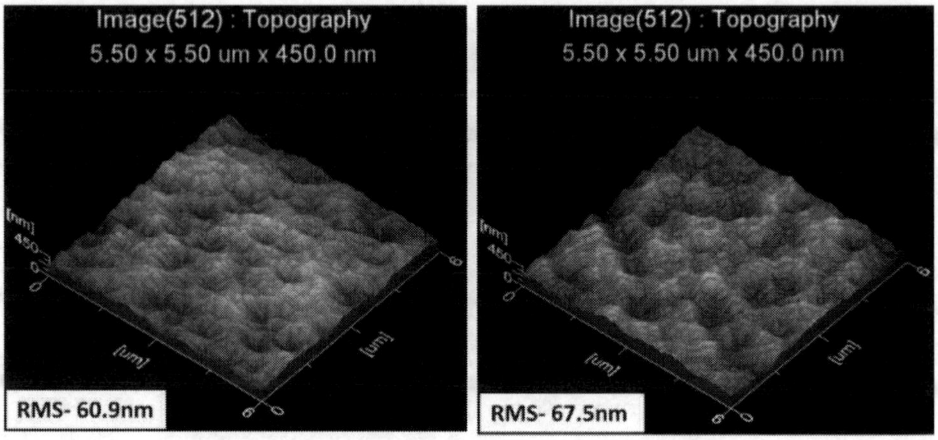

Figure 3: Surface morphology of ZnO thin film (a) before and (b) after annealed in hydrogen at 450°C

3.2 Crystalline quality

Figure 4, 5 and 6 shows the XRD patterns of as-grown ZnO thin film and annealed film for 100°C, 200°C and 450°C, respectively. The dominant peaks from the diffraction pattern shows all thin films have strong (002) orientation along the c-axis preferential growth. The 2θ angle for as-grown thin film is 34.32° and the annealed at 100°C thin film shows at 34.28°. At 200°C, ZnO atoms were recrystallize when the annealing temperature increase to 200deg based on the stronger diffraction peak at 34.36° after the annealing process. This could be happened due to the high flow rate of hydrogen gas has induced self-arrangement of ZnO molecules at the surface of the thin films and improve any dislocation of ZnO atoms.

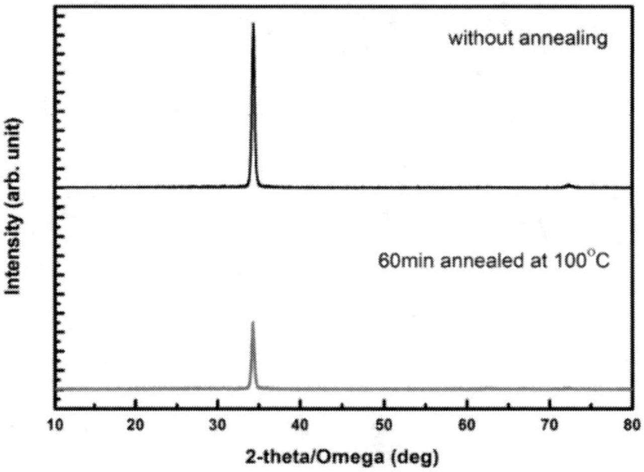

Figure 4: XRD pattern of ZnO thin film annealed at 100°C

ZnO atoms were recrystallized when the annealing temperature increase to 200°C based on the stronger diffraction peak at 34.36° after the annealing process as shown the XRD pattern in figure 5. This could be happened due to the high flow rate of hydrogen gas has induced self-arrangement of ZnO molecules at the surface of the thin films and improve any dislocation of ZnO atoms.

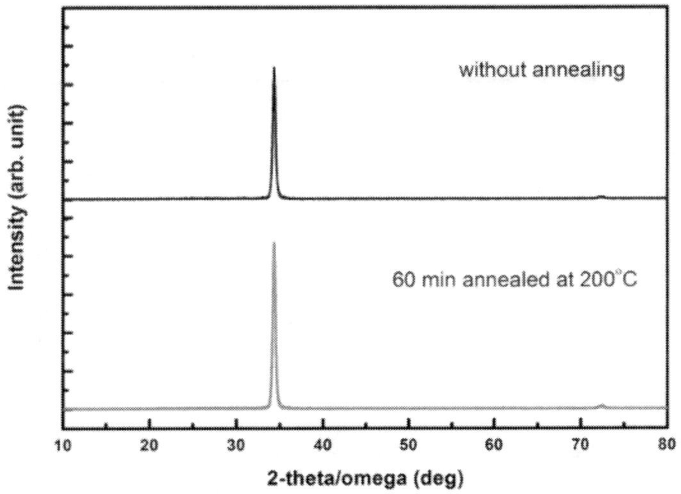

Figure 5: XRD pattern of ZnO thin film annealed at 200°C

When the annealing temperature further increase to 450°C, the diffraction peak at 34.38° become weak. The reason can be assumed due to the effect of atoms in the glass substrate. At around 600deg, the glass substrate generally start to become soft, thus at the higher temperature, movement of the atoms in glass substrate may increase and affect the crystallinity of ZnO thin film.

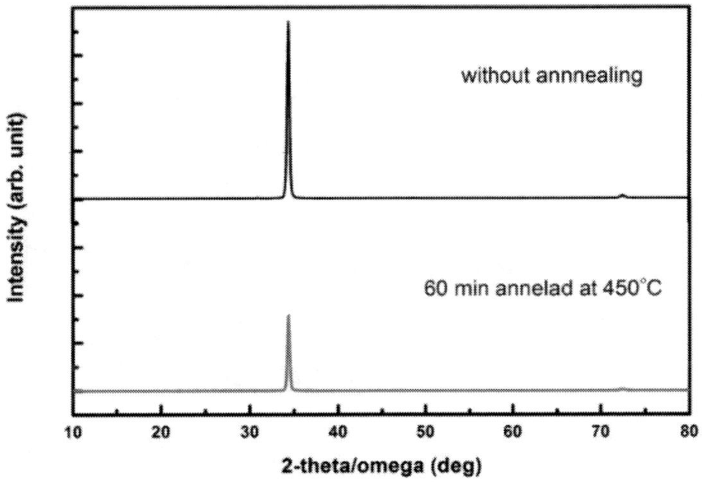

Figure 6: XRD pattern of ZnO thin film annealed at 450°C

Thermal activation will occur at low temperature. The charge transfer was normally dominated by the tunneling effect [12]. However, if the barrier is larger enough, the mobility of electron will be poor. The barrier can be in terms of moisture, OH species and others. Table 1 shows the effect of annealing on sheet resistance of the ZnO thin films. Hydrogen annealing at 100 °C showed sheet resistance improved in 49% of compared to the as-grown thin film. The hydrogen diffused in the grain boundary on the thin film thus removed the moisture at that particular area. The increasing of sheet resistance at higher temperature is due to the formation of voids which

IOP Conf. Series: Materials Science and Engineering **99** (2015) 012030 doi:10.1088/1757-899X/99/1/012030

could adsorb oxygen-related species and act as an electron trap. Besides that, OH species also may dissociate and form a rich-oxygen surface of ZnO thin film.

Table 1: ZnO thin film properties

Sample	Annealing temperature (°C)	Annealing time (min)	Sheet resistance Ω/□		Transmittance	
			Before annealing	After annealing	Before annealing	After annealing
A	100	60	29.6k	15k	78%	79%
B	200	60	1.08k	1.05k	64%	64%
C	450	60	196	1.8k	88%	89%

3.3 Transmittance properties

Table 1 shows the transmittance properties of annealed ZnO films. From the results, there are no significance effect of annealing can be observed at 550 nm visible wavelength. Figure 7 (a) – (c) shows the transmittance properties of ZnO film annealed at 100°C, 200°C and 450°C, respectively.

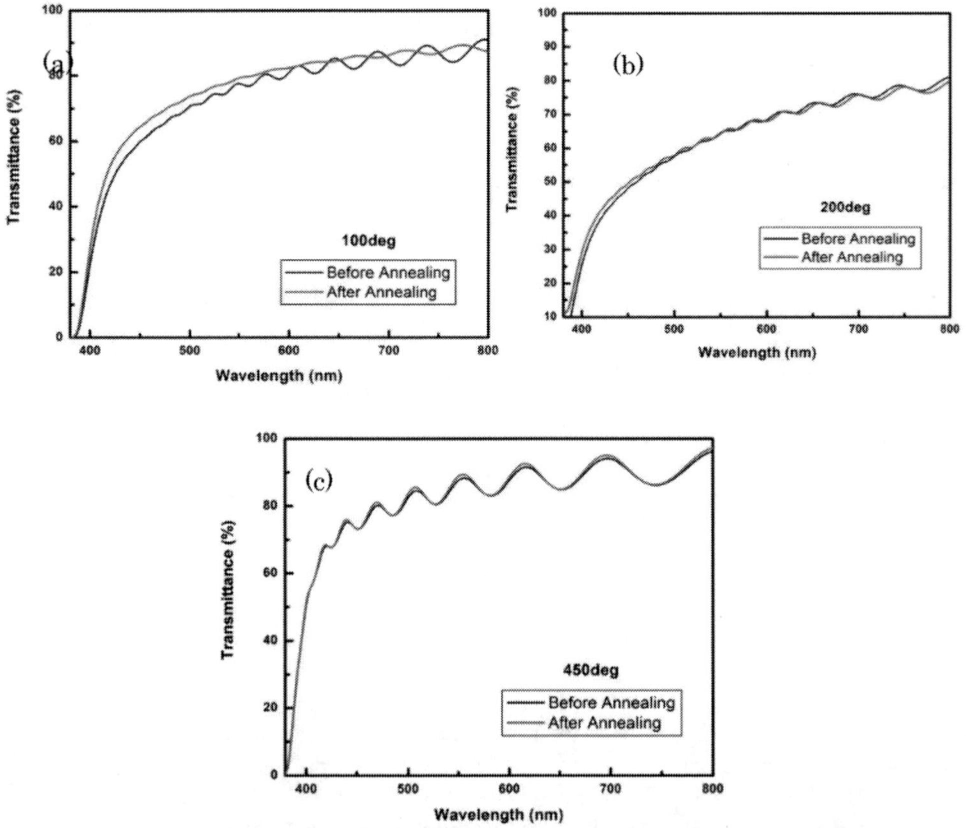

Figure 7 (a) – (c): Transmittance properties of ZnO film before and after annealing

4. Field emission (FE) properties

Annealing was done for nanostructured ZnO thin film at 100°C to investigate the effect on morphological structure and field emission properties. After annealed for 60 minutes in hydrogen, no significance change on the surface morphology of the thin film could be observed using Scanning Electron Microscopy (SEM). However, from the FE result, it can be assumed that the surface was cleaned by the hydrogen gas. Figure 8 shows the FE properties of annealed nanostructured ZnO film. The details on the measurement can be found elsewhere [13]. Higher current density was achieved compared to before annealing (as-grown nanostructured film). The electric field was achieved at 7.5 V/μm for 1 μA/cm^2 current density. By using Fowler-Nordheim equation [14], the calculated field enhancement factor (β) is 1465. Before annealing, the FE result showed non-linear behavior which does not give any result for β. Thus, the FE properties can be improved with annealing under hydrogen ambient.

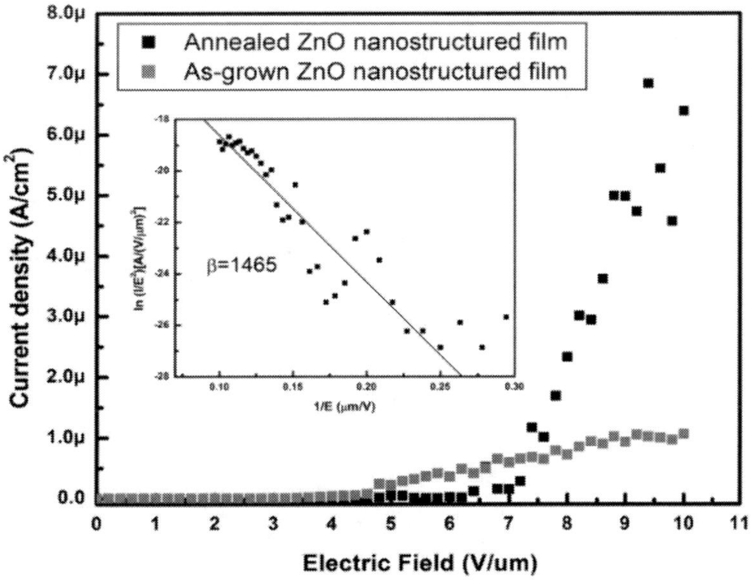

Figure 8: Field emission properties of ZnO thin film (inset figure: F-N plot for annealed ZnO nanostructured film)

5. Conclusion

Due to the hydrogen behavior, improvement of sheet resistance and field emission on thin film and nanostructured film, respectively were achieved at low annealing temperature compared to other reported annealing ambient which higher annealing temperature was required to change the properties of surface thin film and the crystallinity. The ZnO nanostructure film was not affected with the high flow rate of hydrogen. The annealing process also does not affect the optical properties of ZnO thin film regardless the thickness of ZnO thin film. The achievable low annealing temperature of ZnO film can be employed on flexible ZnO thin film in the future work.

References

[1] Zhu, B. L. *et al.* Influence of hydrogen introduction on structure and properties of ZnO thin films during sputtering and post-annealing. *Thin Solid Films* **519**, 3809–3815 (2011).

[2] Xue, S. W. *et al.* Effects of post-thermal annealing on the optical constants of ZnO thin film. *Journal of Alloys and Compounds* **448**, 21–26 (2008).

[3] Fang, Z. B., Yan, Z. J., Tan, Y. S., Liu, X. Q. & Wang, Y. Y. Influence of post-annealing treatment on the structure properties of ZnO films. *Applied Surface Science* **241**, 303–308 (2005).

[4] Guo, D., Ikeda, S., Saiki, K., Miyazoe, H. & Terashima, K. Effect of annealing on the mobility and morphology of thermally activated pentacene thin film transistors. *Journal of Applied Physics* **99**, 094502 (2006).

[5] Gupta, V. & Mansingh, A. Influence of postdeposition annealing on the structural and optical properties of sputtered zinc oxide film. *Journal of Applied Physics* **80**, 1063 (1996).

[6] S. Guillemin, E. Appert, H. Roussel, B. Doisneau et al., "Controlling the Structural Properties of Single Step, Dip Coated ZnO Seed Layers for Growing Perfectly Aligned Nanowire Arrays," J. Phys. Chem. C, **119**, 21694-21703 (2015).

[7] A Mahmood, N. Ahmed, Q. Raza, T. Muhammad et al., "Effect of thermal annealing on thestructural and optical properties of ZnO thin films deposited by the reactive e-beam evaporation technique," Phys. Scr. **82** , 065801, (2010).

[8] S. W. Xue, X. T. Zu, W. L. Zhou, H. X. Deng et al., "Effects of post-thermal annealing on the optical constants of ZnO thin film," Journal of Alloys and Compounds, **448**, 21-26, (2008).

[9] Z. B. Fang, Z. J. Yan, Y. S. Tan, et al., "Influence of post-annealing treatment on the structure properties of ZnO films," Applied Surface Science, **241**, 303-308, (2005).

[10] A. Ceylan, A. K. Rumaiz, D. Caliskan, S. Ozcan, E. Ozbay and J. C. Woicik, "Effects of rapid thermal annealing on the structural and local atomic properties of ZnO: Ge nanocomposite thin films", J. Appl. Phys., **117**, 105303, (2015).

[11] S. Kim, J. Seo, H. W. Jang, J. Bang et al., "Effects of H 2 ambient annealing in fully 0 0 2-textured ZnO:Ga thin films grown on glass substrates using RF magnetron co-sputter deposition", Applied Surface Science, **255**, 4616-4622, (2009).

[12] R. Bourguiga, G. Horowitz, F. Garnier et al. "Charge transport limited by grain boundaries in polycrystalline octithiophene thin film transistors", Eur. Phys. J. AP, **19**, 117-122, (2002)

[13] Z. Zulkifli, M. Subramanian, T. Tsuchiya, M. S. Rosmi, et al., "Highly transparent and conducting C:ZnO thin film for field emission displays", RSC Adv., **4**, 64763, (2014)

[14] Z. Zulkifli, M. Subramanian, M. Zamri, G. Kalita and M. Tanemura, " Fabrication of Nanostructured ZnO Films for Transparent Field Emission Displays", Jpn. J. Appl. Phys. **52**, 11NJ07, (2013)

IOP Conf. Series: Materials Science and Engineering **99** (2015) 012031 doi:10.1088/1757-899X/99/1/012031

Study of Reduced Graphene Oxide for Trench Schottky Diode

Nur Samihah Khairir[*1,2] , **Mohd Rofei Mat Hussin**[2,3] , **Iskhandar Md Nasir**[2] , **A.S.M. Mukhter Uz-Zaman**[3], **Wan Fazlida Hanim Abdullah**[1] , **Ahmad Sabirin Zoolfakar**[*1]

[1] Faculty of Electrical Engineering, Universiti Teknologi MARA, 40450 Shah Alam, Malaysia
[2] MIMOS Berhad, Technology Park Malaysia, 57000 Kuala Lumpur, Malaysia
[3] Faculty of Engineering, Multimedia University, 63100 Cyberjaya, Malaysia

Emails: ahmad074@salam.uitm.edu.my

Abstract. This paper presents the study of reduced Graphene Oxide (RGO) for trench Schottky diode by replacing conventional metal layer that forms schottky contact with a nanostructured carbon thin film *via* Reduced Graphene Oxide (RGO) technique. The RGO was synthesis by chemical exfoliation in which modified Hummer's method was approached. It was then deposited on the trench schottky pattern substrate by pressurized spray coating. The sample was then characterized by FESEM, Raman Spectroscopy and I-V test. The results of FESEM and Raman showed good characteristics and well deposited nanostructures of RGO flakes. The two-point I-V test showed that the samples have a low turn-on voltage and a higher break-down voltage, which is better than the conventional schottky diode used in the market.

1. Introduction

A Schottky diode is a semiconductor diode which has a low forward voltage drop and a very fast switching action [1]. In a Schottky diode, a semiconductor–metal junction is formed between a semiconductor and a metal, thus creating a Schottky barrier. Throughout the history, it has many kinds of schottky structure and one of it is the trench approach which had been proven to have a lot of advantages [2].

However, the device is experiencing limitation in conductivity tolerance at higher temperature due to the limitation characteristics of metal layer in the schottky point interface [3]. By using metal, it is hard to achieve tolerance in heat variation since metal is also having high thermal conductivity [4].

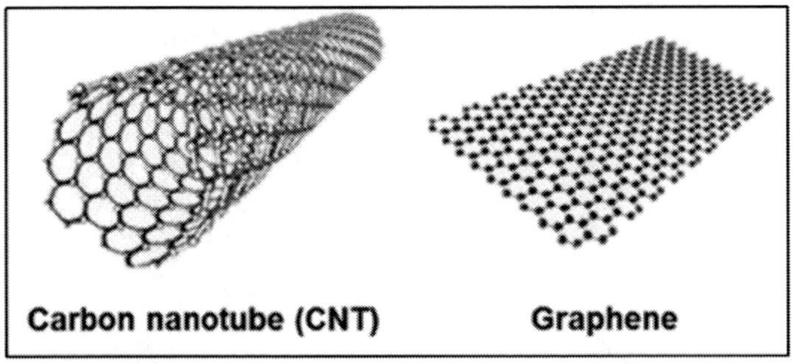

Figure 1. Atomic Structure of Carbon Nanotube (CNT) and Graphene

Thus, nanostructured carbon which is known with superior electronics and thermal properties as well as high carrier mobility is potential to replace conventional metal layer of the schottky diodes [5]. Most of nanoscale carbon materials consist from graphite, the standard form of carbon. In definition, graphite means "an allotropic form of the element carbon consisting of layers of hexagonally arranged carbon atoms in a planar condensed ring system" [6].

A hypothetical material called graphene, which known to be a 'single, one-atom thick, sheets of graphite' was gaining more attention [7]. Many breakthroughs of graphene have been exposed from researches as well as in the mass production system. The atomic structure of the 1D representation of carbon can be seen in Figure 1.

Graphene is the first two-dimensional atomic crystal discovered and has outstanding properties such as extreme mechanical strength and high electronic conductivity of as well as thermal [7]. Graphene and its compounds are increasingly used to make transistors that show extremely good performance which exhibits a definite progress for the raw material [8].

A lot of research on graphene focused on the reduction of Graphene Oxide (RGO) thin films by both chemical and electrochemical methods [9-12] due to the limitation of extracting graphene via mechanical exfoliation. Chemistry of graphene and its related compounds often deals with the exfoliation techniques that could be scaled up for industrial needs. This usually means that exfoliated product should exist in a form of dispersion. Reduction of graphene oxide, as seen in Figure 2, produces graphene-like structures and graphene oxide could be an environmentally friendly way to large-scale production of one to few layer graphitic thin films [13].

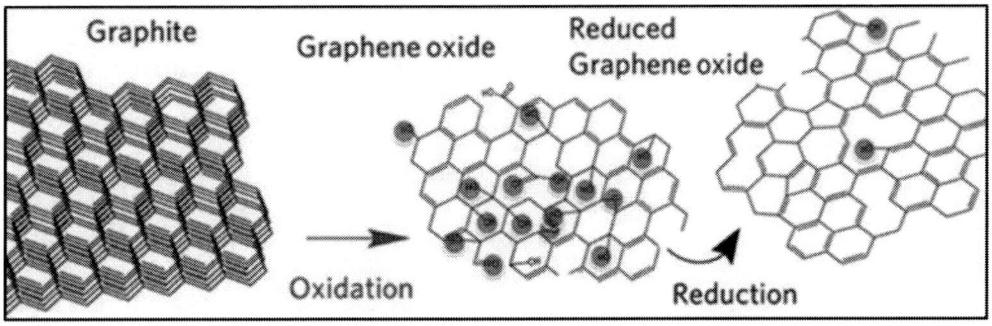

Figure 2. Reduction Process of Graphene Oxide

This paper is to discuss the effect of replacing the conventional metal layer that had been used for creating schottky contact with the nanostructured carbon, which in this case is RGO. It's been proven that a nanostructured carbon has a proper quality of conduction like a metal, which what the current industry is seeking of; a potential material to be incorporated with current transistors and diodes; that can enhance their performances as well as beneficial towards environment.

2. Methodology

Figure 3 shows the process of overall workflow of this research. It started off with benchmarking and literature review of the other researches, to obtain general idea of the current drive and gap for the RGO technology. Then, the experiment started with the fabrication of RGO *via* modified hummus method [14], which is one of the branches of chemical exfoliation.

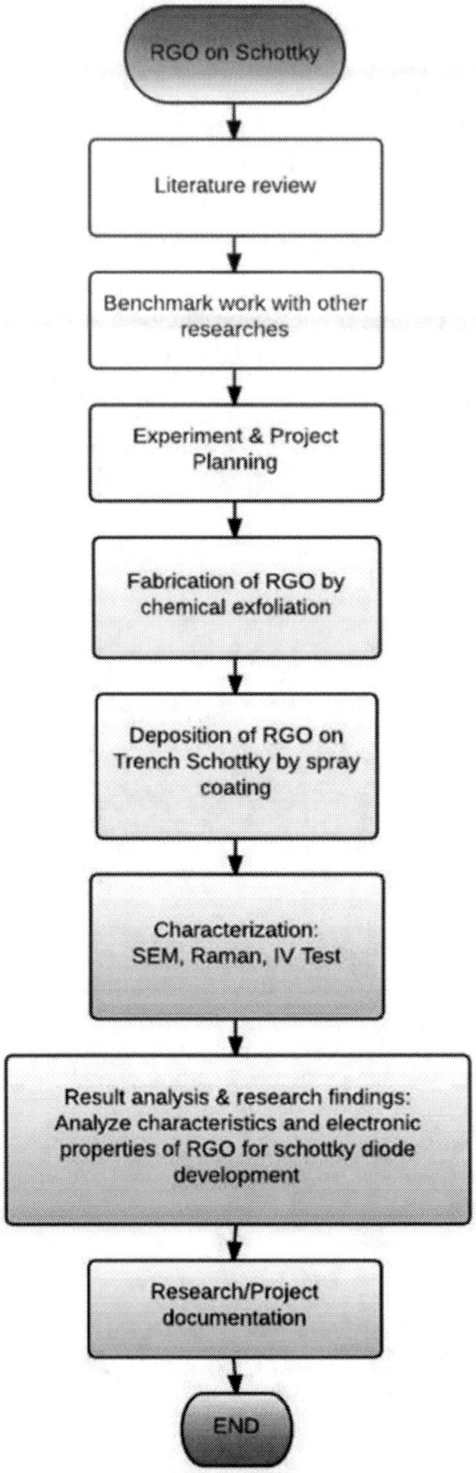

Figure 3. Process Flow of RGO Fabrication and Characterization

2.1. Preparation of Reduced Graphene Oxide (RGO)

2.1.1 Chemical Exfoliation and Reduction Process

The first step in RGO synthetization was the production of graphene oxide (GO) solution via modified Hummer's method by using graphite powder. The powder was mixed with Sodium Nitrate ($NaNO_3$) and concentrated sulphuric acid (H_2SO_4). The solution was stirred at a constant speed. Then, after sometime, while being stirred, potassium permanganate ($KMnO_4$) was gradually added before the solution was being diluted with a portion of distilled water [11].

This caused the formation of active species; diamanganese heptoxide (Mn2O7). It has the ability to selectively oxidize unsaturated double bonds that causes great significant in graphitic structure [15] and reaction pathways during oxidation [16]. The following equation shows the reaction that occurs between the two solutions [17]:

$$KMnO_4 + 3H_2SO_4 \rightarrow K^+ + H_3O^+ + 3HSO_4^-$$
$$MnO_3^+ + MnO_4- \rightarrow Mn_2O^7 \qquad (1)$$

To reduce the graphene oxide solution, aqueous ammonia (NH3.H2O) and hydrazine hydrate was added to form agglomeration of graphene-based nanosheets [17].

2.1.2 Deposition of RGO Thin Film

The RGO thin film was deposited on a substrate (see Figure 4) by pressurized spray coating method. The dispersion was produced by chemical reduction of GO, and the amount of spray volume controls the film thickness [18].

Figure 4 shows the schematic representation of trench-pattern schottky device which was used as a platform to characterize the electrical parameters of schottky behavior with the incorporation of RGO to replace conventional metal films. The RGO thin film sits on the surface of the trench pattern to form the schottky contact between the polysilicon and the RGO sheets itself.

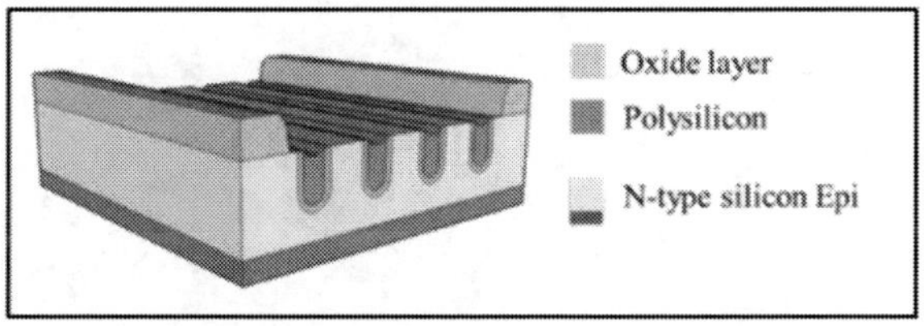

Figure 4. Cross-sectional schematic diagram of trench schottky structure before RGO deposition

2.2. Sample Characterization

2.2.1 Field Effect Scanning Electron Microscopy (FESEM)

By using FESEM, ones can observed the RGO films on the trench patterns, and show contrast between the thickness layers [19]. This is the same type of processes as optical microscope, obtaining views from various sides to be observed. However, it has a resolution ranging from a few nanometers to hundred micrometers, and this suggest that it fills an important gap between the Centre for Microscopy's atomic force microscope and digital 3D optical microscope. The FESEM is used primarily for ultra-high resolution electron imaging of surfaces and interfaces.

The morphology of the sample was observed by looking at the deposited graphene flakes formed at the surface of the trench schottky diode. By this method, we observed the distribution of the RGO coating and also looking to the deposited flakes size range.

2.2.2 Raman Spectroscopy

Carbon properties on the substrate were analyzed by using Raman. The sample was illuminated with a laser beam in the ultraviolet (UV) range. Scattered light was collected with a lens and was sent through interference filter or spectrophotometer to obtain raman spectrum of the sample. A great deal of detail on the fine structure of graphene can be extracted from the raman spectra [20].

2.2.3 I-V Characteristic

I-V characteristic is used to characterize the electrical properties of the device. It can be done either by using two-probe method or four-probe method. The ideal conductivity and resistivity pattern graph should result in a smooth slightly curved diagonal line of proportional I-V. This project used the two-point probe of Keithley 2400 to measure the schottky curve which focused on the turn-on and breakdown voltage.

3. Result and discussion

Figure 5 illustrates the SEM top view image of RGO flakes of different sizes and conditions. From the SEM images, ones can noticed that the RGO flakes synthesized are either multi layered or single layered as shown in figure 5(a) and 5(b) respectively.

3.1 FESEM Images

It is been observed the flakes are scattered almost thoroughly and obtained high coverage on the substrate surface. The average size of the flakes is found to be around 1um to 13um in diameter, though there are some nano-sized flakes spotted on some area. There are also more folded multi-layered of graphene sheets can be found compared to single layer sheets. A single layer flake or sheet is one-atom thick, and considered to have the highest quality of graphitic properties.

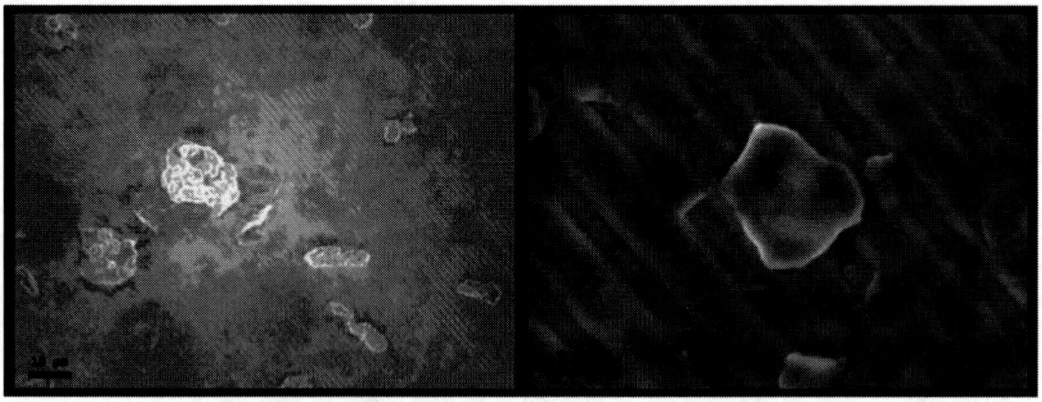

Figure 5. FESEM surface images of RGO deposition on top of silicon trench patterned
a) A few multi-layered RGO flakes on silicon
b) One layer of RGO flake on silicon

Then, the devices undergone Raman Spectroscopy to obtain the quality of carbon flakes and show the characteristics of graphene that had been deposited (Figure 6). The Raman peaks shows good production of RGO which the D peak shows Raman shift at around 1360 cm^{-1} while G peak shows Raman shift at around 1600 cm^{-1}.

This is at consistency, not considering minor differences, with other researchers [21,22], with slight shift of 10 cm^{-1} in G band when being compared to GO G band (1590 cm^{-1}). This is mostly due to the effect of electron-phonon coupling that caused an upshift to the band. G band represents the relative degree of graphitization, which assigns the E_{2g} phonon of sp_2 Carbon atoms while D band can define the defects in structures and disorders induced breathing mode of A_{1g} symmetry [22].

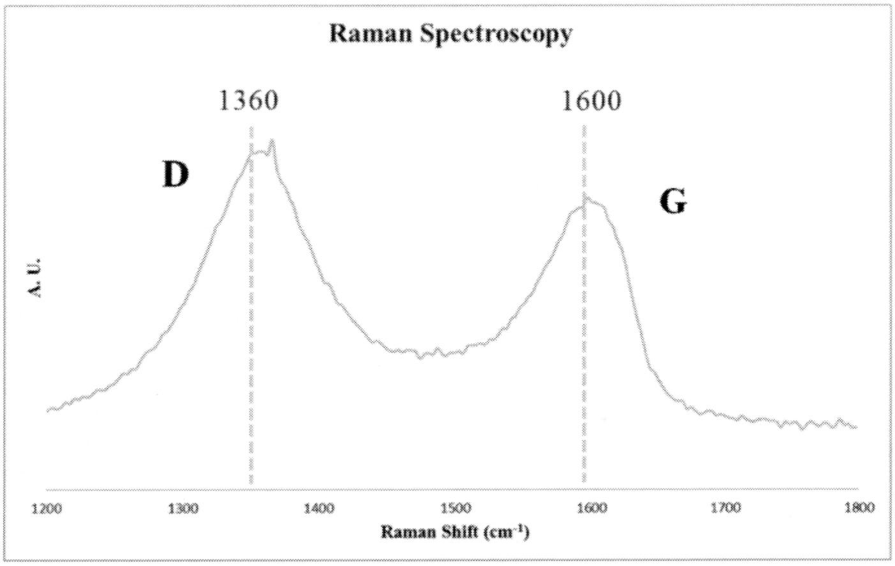

Figure 6. Raman Spectroscopy tested on Schottky surface with RGO deposition

The sample was then tested electrically using a two-point probe to test the schottky contact I-V curve. The position of the probe can be seen at the inset of Figure 7, where the first probe was located on the RGO thin film flake and the other was grounded.

Based on the curve shown, it exhibits a schottky curve flow with low turn-on and high breakdown voltage which are 1V and -10V respectively. The turn-on voltage is reasonable for schottky devices and their breakdown voltage is superior [23] as the schottky devices are required to have larger breakdown voltage. This is due to the known high conductivity of graphene sheet; producing great performance in rectifying the voltage to form schottky curve. The test was conducted on different dies multiple times and based on the observations; the curve produced was similar to the same pattern.

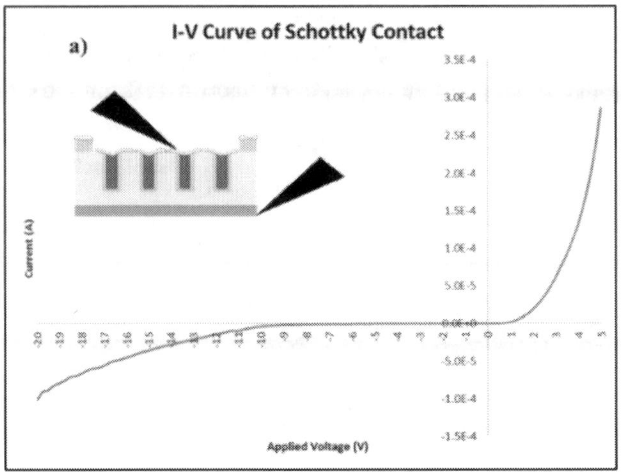

Figure 7. I-VTest Characteristics of Schottky Curve on two-point probing

4. Conclusion

The trench schottky diode is widely used by the industry in various applications as it is one of the most basic electronic devices in the market. RGO has the potential of replacing the conventional metal layer of schottky pattern as it has characteristics of better conductivity as well as a one-atom layer in thickness. The fabrication of RGO is achieved by modified Hummers method and deposited by pressurized spray coating onto the trench schottky substrate. When the sample had being tested for characterization, results show that by incorporating and replacing the metal layer with RGO on the schottky diode, the device exhibits better performance and have great potential and beneficial in the industry of nano-scaled transistors and diodes.

Acknowledgement

The author would like to acknowledge with gratitude the Semiconductor Foundry Operation of MIMOS Berhad and Faculty of Electrical Engineering UiTM for the encouragement. High appreciation conveyed specifically Dr. Lee's team from Nano Semiconductor Technology Division (NSTD) and also to the staffs of Nanoelectronics Lab and Advance Analytical Services Lab. This work is partially supported by Ministry of Science, Technology and Innovation Malaysia (MOSTI).

References

[1] C. Blake, D. Kinzer, and P. Wood, "Synchronous rectifiers versus Schottky diodes: a comparison of the losses of a synchronous rectifier versus the losses of a Schottky diode rectifier," *Proc. 1994 IEEE Appl. Power Electron. Conf. Expo. - ASPEC'94*, 1994.

[2] J. Raymond and E. Hueting, "Trenched Schottky Rectifiers," US 6441454 B2, 2002.

[3] T. Munir, A. AbdulAziz, M. J. Abdullah, and N. M. Ahmed, "Temperature variation efects on current-voltage (I-V) characteristics of n-GaN schottky diode," *Funct. Mater. Devices*, vol. 517, pp. 141–146, 2006.

[4] Y. P. Mamunya, V. V. Davydenko, P. Pissis, and E. V. Lebedev, "Electrical and thermal conductivity of polymers filled with metal powders," *Eur. Polym. J.*, vol. 38, no. 9, pp. 1887–1897, 2002.

[5] O. a. Shenderova, V. V. Zhirnov, and D. W. Brenner, "Carbon Nanostructures," *Crit. Rev. Solid State Mater. Sci.*, vol. 27, no. 3–4, pp. 227–356, 2002.

[6] M. Nič, "IUPAC Compendium of Chemical Terminology 2nd Edition (1997)," *IUPAC Compend. Chem. Terminol. Gold B.*, vol. 2, pp. 1997–1997, 1997.

[7] K. S. Novoselov, V. I. Fal'ko, L. Colombo, P. R. Gellert, M. G. Schwab, and K. Kim, "A roadmap for graphene," *Nature*, vol. 490, no. 7419, pp. 192–200, 2012.

[8] M. Lemme, "Current Status of Graphene Transistors," *Diffus. defect data. Solid state data. Part B*, pp. 1–12, 2010.

[9] H. J. Shin, K. K. Kim, A. Benayad, S. M. Yoon, H. K. Park, I. S. Jung, M. H. Jin, H. K. Jeong, J. M. Kim, J. Y. Choi, and Y. H. Lee, "Efficient reduction of graphite oxide by sodium borohydride and its effect on electrical conductance," *Adv. Funct. Mater.*, vol. 19, no. 12, pp. 1987–1992, 2009.

[10] W. Park, J. Hu, L. a. Jauregui, X. Ruan, and Y. P. Chen, "Electrical and thermal conductivities of reduced graphene oxide/polystyrene composites," *Appl. Phys. Lett.*, vol. 104, pp. 1–13, 2014.

[11] L. Meng and S. Park, "Preparation and Characterization of Reduced Graphene Nanosheets via Pre-exfoliation of Graphite Flakes," vol. 33, no. 1, pp. 209–214, 2012.

[12] S. Pei and H. M. Cheng, "The reduction of graphene oxide," *Carbon N. Y.*, vol. 50, no. 9, pp. 3210–3228, 2012.

[13] H. Feng, R. Cheng, X. Zhao, X. Duan, and J. Li, "A low-temperature method to produce highly reduced graphene oxide.," *Nat. Commun.*, vol. 4, p. 1539, 2013.

[14] L. Shahriary and A. a. Athawale, "Graphene Oxide Synthesized by using Modified Hummers Approach," *Int. J. Renew. Energy Environ. Eng.*, vol. 02, no. 01, pp. 58–63, 2014.

[15] M. Tro¨mel and M. Russ, Angew. Chem., 1987, 99, 1037–1038.

[16] D. R. Dreyer, S. Park, C. W. Bielawski, and R. S. Ruoff, "The chemistry of graphene oxide.," *Chem. Soc. Rev.*, vol. 39, no. 1, pp. 228–240, 2010.

[17] N. Cao and Y. Zhang, "Study of Reduced Graphene Oxide Preparation by Hummers ' Method and Related Characterization," vol. 2015, 2015.

[18] H. Shi, C. Wang, Z. Sun, Y. Zhou, K. Jin, and G. Yang, "Transparent conductive reduced graphene oxide thin films produced by spray coating," *Sci. China Physics, Mech. Astron.*, vol. 58, no. 1, pp. 1–5, 2015.

[19] D. a Dikin, S. Stankovich, E. J. Zimney, R. D. Piner, G. H. B. Dommett, G. Evmenenko, and R. S. Ruoff, "Preparation and characterization of graphene oxide paper.," *Nature*, vol. 448, pp. 457–460, 2007.

[20] A. C. Ferrari, "Raman spectroscopy of graphene and graphite: Disorder, electron-phonon coupling, doping and nonadiabatic effects," *Solid State Commun.*, vol. 143, pp. 47–57, 2007.

[21] P. Wang, Z.-G. Liu, X. Chen, F.-L. Meng, J.-H. Liu, and X.-J. Huang, "UV irradiation synthesis of an Au–graphene nanocomposite with enhanced electrochemical sensing properties," *J. Mater. Chem. A*, vol. 1, no. 32, p. 9189, 2013.

[22] Z. Ji, X. Shen, M. Li, H. Zhou, G. Zhu, and K. Chen, "Synthesis of reduced graphene oxide/CeO2 nanocomposites and their photocatalytic properties.," *Nanotechnology*, vol. 24, no. 11, p. 115603, 2013.

[23] D. Sinha and J. U. Lee, "Ideal graphene/silicon schottky junction diodes," *Nano Lett.*, vol. 14, pp. 4660–4664, 2014.

Electrical and Optical Properties of Copper Oxide Thin Films by Sol-Gel Technique

H. Hashim[1], S. S. Shariffudin, P. S. M. Saad and H. A. M. Ridah

Faculty of Electrical Engineering, Universiti Teknologi Mara,
40450 Shah Alam, Selangor, MALAYSIA

E-mail: hashimah655@salam.uitm.edu.my

Abstract. Copper oxide were prepared by sol-gel technique and deposited onto quartz substrates as thin films using spin coating method. The aim of this research was to study the effects of different spin coating speeds of copper oxide thin films on the electrical and optical properties of the thin films. Five samples of copper oxide thin films with different spin coating speeds of 1000, 1500, 2000, 2500 and 3000 rpm were annealed at 600°C for 30 minutes. UV-Vis spectrophotometer and two-point probe technique were used to characterize the optical and electrical properties of the deposited films. Based on the results obtained, it revealed that the electrical conductivity of the copper oxide thin films reduce as the spin coating speeds increase. The calculated optical band gap and the resistivity of the copper oxide thin films also decrease when the spin coating speeds are increased.

1. Introduction

Copper oxide was one of the oldest and first semiconductor material that discovered and characterized, but then was overtaken by the fast development of silicon [1, 2]. Copper oxide thin films could be used for several applications in catalyst, semiconductor, solar cells [6] and electronics [7-9]. It has direct optical band gap energy in the range of 2.1 to 2.6 eV [10]. The transparent conductive copper oxide thin films had been developed by several techniques such as electrodeposition [11], thermal evaporation, sol-gel technique [12] and wet chemical method [13] which could be deposited on various substrates [14]. However, the common technique used is sol-gel method due to its simplicity, lower cost, reproducibility, ease of composition control and large area deposition [15]. In addition, sol-gel method also easily to deposit on different type of substrates that could be performed under non-vacuum environment.

Therefore in this research sol-gel method was used to form the solution of copper oxide which then deposited on quartz substrates using spin coating technique. The main goal of this paper is to study the effects of different spin coating speeds on the electrical and optical properties of the copper oxide thin films on quartz substrates. In this work, the samples were characterized using UV-Vis spectrophotometer and two-point probe method for the optical and electrical properties respectively.

[1] To whom any correspondence should be addressed.

IOP Conf. Series: Materials Science and Engineering **99** (2015) 012032 doi:10.1088/1757-899X/99/1/012032

2. Experimental Details

The procedures for preparing copper oxide thin films were summarized in this section.

2.1. Preparation of substrates

In this work, 2 x 2 cm^2 quartz substrates were used with the thickness of 0.1 cm. All substrates were cleaned using acetone, methanol and deionized water in the ultrasonic bath in order to obtain a good and clean surface contact for the next process.

2.2. Preparation of copper oxide solution using sol-gel method

In the sol-gel process, the materials used were copper acetate powder [17], diethanolamine, isopropanol and polyethylene glycol without any further purification. 0.5 molar concentration of solution was formed by dissolving 0.5 gram copper acetate powder in 9 ml isopropyl alcohol and 0.5 ml diethanolamine which act as precursor, solvent and complexing agent respectively. Then, 0.5 ml polyethylene glycol was added into the solution. Next, the copper oxide solution was stirred for 10 minutes at room temperature to achieve a transparent solution. The solution with concentration of 0.5 molar was a clear and dark blue without any suspension of particles.

2.3. Preparation of copper oxide thin films deposition by spin coating technique

Prior to the thin films deposition process, quarter area of the substrate was covered with capton tape while the other area of the substrate was left uncovered for the layer thickness measurements. After 10 minutes of stirring process, a double step spinning program was applied to obtain the homogenous precursor films. In this step, the solution was spread onto the substrates at 100 rpm for 10 seconds. The solution was spin coated with spinning speed of 1000, 1500, 2000, 2500 and 3000 rpm for 5 minutes each. Immediately after coating process, the sample was dried for 5 minutes at 250°C for each layer. Five layers of copper oxide thin films were coated one over the above. After the drying of the last layer, sample was annealed for 30 minutes at 600°C.

2.4. Samples Characterization

The absorbance and transmittance of copper oxide thin films were observed using UV-Vis spectrophotometer. I-V measurements were performed by using two-point probe technique. Prior to the I-V measurements, the thin films were coated with platinum (Pt) as the mental contacts. The sample cross section of copper oxide thin film was shown in Figure 1. The thickness of metal contacts deposited onto copper oxide thin films were 0.00006 mm while their width were 3 mm.

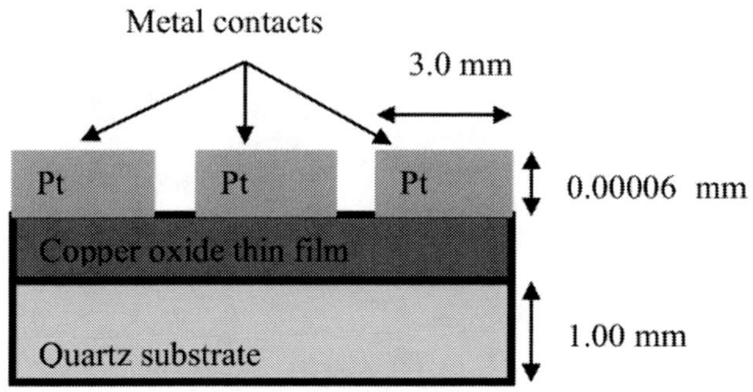

Figure 1. Sample cross section of copper oxide thin film deposited by metal contacts.

3. Results and Discussion

The results obtained from the experiment are discussed in this section.

3.1. Optical properties of copper oxide thin films

Figure 2 shows the absorbance spectra of copper oxide thin films at five different spinning rates. It could be observed that the absorbance of each film decreased with increasing of wavelength and spin coating speeds. The absorbance of the copper oxide thin film for 1000 rpm spinning rate was higher compare to the others. This due to the thin film layer thickness. The highest spin coating speed produced the thinnest film with the lowest absorbance while the lowest spinning rate produced the thickest film with the highest absorbance [20].

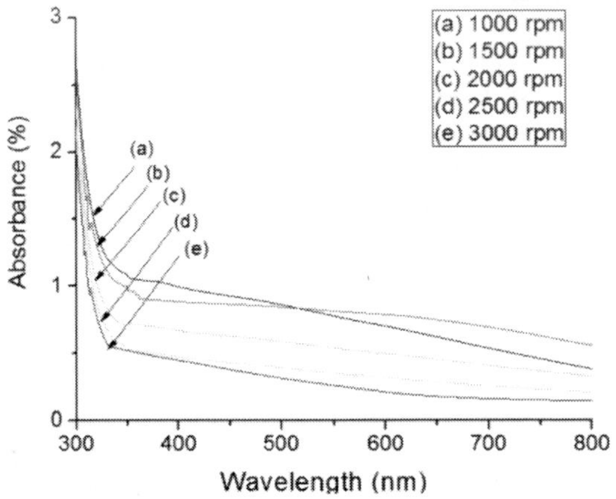

Figure 2. The plot of absorbance against wavelength.

The trend show here is similar with Halin et al [20]. The absorption was occurred when electron made the transition from a lower energy state in valence band to high energy state in conduction band while absorbing the excess energy as photon. It could be observed that almost absorption edge at wavelength of about 340 nm present in the spectra indicating the absorption properties of copper oxide in the UV region. The optical absorption at absorption edge related to the transition from valence band to conduction band and was closed with band gap energy [21]. Figure 3 shows the transmittances spectra of copper oxide thin films as a function of spin coating speeds.

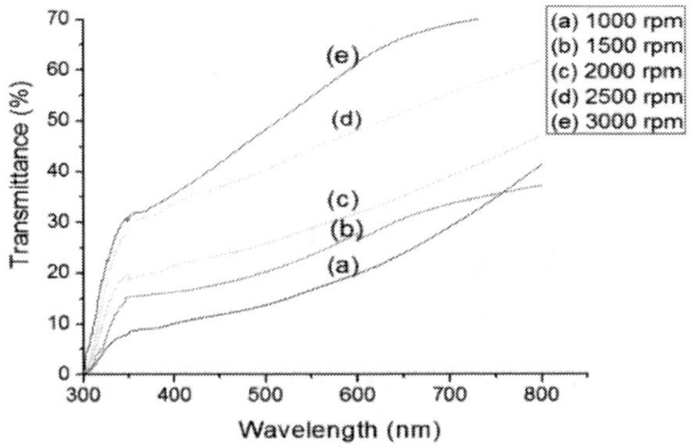

Figure 3. The plot of optical transmittance against wavelength.

From Figure 3, it can be seen that copper oxide thin film at 3000 rpm had high transparency in the visible region (400-800nm) which was profited for transparent electronic device fabrication. While, as the spin coating speeds increased from 1000 to 3000 rpm, the transmittance was below than 70%. It can be observed that the transmittance of the light at visible region reduced when the thickness of the thin films increased [22]. Therefore, the transparency at visible region was higher in copper oxide thin films due to it had low thickness. Transmission is an inverse process of absorption where electron at conduction band moving down to a lower energy state in valence band thus annihilating the electron hole pair. The absorption coefficient, α was obtained using Lambert's law as indicated by equation 1 [23, 24]:

$$\alpha = 1/t \ \ln 1/(T) \tag{1}$$

where t is the thin film thickness and T is the transmittance spectrum of the thin film. Figure 4 shows all thin films exhibited a low absorption in the visible but revealed high absorption properties in the UV region (below 350 nm).

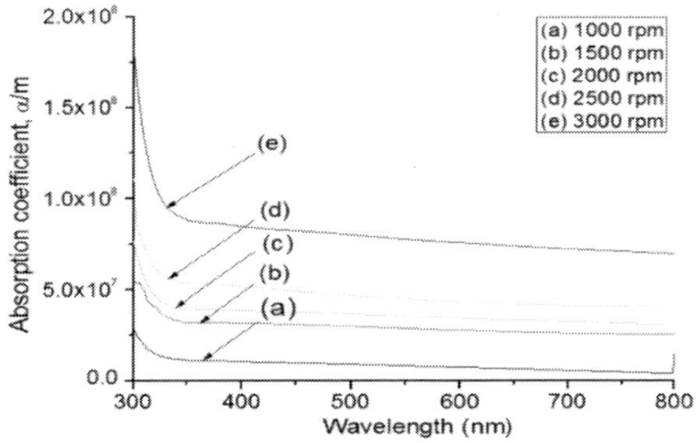

Figure 4. Relationship between absorption coefficient and wavelength.

It could be observed that copper oxide thin film at 3000 rpm had high absorption coefficient that was 6.79×10^{7} m^{-1} while absorption coefficient at 1000 rpm was 15.21×10^{6} m^{-1}. It can be seen that the absorption coefficient reduced when the spin coating speeds were reduced and the thickness of the thin films were increased. Figure 5 shows the optical band gap energy that plotted using absorption coefficient, α data.

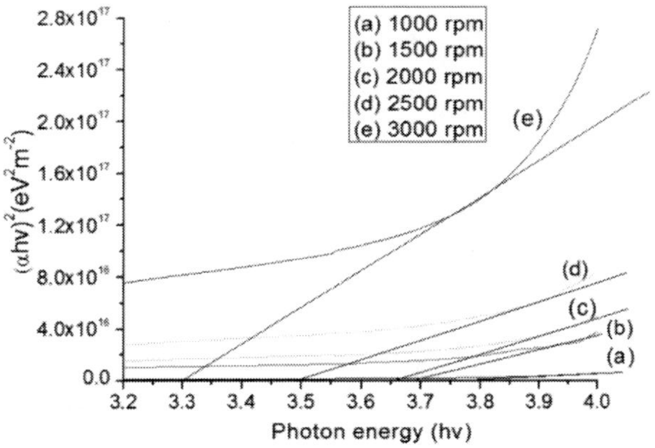

Figure 5. The plot of optical band gap energy versus photon energy.

IOP Conf. Series: Materials Science and Engineering **99** (2015) 012032 doi:10.1088/1757-899X/99/1/012032

The absorption coefficient, α of the copper oxide thin films were related to the photon energy, hv [25], where hv is the photon energy (eV). These optical band gap were obtained by tangent line intersection with energy axis [26] on the photon energy (hv) of the $(\alpha hv)^2$ versus photon energy plot. The photon energy (hv) was calculated using equation 2:

$$hv = hc/\lambda \qquad (2)$$

where h is the Plank constant which is 6.63×10^{-34}, c is the light constant that is 3×10^8 and λ is wavelength. The optical band gap energy for these films were found to be in the range of 3.30 to 3.80 eV which were increased as the spin coating speeds increased and also were found to be affected by film thickness [20].

From the graph, the optical band gap increased when the spin coating speeds were reduced. These results somehow contradict with results reported by Halin et al [21]. Halin at al [20] explained that the decreasing in grain size of films with increasing of spinning rates will be contributed to the reduction in optical transparency trough scattering phenomenon [19, 20]. By referring the results in Table 1, it can be observed that the thickness influenced the optical band gap.

Table 1. Thickness of thin films and optical band gap at different spin coating speeds.

Spin speed (rpm)	Thickness (nm)	Optical band gap (eV)
1000	215.43	3.800
1500	129.92	3.675
2000	99.65	3.650
2500	64.53	3.500
3000	39.97	3.300

3.2. Electrical properties of copper oxide thin films

The I-V measurements plot for copper oxide thin films at different spin coating speeds were shown in Figure 6. In this work, Platinum (Pt) was used as the mental contacts. The I-V measurements were performed using two probe method with applied voltage -10 to 10 V. The needles of the probe must touch the surface firmly to produce reliability data.

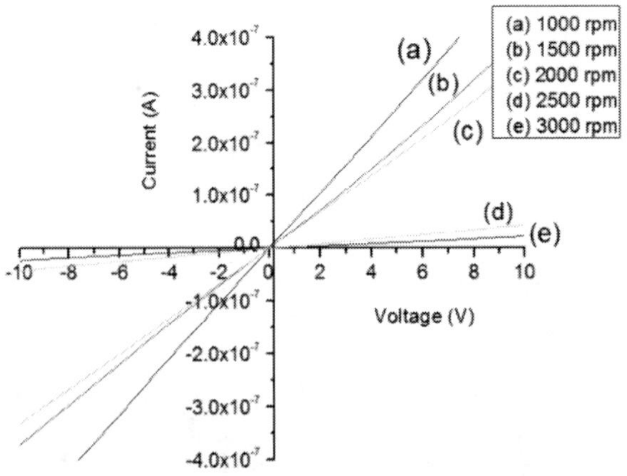

Figure 6. The plot of I-V curve.

Figure 6 shows the I-V curve of copper oxide thin films at 1000, 1500, 2000, 2500 and 3000 rpm which shown that all films indicated the Ohmic behavior with linear I-V curve. It also could be observed that the current values at fixed voltage decreased due to the copper oxide thin films as spin coating speeds were increased. The increased degree of alignment of thin films after undergo the increased of spin speeds caused significant decreased in the electrical conductivity was confirmed by SEM and WAXD studies [27]. Figure 6 also shows the linear graph which obeys Ohm's law. The current through a conductor between two points was directly proportional to the applied voltage across the two points. The current will flow and increase accordingly to the magnitude of the applied voltage as indicated by following equation 3:

$$V = IR \tag{3}$$

where V is voltage, I is current and R is resistance. From the results, it shows that all prepared samples were strongly depending on the speeds of spinning. The spectra revealed that the current at 3000 rpm was the lowest while the current at 1000 rpm was the highest among the other spectrums. As the increased of spin coating speeds, it actually helped to spread the active layer uniformly on the glass substrates [28]. As a result, the recombination process took place effectively as more holes and electrons were transported to the active area, which translated into better electrical characteristics [28].

From Figure 7, the resistivity ρ, of copper oxide thin films were calculated using following equation 4;

$$\rho = (V/I)(wt)/L \tag{4}$$

where V is the supplied voltage, I is the measured current, w is the width of metal contact, t is the film's thickness and L is the length between metal contact. The calculated resistivity obtained in this work to be 2.79 , 7.54 , 11.9 , 43.86 and 114.57 Ω m for thin films spin at 1000, 1500, 2000, 2500 and 3000 rpm respectively. The lowest resistivity was about 2.79 Ω m at 1000 rpm while the highest resistivity was about 114.57 Ω m at 3000 rpm. This was due to increase of electron and hole as carrier concentration and hence decreased the resistivity [29]. Therefore it can be seen that the lowest spin coating speeds with the thickest thin films and the increased carrier concentration produced the lowest resistivity.

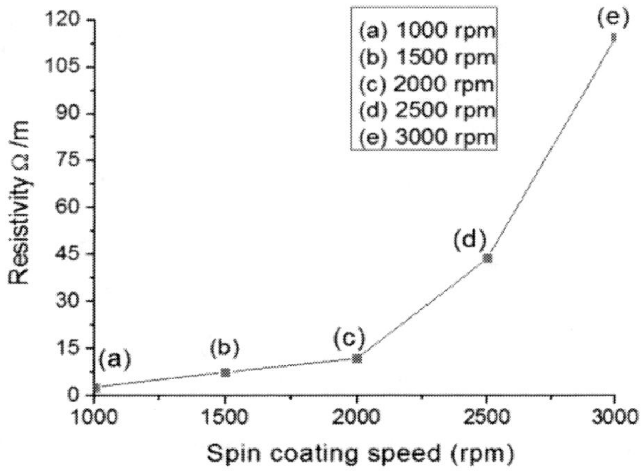

Figure 7. The plot of resistivity of copper oxide thin films versus spin coating speeds.

The relationship between resistivity and conductivity was shown by the following equation 5;

$$\sigma = \frac{1}{\rho} \tag{5}$$

where σ is conductivity and ρ is resistivity values as calculated. The conductivity of the thin films in the range of 0.0087 to 0.3584 Sm^{-1}. From the graph, as the spin coating speeds being increased, the conductivity of the prepared films displaying a slow down trend. This means that, the lowest spin coating speeds can improved the conductivity the thin films. Besides, the conductivity of the thin films was influenced by carrier concentration or density. It can be seen that the highest spinning rates and the highest uniformity of thin films produced the lowest conductivity of copper oxide thin films.

From the results, it can be observed that the conductivity of the thin films decreased from 1000 to 3000 rpm. The results indicated that the spin coating speed at 1000 rpm was the optimum parameter which gives the highest conductivity at 0.3584 Sm^{-1} compared to the other spin coating speeds which give lower conductivity. There were many researchers reported that the different conductivity will be affected the electron carrier mobility.

As slowest spin coating speeds were introduced, the thickest of the thin films will be generated and the carrier concentration will be higher in the thin films. This will be resulted in the increasing of the conductivity of the thin films. There were always related the conductivity, σ of a material to microscopic parameters that described the motion of the electrons or other charge carrying particles such as holes or ions. Besides that, the thickness of copper oxide thin films which acted as active layer were believed to be important factor to study the effect on the electrical properties [29]. The results of conductivity and resistivity with the thickness of thin films at different spin coating speeds were summarized in table 2.

Table 2. Thickness of thin films, resistivity and conductivity at different spin coating speeds.

Spin speed (rpm)	Thickness (nm)	Resistivity, ρ (Ωm)	Conductivity, σ (Sm^{-1})
1000	215.43	2.79	0.3584
1500	129.92	7.54	0.1326
2000	99.65	11.9	0.0840
2500	64.53	43.86	0.0228
3000	39.97	114.57	0.0087

4. Conclusion

Copper oxide thin films were deposited on quartz substrates by spin coating technique. The thicknesses of thin films, surface topology, optical and electrical properties were investigated. In this work, the goal has been achieved to study the effects of varying the spin coating speeds of copper oxide thin films. From the results obtained, copper oxide thin film that deposited at 1000 rpm shows the highest thickness and the lowest of surface roughness. At 1000 rpm of spinning rate also shows the highest of transparency. However, at copper oxide thin film that deposited at 3000 rpm shows the lowest optical band gap. Besides, it was found that at 1000 rpm of spinning rate shows the best of electrical conductivity. Future investigation should be done on the crystallinity of the thin film by using XRD. Besides that, the investigation on the electrical and optical properties of the copper oxide thin films could be performed using doping technique.

Acknowledgments
The author would like to thank to the NANO-Electronic Centre (NET), Faculty of Electrical Engineering, Universiti Teknologi MARA and NANO-SciTech Centre, Institute of Science, Universiti Teknologi MARA for providing the laboratory facilities.

References

[1] Michael N and Simon D. E 2008 *J. Thin Solid Films* **516** 1468-1472
[2] Davis S. D and Harry A. A 2009 *California Institute of Technology*
[3] Habibi M. H, Kamrani R and Mokhtari R 2010
[4] Rehman S, Mumtaz A and Hasanain S. K 2011 *J. Nanopart. Res.*
[5] Alikhanzadeh-Arani S, Salavati-Niasari M and Almasi-Kashi M 2012 *J. Inorg. Organomet. Polym.*
[6] D. Arun Kumar, Francis P. Xavier and J. Merline Shyla 2012 *Applied Science Research*, 2174-2183
[7] Chun-Lung Chu, Hsin-Chun Lu, Chen-Yang Lo, Chi-You Lai and Yu-Hsiang Wan 2009 *Physical B. Condensed Matter* 404 4831-4834
[8] A. Parreta 1996 *Phys. Stat. Sol.* **155** 399
[9] Kawazoe, Yasukawa, Hyodo, Kurita, Yanagi and Hosono 1997 **389** 939-942
[10] A. A. Ogwu, E. Bouquerel, O. Ademosu, S. Moh, E. Crossan and F. Placido 2005 *Acta Materialia* **53** 5151-5159
[11] T. Mahalingam, J. S. P. Chitra, S. Rajendran, M. Jayachandran, M. J. Chockalingam 2000 *J. Cryt. Growth* 216 304-310
[12] L. Armelao, D. Barreca, M. Bertapelle, G. Bottaro, C. Sada and E. Tondello 2003 *Thin Solid Films* **442** 48-52
[13] Y. Qu, X. Li, G. Chen, H. Zhang and Y. Chen 2008 *Mater. Lett.* **62** 886-888
[14] M. F. Al-Kuhaili 2008 *Vacuum* **82** 623-629
[15] B. Roy, J. D Perkins, T. Kaydanova, D. L Young, M. Taylor, A. Miedaner, C. Curtis, H. J Kleebe, D. W Readey, D. S Ginley 2008 *Thin Solid Films* **516** 4093-4101
[16] B. K Mohammad Hossein Habibi 2013 **Springer**
[17] E. M. A.Y. Oral, E. Mensur, M.H. Aslan and E. Basaran 2013 *Elsevier Ltd.*
[18] R. Dewi, N. I. Baa'yah, and I. A. Talib 2007 *Materials Science-Poland* **25** 3
[19] M. M. Zahidi, M. F. Malek, M. H. Mamat, N. A. Rashied, U. M. Noor, M. R. Mahmood 2010 *International Conference on Enabling Science and Nanotechnology*
[20] D. S. C. Halin, I. A. Talib A. R. Daud and M. A. A. Hami 2009 *Solid State Science and Technology* **17**
[21] M. H. Mamat, Z. Khusaimi, M. Z. Musa, and M. Rusop, Electronic Devices, Systems and Applications (ICEDSA),Intl Conf on, 2010.
[22] S. A. Kamaruddin, M. Z. Sahdan, K. Y. Chan, Y. Ho-Kwang, H. Jian-Wei, M. R. Mahmood, Semiconductor Electronics (ICSE), 2010 IEEE International Conference on 2010, pp. 171-174.
[23] M. Z. S. M.H. Mamat, Z. Khusaimi, A.Z. Ahmed, S. Abdullah, M. Rusop 2010 *Optical Materials* 696–699
[24] C.-H. L. C.C. Ting, C.Y. Kuo, C.C. Hsu, H.C. Wang, M.-H.Yang 2010 *Thin Solid Films* 4156–4162
[25] Ray S. C 2001 *Solar Energy Material & Solar Cells* **68** 307-312
[26] W.P. T. Young Sung Kim 2007 *Applied Surface Science* **253** 4911-4916
[27] C. Zhu, J. Chen, K. K. Koziol, J. W. Gilman, P. C. Trulove, S. S. Rahatekar 2014 *Express Polymer Letters* **8** 154–163
[28] A. H Reshak, M. M Shahimin, N. Juhari and S. Suppiah 2013 *Progress in Biophysics and Molecular Biology* **113** 289-294
[29] Betty Lise Anderson and Richard L.Anderson 2005 *Mc Graw Hill International Edition*

Institute of Physics
Dirac House, Temple Back
Bristol BS1 6BE UK

ISSN: 1757-8981
ISBN 978-1-5108-1636-7

9 781510 816367

CURRAN ASSOCIATES INC.
proceedings.com